Editors

Lecture Notes in Chemistry

Edited by G. Berthier M. J. S. Dewar H. Fischer
K. Fukui G. G. Hall J. Hinze H. H. Jaffé J. Jortner
W. Kutzelnigg K. Ruedenberg J. Tomasi

38

E. Lindholm
L. Åsbrink

Molecular Orbitals and their Energies, Studied by the Semiempirical HAM Method

Springer-Verlag
Berlin Heidelberg New York Tokyo 1985

Authors

E. Lindholm
L. Åsbrink
Physics Department, Royal Institute of Technology
S-100 44 Stockholm

ISBN-13: 978-3-540-15659-8 e-ISBN-13: 978-3-642-45595-7
DOI: 10.1007/978-3-642-45595-7

Library of Congress Cataloging in Publication Data. Lindholm, E. (Einar), 1913- Molecular orbitals and their energies, studied by the semiempirical HAM method. (Lecture notes in chemistry; 38) 1. Molecular orbitals. I. Asbrink, L., 1944- II. Title. III. Series. QD461.L724 1985 541.2'2 85-14797
ISBN-13: 978-3-540-15659-8

Preface

This treatment of molecular and atomic physics is primarily meant as a textbook. It is intended for both chemists and physicists. It can be read without much knowledge of quantum mechanics or mathematics, since all such details are explained. It has developed through a series of lectures at the Royal Institute of Technology.

The content is to about 50 % theoretical and to 50 % experimental. The reason why the authors, who are experimentalists, went into theory is the following.

When we during the beginning of the 1970's measured photoelectron spectra of organic molecules, it appeared to be impossible to understand them by use of available theoretical calculations. To handle hydrocarbons we (together with C. Fridh) constructed in 1972 a purely empirical procedure, SPINDO [1] , which has proved to be useful, but the extension to molecules with heteroatoms appeared to be difficult.

One of us (L.Å.) proposed then another purely empirical procedure (Hydrogenic Atoms in Molecules, HAM/1, unpublished), in which the Fock matrix elements $F_{\mu\nu}$ were parametrized using Slater's shielding concept. The self-repulsion was compensated by a term "-1".

The second effort, HAM/2 [2] , started from the total energy E of the molecule. The atomic parts of E used the Slater shielding constants, and the bond parts of E were taken from SPINDO. The Fock matrix elements $F_{\mu\nu}$ were then obtained from E in a conventional way.

The third effort, HAM/3 [3] , started from atomic spectroscopy instead. Detailed studies by one of us (L.Å.) produced expressions for the atomic shielding efficiencies, which gave high accuracy in atomic calculations. HAM/3 has been used to calculate ionization energies, excitation energies and electron affinities, and many examples are found in the papers listed in Sec.H.6.

The HAM/3 computer program has been submitted to Quantum Chemistry Program Exchange (QCPE) [4] and all calculations described in this treatment can therefore easily be reproduced using the QCPE program.

The method obtained a very important support when D.P. Chong in Vancouver, Canada, compared the HAM/3 results with his very reliable calculations of ionization energies [5] and found good agreement. He has afterwards contributed very much to the development of the method, and he has recently adapted the QCPE program to a personal computer [6].

In spite of its results the HAM method had been criticized [7.8]. It had never been shown that the exchange integral in the Hartree-Fock method can be replaced by a term "-1". In 1979 we (together with C. Fridh) could show that exploitation of the idempotency of density matrices can give the desired proof. The paper was published together with the critic [9].

It remained then only to deduce the HAM method from the Hartree-Fock method. After many discussions this was solved in 1979 (by L.Å.) as described in Sec.D. We found later (in 1980) that pair-correlation energies can be included in this proof. This deduction has recently (1985) been improved when it appeared to be possible to deduce the treatment of correlation from density functional theory.

The _fourth effort_, HAM/4, could now start. A large number of atoms have been studied as described in Sec.E. and new theoretical methods to handle the multiplet splitting in atomic spectroscopy have been presented.

A determination of the molecular parameters would certainly now result in a theoretical procedure, capable of high accuracy in calculation of ionization energies, excitation energies, electron affinities and other properties of small and large molecules.

Outside Sweden, the HAM/3 method has been used in several laboratories after it had been submitted to QCPE in 1980. Its use has predominantly been restricted to interpretations of photoelectron spectra. We believe, however, that applications to UV spectroscopy and electron affinities will be more important for the future, since these fields are less developed and more important for a further application on chemical reactions. The study of electron affinities is therefore stressed in two chapters.

It is a pleasure to acknowledge our gratitude to our coworker C. Fridh and to G. Ahlgren, G. Bieri, S. de Bruijn, J.-L. Calais, D.P. Chong, O. Edqvist, K.F. Freed, O. Goscinski, I. Lindgren, S. Ljunggren, B.I. Lundqvist, S. Lundqvist, R. Manne, R.G. Parr, P. Sand, L.E. Selin and A. Svensson for cooperation or discussions, and to the Swedish Natural Science Research Council, the Bank of Sweden Tercentenary Foundation and Knut och Alice Wallenbergs Stiftelse for economic support.

Stockholm February 1985 E. Lindholm L. Åsbrink

References

1. C. Fridh, L. Åsbrink and E. Lindholm, Chem.Phys.Letters 15, 282 (1972).
2. L. Åsbrink, C. Fridh and E. Lindholm, in: Chemical Spectroscopy and Photochemistry in the Vacuum-Ultraviolet, C. Sandorfy, P. Ausloos and M.B. Robin (eds.), Reidel, Dordrecht (1974).
3. L. Åsbrink, C. Fridh and E. Lindholm, Chem.Phys.Letters 52, 63, 69, 72 (1977).
4. L. Åsbrink, C. Fridh and E. Lindholm, QCPE 12, 393 (1980) (Quantum Chemistry Program Exchange, Indiana University, Bloomington, Indiana).
5. D.P. Chong, Theoret.Chim.Acta 51, 55 (1979).
6. D.P. Chong, QCPE QCMP005 (1985).
7. S. de Bruijn, Chem.Phys.Letters 52, 76 (1977).
8. S. de Bruijn, Theoret.Chim.Acta 50, 313 (1979).
9. L. Åsbrink, C. Fridh, E. Lindholm and S. de Bruijn, Chem.Phys.Letters 66, 411 (1979).

Contents

Contents

The HAM method is obtained in the following way.

Density functional theory starts from the total electronic energy E of a molecule, which is the sum of kinetic energy, attraction to nuclei, electron-electron repulsion, exchange energy (E_x) and correlation energy (E_c). All these terms are functionals of the total electronic density ϱ.

Kohn and Sham introduced orbitals ψ_i , defined by

$$\varrho = \sum_i \psi_i^* \psi_i \tag{1}$$

and determined them from E by variational calculus. It appears that E_c can be written

$$E_c = \frac{1}{2} \sum_{\mu\nu\lambda\sigma} P_{\mu\nu} P_{\lambda\sigma} \, \varepsilon^{corr}_{\mu\nu\lambda\sigma} \tag{2}$$

where the pair-correlation energy is denoted $\varepsilon^{corr}_{\mu\nu\lambda\sigma}$.

If this is combined with the Hartree-Fock-LCAO total energy expression, we find

$$E = -\frac{1}{2} \sum_{\mu\nu} P_{\mu\nu} \int \phi_\mu^* \nabla^2 \phi_\nu \, d\tau \tag{3}$$

$$- \sum_{\mu\nu} P_{\mu\nu} \int \phi_\mu^* \phi_\nu \sum_B Z_B r_B^{-1} \, d\tau \tag{4}$$

$$+ \frac{1}{2} \sum_{\mu\nu\lambda\sigma} P_{\mu\nu} P_{\lambda\sigma} \left[(\mu\nu|\lambda\sigma) - \frac{1}{2}(\mu\sigma|\lambda\nu) + \varepsilon^{corr}_{\mu\nu\lambda\sigma} \right] \tag{5}$$

$$+ \sum_{A>B} Z_A Z_B R_{AB}^{-1} \tag{6}$$

The solutions of the corresponding Kohn-Sham one-particle self-consistent equation are the Kohn-Sham orbitals. These orbitals give the correct energy E , they are orthonormal and their density matrix is idempotent.

In eq.(5) the self-repulsion is eliminated by the exchange integral. This is handled in HAM by adding the idempotency relation

$$\mathbb{P} \mathbb{S} \mathbb{P} - 2 \mathbb{P} = 0 \tag{7}$$

after multiplication with a factor. This trick does not change the total energy E since the relation is zero, but after some algebra the HAM total energy appears in a new formulation

$$E = -\frac{1}{2}\sum_{A}\sum_{\mu} N_{\mu}\zeta_{\mu}^{2} \quad (8)$$

$$-\sum_{\mu_A \nu_B} P_{\mu\nu}\beta_{\mu\nu} \quad (9)$$

+ electrostatic repulsion between atoms in molecule (10)

+ two small terms (11)

Since the atomic terms in eq.(8) and the molecular terms in eqs.(9-11) are well separated, the one-center parameters in eq.(8) can be determined from atomic spectroscopy with full accuracy for molecular calculations.

The separation of atomic (one-center) and molecular (two-center) terms is important. In ZDO type methods this goal is achieved only after neglect of a large number of integrals.

Let us study the atomic terms.

Before the final expression, eq.(8), is obtained we have

$$(8) = -\frac{1}{2}\sum_{A}\sum_{\mu} N_{\mu}\cdot\Big[\int \phi_{\mu}^{*}\nabla^{2}\phi_{\mu}\,d\tau + 2Z_A\int\phi_{\mu}^{2}r^{-1}\,d\tau \quad (12)$$

$$-(N_{\mu}-1)\gamma_{\mu\mu} - \sum_{\nu\neq\mu} N_{\nu}\gamma_{\nu\mu}\Big] \quad (13)$$

In eq.(12) we observe the kinetic energy of electron μ and its attraction to nucleus A. In eq.(13) we have the repulsion between μ and the other electrons on A. The term (-1) here appears directly when the idempotency relation eq.(7) is combined with eq.(3-6) and has its origin in the exchange integral. It eliminates the self-repulsion.

The repulsions $\gamma_{\nu\mu}$ are not suitable for parametrization, since they depend upon Z_A, but if we divide by the second term in eq.(12) we obtain shielding efficiencies $\sigma_{\nu\mu}$ which are independent of Z_A. Many atoms can now be used in the parametrization, and the success of HAM is simply due to the fact that the mathematical framework in eqs.(8-11) is suitable for parametrization.

Since HAM is a density functional method, it exhibits the same pecularities with respect to the Hartree-Fock rules as other density functional theories. For instance, Koopmans' theorem has no meaning.

Due to the unique way in which the self-repulsion is handled in HAM, the formulas for ionization energies, excitation energies and electron affinities must be deduced in special ways.

Hartree-Fock schemes can therefore not be used.

A. The LCAO model

A.1. Molecular orbitals ψ_i

There are many orbitals in a molecule. The simplest way to describe the molecule is to assume that in a certain meaning the electrons move independently of each other. This is called the "orbital approximation".

The total wavefunction Ψ of the molecule can then be written as a product of molecular orbitals ψ_i

$$\Psi = \psi_1(1) \cdot \psi_2(2) \cdot \psi_3(3) \cdot \psi_4(4) \cdot \cdot \cdot \cdot$$

Here, the subscript denotes the number of the orbital and the parenthesis denotes the number of the electron. The product is known as a Hartree product.

The concept "molecular orbital ψ_i" means that the electron "belongs" to the whole molecule. It can move from one part of the molecule to another part without "belonging" more to a certain part of the molecule than to another.

As an illustration we show a molecular orbital in benzene (with energy 17 eV). It is negative inside the molecule and positive outside the molecule.

In this orbital there are normally two electrons which in a certain meaning move independently of the many electrons in the other orbitals in the molecule.

Since the two electrons have opposite spins, α and β, respectively, one can partition the molecular orbital ψ_i, which contains two electrons, into two orbitals, ψ_i^{α} and ψ_i^{β}, which contain one electron each.

We have written above that in a certain meaning the electrons move independently of each other. This means that we assume that an electron moves under the influence of the electric field from the nuclei and from the average fields due to the other electrons. This assumption, which is the "orbital approximation", is a very good approximation. If it is taken as a basis for quantum-mechanical calculations, the orbitals and energies, obtained in this way, are in very reasonable agreement with nature.

In the orbital approximation the instantaneous repulsions between the electrons are neglected since only the average repulsion is considered. In this approximation, therefore, two electrons may come very near each other. In nature, on the other hand, this never happens due to their repulsion. Therefore, in nature, both the orbitals and their energies differ slightly from them in the orbital approximation. Although the effects of this "correlation" are small, they often spoil the possibilities to attain agreement between calculation and nature. The correlation will be discussed in detail below.

A.2. The LCAO formalism

It is not easy to solve the Schrödinger equation for a molecule. Usually, an approximate method, the variational method, is used. This approach requires, however, that we first are able to guess Ψ and ψ_i .

Every molecule is composed of atoms, and in a free atom its electrons move in atomic orbitals ϕ_μ . If we restrict our atoms to those in the first row of the periodic table, we have on each atom the following atomic orbitals:

$1s$, $2s$, $2px$, $2py$ and $2pz$.

The electrons in the $1s$ orbital move so near the nucleus and have so high binding energies that they do not take part in the chemical bonding in the molecule. They are even not much influenced by the formation of a molecule from its atoms. We will therefore treat them together with the nucleus and say that the nucleus together with the two $1s$ electrons constitute the "core" with positive charge $Z-2$, where Z is the nuclear charge.

In our discussions here we will therefore consider mainly the "valence electrons"

$2s$, $2px$, $2py$ and $2pz$

except for hydrogen, where the valence electron is $1s$.

It is convenient to illustrate the atomic orbitals ϕ_μ in the following schematic way:

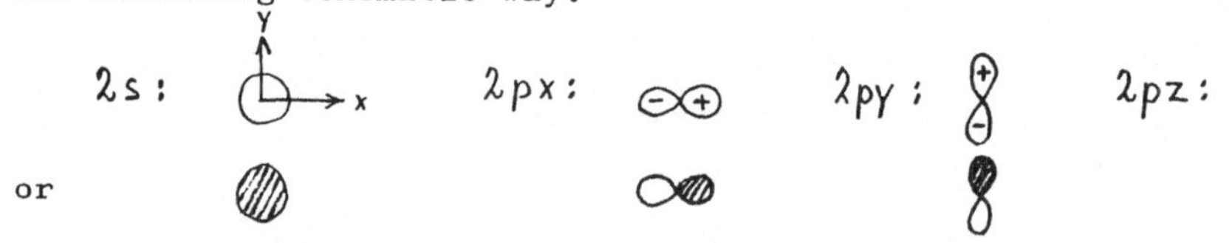

We observe that the $2p$ orbitals in the normal situation have their positive part in the direction of the positive coordinate axis.

These four orbitals are numbered μ = 1, 2, 3 and 4:

$$\phi_1 = \phi_{2s} \qquad \phi_2 = \phi_{2px} \qquad \phi_3 = \phi_{2py} \quad \text{and} \quad \phi_4 = \phi_{2pz}$$

This concerns the first atom in the molecule, often denoted as A. For the second atom, B, the numbers μ of the atomic orbitals are 5, 6, 7 and 8, and so on.

In the LCAO formalism we assume that every molecular orbital ψ_i can be written as a Linear Combination of Atomic Orbitals

$$\psi_i = \sum_\mu c_{\mu i} \phi_\mu \tag{A.1}$$

where $c_{\mu i}$ are coefficients. Since this is a guess to be used in the calculus of variations, we do not know the coefficients $c_{\mu i}$ but hope to be able to determine them.

All this can be illustrated for the molecular orbital in benzene at -17 eV which was discussed above. We see how the molecular orbital ψ_i is formed from atomic orbitals on the carbons, which all are directed outwards.

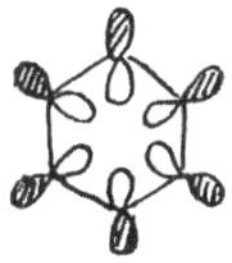

Some of the $2p$ orbitals are not directed along a coordinate axis. Such an atomic orbital must then be formed from a superposition of $2px$ and $2py$, which is easy, since the orbitals can be added like vectors.

A.3. The normalization and orthogonality of orbitals

All orbitals are assumed to be normalized, i.e.

$$\int \psi_i^*(1)\ \psi_i(1)\ d\tau_1 = 1$$

and

$$\int \phi_\mu^*(1)\ \phi_\mu(1)\ d\tau_1 = 1$$

where the parenthesis denotes one of the electrons.

All molecular orbitals must be orthogonal to each other

$$\int \psi_i^*(1)\ \psi_j(1)\ d\tau_1 = \delta_{ij} \tag{A.2}$$

where δ_{ij} is the Kronecker delta, which is = 1 for $i = j$ and = 0 for $i \neq j$.

For the atomic orbitals this is true only if μ and ν denote atomic orbitals on the same atom

$$\int \phi^*_{\mu_A}(1)\ \phi_{\nu_A}(1)\ d\tau_1 = \delta_{\mu\nu} \qquad (A.3)$$

but when μ_A and ν_B are on different atoms the integral is called overlap integral $S_{\mu\nu}$

$$S_{\mu\nu} = \int \phi^*_{\mu}(1)\ \phi_{\nu}(1)\ d\tau_1 \qquad (A.4)$$

The overlap integral takes the following values:

$S_{\mu\mu} = 1$ since ϕ_μ normalized

$S_{\mu_A\nu_A} = 0$ since ϕ_{μ_A} and ϕ_{ν_A} are orthogonal $(\mu \neq \nu)$

$S_{\mu_A\nu_B}$ = usually between -0.5 and +0.5 :

ETHYLENE OVERLAP MATRIX

		C1 1	C1 2	C1 3	C1 4	C2 5	C2 6	C2 7	C2 8
1	S	1.000000	0.0	0.0	0.0	0.160110	0.335444	0.0	0.0
2 C1	X	0.0	1.000000	0.0	0.0	-0.335444	-0.332302	0.0	0.0
3	Y	0.0	0.0	1.000000	0.0	0.0	0.0	0.213173	0.0
4	Z	0.0	0.0	0.0	1.000000	0.0	0.0	0.0	0.213173
5	S	0.160110	-0.335444	0.0	0.0	1.000000	0.0	0.0	0.0
6 C2	X	0.335444	-0.332302	0.0	0.0	0.0	1.000000	0.0	0.0
7	Y	0.0	0.0	0.213173	0.0	0.0	0.0	1.000000	0.0
8	Z	0.0	0.0	0.0	0.213173	0.0	0.0	0.0	1.000000
9	H3	0.379850	0.228471	0.373789	0.0	0.066128	0.119622	0.058238	0.0
10	H4	0.379850	0.228471	-0.373789	0.0	0.066128	0.119622	-0.058238	0.0
11	H5	0.066128	-0.119622	0.058238	0.0	0.379850	-0.228471	0.373789	0.0
12	H6	0.066128	-0.119622	-0.058238	0.0	0.379850	-0.228471	-0.373789	0.0

		H3 9	H4 10	H5 11	H6 12
1	S	0.379850	0.379850	0.066128	0.066128
2 C1	X	0.228471	0.228471	-0.119622	-0.119622
3	Y	0.373789	-0.373789	0.058238	-0.058238
4	Z	0.0	0.0	0.0	0.0
5	S	0.066128	0.066128	0.379850	0.379850
6 C2	X	0.119622	0.119622	-0.228471	-0.228471
7	Y	0.058238	-0.058238	0.373789	-0.373789
8	Z	0.0	0.0	0.0	0.0
9	H3	1.000000	0.208369	0.088150	0.034612
10	H4	0.208369	1.000000	0.034612	0.088150
11	H5	0.088150	0.034612	1.000000	0.208369
12	H6	0.034612	0.088150	0.208369	1.000000

It is interesting to introduce the LCAO expansion into the orthonormality expression for ψ_i . We obtain

$$\int \psi^*_i(1)\ \psi_j(1)\ d\tau_1 = \int \sum_\mu c^*_{\mu i}\ \phi^*_\mu \cdot \sum_\nu c_{\nu j}\ \phi_\nu \cdot d\tau_1 =$$

$$= \sum_{\mu\nu} c^*_{\mu i}\ c_{\nu j}\ S_{\mu\nu} = \delta_{ij} \qquad (A.5)$$

A.4. How to interpret the print-out from a quantum-chemical calculation

The central part of the print-out from a quantum-chemical calculation presents the eigenvalues and the eigenvectors from the solution of the Schrödinger equation. We will illustrate this with ethylene as an example since benzene, which was the example above, is too large.

We must first give the coordinates of the different atoms in the ethylene molecule and can then make a drawing of the molecule.

ETHYLENE PLANAR

COORDINATES IN ANGSTROM UNITS

	ATOM	X	Y	Z
1	C	0.6680	0.0	0.0
2	C	-0.6680	0.0	0.0
3	H	1.2340	0.9260	0.0
4	H	1.2340	-0.9260	0.0
5	H	-1.2340	0.9260	0.0
6	H	-1.2340	-0.9260	0.0

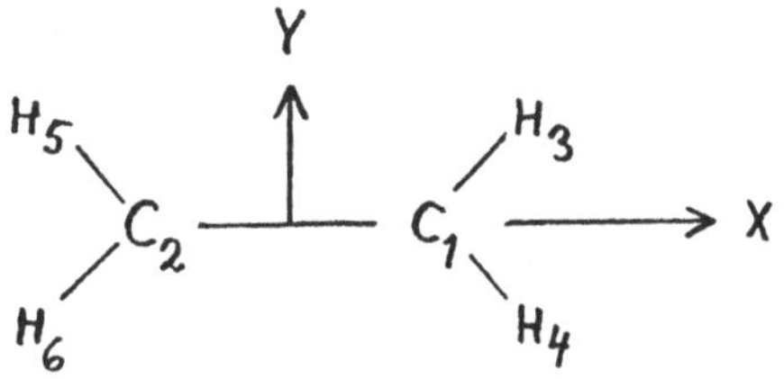

The eigenvalues and the eigenvectors are then printed as shown in the table below.

At the top the numbers i of the orbitals are given. Below them the eigenvalues ε_i are shown. Due to the special type of calculation, used here, the eigenvalues in the table mean the ionization energies of the electrons in the different orbitals. This is, however, true only up to orbital 6, since only the first six orbitals are filled with electrons (two in each orbital). The orbitals 7,8 --- 12 are empty and are called "excited".

Below the eigenvalues ε_i the eigenvectors are given. They give the coefficients $c_{\mu i}$ for each orbital. The indices μ are given to the left. For the first carbon atom C 1 the indices μ = 1, 2, 3 and 4 mean 2s, 2px , 2py and 2pz and so on.

Using the coefficients $c_{\mu i}$ the orbitals can be plotted, which is done below the table. For each orbital we begin by drawing the molecule, correctly positioned in the coordinate system. We include only those atomic orbitals for which $c_{\mu i}$ is large.

$\mu \downarrow$	$i \rightarrow$	1	2	3	4	5	6	7	8	9	10	11	12
	$\varepsilon_i \rightarrow$	-24.292	-19.848	-16.256	-14.883	-13.173	-10.538	-4.269	15.420	19.620	25.496	29.969	36.434
1	C 1	0.466	0.334	-0.000	0.081	0.000	0.000	-0.000	0.676	0.835	0.000	0.003	0.000
2	C 1	-0.055	0.213	0.000	-0.522	0.000	-0.000	-0.000	0.413	-0.381	-0.000	1.004	0.000
3	C 1	-0.000	0.000	-0.420	-0.000	-0.482	0.000	-0.000	-0.000	-0.000	0.687	0.000	-0.857
4	C 1	-0.000	-0.000	-0.000	0.000	-0.000	-0.642	-0.797	0.000	0.000	-0.000	0.000	-0.000
5	C 2	0.466	-0.334	-0.000	0.081	-0.000	0.000	-0.000	0.676	-0.835	-0.000	-0.003	-0.000
6	C 2	0.055	0.213	0.000	0.522	-0.000	0.000	0.000	-0.413	-0.381	-0.000	1.004	0.000
7	C 2	0.000	0.0	-0.420	0.000	0.482	-0.000	-0.000	0.000	-0.000	0.687	0.000	0.857
8	C 2	0.000	0.000	-0.000	0.000	-0.000	-0.642	0.797	-0.000	-0.000	-0.000	0.000	-0.000
9	H 3	0.177	0.257	-0.249	-0.180	-0.301	0.000	-0.000	-0.464	-0.288	-0.635	-0.476	0.652
10	H 4	0.177	0.257	0.249	-0.180	0.301	-0.000	-0.000	-0.464	-0.288	0.635	-0.476	-0.652
11	H 5	0.177	-0.257	-0.249	-0.180	0.301	-0.000	0.000	-0.464	0.288	-0.635	0.476	-0.652
12	H 6	0.177	-0.257	0.249	-0.180	-0.301	0.000	-0.000	-0.464	0.288	0.635	0.476	0.652

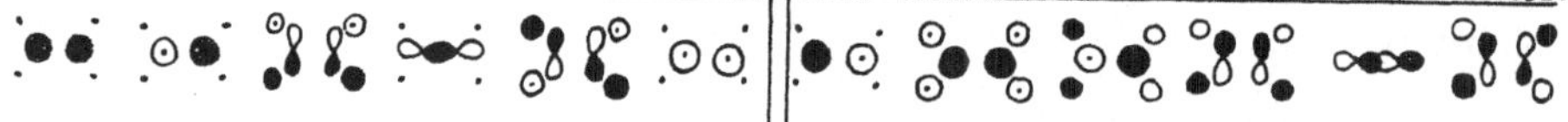

s+s	s−s	π_{CH}	σ	π^*_{CH}	π	π^*	σ^*_{CH}	σ^*_{CH}	σ^*_{CH}	σ^*_{CC}	σ^*_{CH}
a_g	b_{1u}	b_{2u}	a_g	b_{3g}	b_{3u}	b_{2g}	a_g	b_{1u}	b_{2u}	b_{1u}	b_{3g}

It is generally better to omit atomic orbitals than to include too many. The $2p$ orbitals are directed along the positive coordinate axis for positive $c_{\mu i}$.

The pictures of the six filled orbitals give a qualitative explanation of the chemical bonding. We see directly that orbital 1 is bonding between the carbon atoms but orbital 2 is antibonding. The two orbitals together are therefore non-bonding. The same is true for orbitals 3 and 5. Finally, the orbitals 4 and 6 are bonding, which explains the double bond in ethylene.

For a free atom the situation is simple. The eigenvalues and eigenvectors are:

$\mu \downarrow$	$i \rightarrow$	1	2	3	4
	$\varepsilon_i \rightarrow$	ε_1	ε_2	ε_3	ε_4
1		1	0	0	0
2		0	1	0	0
3		0	0	1	0
4		0	0	0	1

A.5. The charge in $\mu\nu$

We have already mentioned that a molecular orbital ψ_i generally accomodates 2 electrons, but that a spinorbital ψ_i^α accomodates only 1 electron. We will denote the number of electrons as the electronic charge q_i in the molecular orbital ψ_i.

The charge in the volume element $d\tau$ is then

$$dq_i = q_i \cdot \psi_i^* \psi_i \, d\tau$$

since the molecular orbitals are normalized. Integration over the whole space gives

$$q_i = q_i \int \psi_i^* \psi_i \, d\tau = q_i \cdot \sum_{\mu\nu} c_{\mu i}^* c_{\nu i} \cdot \int \phi_\mu^* \phi_\nu \, d\tau =$$

$$= q_i \sum_{\mu\nu} c_{\mu i}^* c_{\nu i} S_{\mu\nu}$$

If we partition the expression above, we find that

$$q_i c_{\mu i}^* c_{\nu i} S_{\mu\nu}$$

can be interpreted as the charge in the overlap region between ϕ_μ and ϕ_ν, which is derived from the charge in the molecular orbital ψ_i.

But there are also other molecular orbitals which can contribute to the charge in this region, and therefore the total charge in the region $\mu\nu$ is

$$q_{\mu\nu} = \sum_i q_i c_{\mu i}^* c_{\nu i} S_{\mu\nu} \qquad \text{(A.6)}$$

We sum here over all molecular orbitals ψ_i. Since $q_i = 0$ for the unoccupied orbitals, this means that we sum over all occupied orbitals.

In the region $\nu\mu$ there is another charge $q_{\nu\mu}$ of the same size since $S_{\mu\nu} = S_{\nu\mu}$.

The density matrix element $P_{\mu\nu}$ will now be introduced

$$P_{\mu\nu} = \sum_i q_i c^*_{\mu i} c_{\nu i} \tag{A.7}$$

The charge in $\mu\nu$ is then simplified to

$$q_{\mu\nu} = P_{\mu\nu} S_{\mu\nu} \tag{A.8}$$

and the total number of electrons in the molecule will be given by

$$\sum_{\mu\nu} q_{\mu\nu} = \sum_{\mu\nu} P_{\mu\nu} S_{\mu\nu} \tag{A.9}$$

As an example we will calculate P_{15} for ethylene. P_{15} means $P_{2s_A 2s_B}$ where A and B are the two carbon atoms which previously were denoted $C\,1$ and $C\,2$. We use the print-out in Sec. A.4. and obtain

$$P_{\mu\nu} = 2 \cdot \sum_i c_{\mu i} c_{\nu i}$$ summing over six occupied orbitals

$$P_{15} = 2 \cdot \left[0.466 \cdot 0.466 + 0.334 \cdot (-0.334) + 0 \cdot 0 + 0.081 \cdot 0.081 + 0 \cdot 0 + 0 \cdot 0\right] = 0.2243$$

The P-matrix is shown below. It differs slightly from our calculation since we used only 3 decimals but the computer has access to 7 decimals.

ETHYLENE PLANAR
P-MATRIX

			C1				C2			
			1	2	3	4	5	6	7	8
1	C1	S	0.643355	0.006454	0.0	0.0	0.215235	0.266282	0.0	0.0
2		X	0.006454	0.614597	0.0	0.000048	-0.266281	-0.440852	0.0	-0.000035
3		Y	0.0	0.0	0.783836	0.0	0.0	0.0	-0.106265	0.0
4		Z	0.0	0.000048	0.0	0.789934	0.0	-0.000037	0.000024	0.789938
5	C2	S	0.215235	-0.266281	0.0	0.0	0.643357	-0.006452	0.0	0.0
6		X	0.266282	-0.440852	0.0	-0.000037	-0.006452	0.614600	0.0	0.000051
7		Y	0.0	0.0	-0.106265	0.000024	0.0	0.0	0.783839	-0.000027
8		Z	0.0	-0.000035	0.0	0.789938	0.0	0.000051	-0.000027	0.789942
9		H3	0.295271	0.266562	0.478867	0.0	-0.034547	-0.056451	-0.077725	0.0
10		H4	0.295270	0.266561	-0.478867	0.0	-0.034545	-0.056452	0.077725	0.0
11		H5	-0.034545	0.056451	-0.077724	0.000035	0.295272	-0.266561	0.478866	-0.000027
12		H6	-0.034547	0.056451	0.077725	0.0	0.295269	-0.266560	-0.478867	0.0

			H 3	H 4	H 5	H 6
			9	10	11	12
1	C1	S	0.295271	0.295270	-0.034545	-0.034547
2		X	0.266562	0.266561	0.056451	0.056451
3		Y	0.478867	-0.478867	-0.077724	0.077725
4		Z	0.0	0.0	0.000035	0.0
5	C2	S	-0.034547	-0.034545	0.295272	0.295269
6		X	-0.056451	-0.056452	-0.266561	-0.266560
7		Y	-0.077725	0.077725	0.478866	-0.478867
8		Z	0.0	0.0	-0.000027	0.0
9		H3	0.542079	-0.043453	-0.060050	0.050506
10		H4	-0.043453	0.542080	0.050505	-0.060049
11		H5	-0.060050	0.050505	0.542077	-0.043452
12		H6	0.050506	-0.060049	-0.043452	0.542076

For a free atom we take the coefficients $c_{\mu i}$ from Sec.A.4. If, for instance, we study an atom with 2 electrons in the 2s orbital and 2 electrons in 2px and no electrons in the other orbitals, we have

ATOM WITH q_S = 2 AND q_X = 2 AND q_Y = q_Z = 0
P-MATRIX

		1	2	3	4
1	S	2.000	0.0	0.0	0.0
2	X	0.0	2.000	0.0	0.0
3	Y	0.0	0.0	0.0	0.0
4	Z	0.0	0.0	0.0	0.0

A.6. The charges on atoms and in bonds

In a free atom, e.g. a carbon atom, the electronic charge in the $2s$ orbital is given by

$$q_{2s2s} = P_{2s2s} \cdot S_{2s2s} = P_{2s2s} = 2$$

if $q_i = 2$ for this orbital.

If now this carbon atom, called A, becomes part of a molecule, a lot of the electronic charge will be localized in the bonds, described by $\mu_A \nu_B$. The charge in the $2s$ orbital is therefore smaller now.

It is in this situation not obvious how to calculate or even to define the total charge of the carbon atom (= nuclear charge Z_A - electronic charges), since it is not clear how to handle the charges in the bonds.

Mulliken proposed 1955 that one simply partitions the "overlap population" (the electronic charge in the bonds)

$P_{\mu_A \nu_B} S_{\mu_A \nu_B} + P_{\nu_B \mu_A} S_{\nu_B \mu_A}$ (where the two terms are of equal size)

into two equal parts, and that one assumes that one part belongs to atom A and the other to atom B, or, to be more precise, one part belongs to ϕ_μ on A and the other to ϕ_ν on B.

The total electronic charge (total gross population) belonging to ϕ_μ on A is then

$$N_\mu = P_{\mu\mu} + \sum_{B \neq A} \sum_{\nu_B} \tfrac{1}{2} \left(P_{\mu\nu} S_{\mu\nu} + P_{\nu\mu} S_{\nu\mu} \right) =$$

$$= \sum_{\nu} \tfrac{1}{2} \left(P_{\mu\nu} S_{\mu\nu} + P_{\nu\mu} S_{\nu\mu} \right) \qquad (A.10)$$

The total electronic charge (total gross population) on atom A is then

$$P_{AA} = \sum_{\mu}^{A} N_{\mu} = \sum_{\mu}^{A} \sum_{\nu} \tfrac{1}{2} \left(P_{\mu\nu} S_{\mu\nu} + P_{\nu\mu} S_{\nu\mu} \right) \qquad (A.11)$$

and the resulting charge on atom A (gross atomic charge) is finally

$$Q_A = Z_A - P_{AA} \qquad (A.12)$$

A.7. The idempotency of density matrices

The density matrix elements $P_{\mu\nu}$ constitute the density matrix $\mathbb{P}$. If all $S_{\mu\nu}$ are = 0 for $\mu \neq \nu$, this matrix has an interesting property

$$2\,\mathbb{P} = \mathbb{P} \cdot \mathbb{P}$$

The matrix is said to be idempotent.

For proof we study the $\mu\nu$ element

$$2\,P_{\mu\nu} = \left(\mathbb{P} \cdot \mathbb{P} \right)_{\mu\nu}$$

Using the rules for matrix multiplication we find

$$\left(\mathbb{P} \cdot \mathbb{P} \right)_{\mu\nu} = \sum_{\lambda} P_{\mu\lambda} \cdot P_{\lambda\nu} = \sum_{\lambda} \sum_{i} \sum_{j} 2 c_{\mu i}^{*} c_{\lambda i} \cdot 2 c_{\lambda j}^{*} c_{\nu j} =$$

$$= 4 \sum_{i} \sum_{j} c_{\mu i}^{*} c_{\nu j} \cdot \underbrace{\sum_{\lambda} c_{\lambda j}^{*} c_{\lambda i}}_{\delta_{ij}} = 2\,P_{\mu\nu} \qquad (A.13)$$

if equ. ($A.5$) is used.

In full overlap basis we have

$$2\,\mathbb{P} = \mathbb{P} S \mathbb{P}$$

which means

$$\left(\mathbb{P} S \mathbb{P} \right)_{\mu\nu} = \sum_{\lambda\sigma} P_{\mu\lambda} S_{\lambda\sigma} P_{\sigma\nu} = \sum_{\lambda\sigma} \sum_{i} \sum_{j} 2 c_{\mu i}^{*} c_{\lambda i} \cdot S_{\lambda\sigma} \cdot 2 c_{\sigma j}^{*} c_{\nu j} =$$

$$= 4 \sum_{i} \sum_{j} c_{\mu i}^{*} c_{\nu j} \cdot \underbrace{\sum_{\lambda\sigma} c_{\sigma j}^{*} c_{\lambda i} S_{\lambda\sigma}}_{\delta_{ij}} = 2\,P_{\mu\nu} \qquad (A.14)$$

again by use of equ. ($A.5$).

B. Hartree-Fock total energy

Our deduction of the expression for the total energy in Hartree-Fock theory will be very brief. For details we make reference to the book: "Approximate Molecular Orbital Theory" by Pople and Beveridge [1].

B.1. The Hamilton operator

In the hamiltonian operator for a molecule there are terms corresponding to the different energies in the molecule, namely
kinetic energy of the electrons,
attraction between the electrons and the nuclei,
repulsion between the electrons,
repulsion between the nuclei and
kinetic energy of the nuclei.

Due to the large masses of the nuclei we will completely neglect the kinetic energy of the nuclei. We will thus study molecules with a fixed geometry. The total energy of such a molecule is then the sum of the first four terms above.

The fourth term, the repulsion between the nuclei, does not constitute any problem, since it is

$$\text{repulsion between nuclei} = \sum_{A>B} Z_A Z_B R_{AB}^{-1} \tag{B.1}$$

supposed that atomic units (see below) are used.

The electronic energy is thus the sum of the first three terms above.

It is convenient to introduce atomic units. The atomic unit of length is the radius of the first orbit in the Bohr theory of the hydrogen atom. The atomic unit of charge is the charge of the electron. The atomic unit of energy is twice the ionization energy of the hydrogen atom, i.e. 27.2 eV. The atomic unit of mass is the electron mass.

Using atomic units the electronic hamiltonian operator is

$$\mathcal{H} = -\frac{1}{2}\sum_p \nabla_p^2 - \sum_B \sum_p Z_B r_{Bp}^{-1} + \sum_{p<q} r_{pq}^{-1} \tag{B.2}$$

where p and q denote electrons, B nuclei and

$$\nabla^2 = \frac{\partial^2}{\partial x^2} + \frac{\partial^2}{\partial y^2} + \frac{\partial^2}{\partial z^2}$$

B.2. The wavefunctions in Hartree-Fock theory

We have assumed above that the total wavefunction of a molecule can be written as a Hartree product

$$\Psi = \psi_1(1) \cdot \psi_2(2) \cdot \psi_3(3) \cdot \psi_4(4) \cdot\cdot\cdot\cdot\cdot\cdot$$

Since there are two electrons in each orbital with different spins, α and β, respectively, it is sometimes better to write the Hartree product as

$$\Psi = \psi_1(1)\alpha(1) \cdot \psi_1(2)\beta(2) \cdot \psi_2(3)\alpha(3) \cdot \psi_2(4)\beta(4) \cdot\cdot\cdot\cdot$$

where $\psi_i(n)\,\alpha(n)$ is called spin orbital.

According to Pauli's exclusion principle the total wavefunction must, however, be antisymmetric with respect to exchange of two electrons. It is therefore necessary to form a Slater determinant from the Hartree product according to well-known rules which are discussed at length by Pople and Beveridge [1].

B.3. The total energy in Hartree-Fock theory

In quantum mechanics the electronic energy is calculated from the electronic hamiltonian operator as

$$E = \int \Psi^* \mathcal{H} \Psi \, d\tau \qquad (B.3)$$

Since we have given $\mathcal{H}$ above and pointed out that Ψ is a determinant, it is a straight-forward but lengthly procedure to calculate E. This is done by Pople and Beveridge (ref. [1], page 34 and 35) and the result is given in equ. (2.22) there.

$$E = 2\sum_j H_{jj} + \sum_j \sum_k \left(2J_{jk} - K_{jk}\right) \qquad (B.4)$$

Here, H_{jj} describes the kinetic energy of one electron and its attraction towards the nuclei

$$H_{jj} = \int \psi_j^* \left[-\tfrac{1}{2}\nabla^2 - \sum_B Z_B r_B^{-1}\right] \psi_j \, d\tau \qquad (B.5)$$

J_{jk} describes the repulsion between two electrons, which are denoted as (1) and (2)

$$J_{jk} = \int \psi_j^*(1)\ \psi_j(1)\ \frac{1}{r_{12}}\ \psi_k^*(2)\ \psi_k(2)\ d\tau_1 d\tau_2 = \qquad (B.6)$$

$$= (jj\,|\,kk) \qquad (B.7)$$

and K_{jk} is the corresponding exchange integral

$$K_{jk} = \int \psi_j^*(1)\ \psi_k(1)\ \frac{1}{r_{12}}\ \psi_j^*(2)\ \psi_k(2)\ d\tau_1 d\tau_2 = \qquad (B.8)$$

$$= (jk\,|\,jk) \qquad (B.9)$$

It is assumed in eq.(B.4) that there are exactly two electrons in each molecular orbital ψ_i. This is natural since the proof for eq.(B.4) depends upon permutation of the electrons.

If we assume that we have q_i^{α} electrons in the spinorbital ψ_i^{α} eq.(B.4) will be changed:

$$E = \sum_j \left[q_j^{\alpha} H_{jj}^{\alpha} + q_j^{\beta} H_{jj}^{\beta} \right] + \qquad (B.10)$$

$$+ \frac{1}{2} \sum_{jk} q_j q_k J_{jk} - \frac{1}{2} \sum_{jk} q_j^{\alpha} q_k^{\alpha} K_{jk}^{\alpha} - \frac{1}{2} \sum_{jk} q_j^{\beta} q_k^{\beta} K_{jk}^{\beta} \qquad (B.11)$$

In Sec.G. we will have use for the derivatives of E, which can be obtained easily

$$\varepsilon_j^{HF} = \frac{\partial E}{\partial q_j} = H_{jj} + \sum_k \left(2 J_{jk} - K_{jk} \right) \qquad (B.12)$$

$$\varepsilon_j^{HF^{\alpha}} = \frac{\partial E}{\partial q_j^{\alpha}} = H_{jj}^{\alpha} + \sum_k q_k J_{jk} - \sum_k q_k^{\alpha} K_{jk}^{\alpha} \qquad (B.13)$$

B.4. The total energy in LCAO Hartree-Fock theory

We introduce now the LCAO expression

$$\psi_i = \sum_{\mu} c_{\mu i} \, \phi_\mu \tag{B.14}$$

into eq. (B.4). Further, we add the nuclear repulsion to the electronic energy and obtain then the LCAO Hartree-Fock total energy as

$$E = -\frac{1}{2} \sum_{\mu\nu} P_{\mu\nu} \int \phi_\mu^* \nabla^2 \phi_\nu \, d\tau - \tag{B.15}$$

$$- \sum_{\mu\nu} P_{\mu\nu} \int \phi_\mu^* \phi_\nu \sum_B Z_B \, r_B^{-1} \, d\tau + \tag{B.16}$$

$$+ \frac{1}{2} \sum_{\mu\nu\lambda\sigma} P_{\mu\nu} P_{\lambda\sigma} \left[(\mu\nu|\lambda\sigma) - \frac{1}{2} (\mu\sigma|\lambda\nu) \right] + \tag{B.17}$$

$$+ \sum_{A>B} Z_A Z_B R_{AB}^{-1} \tag{B.18}$$

where

$$(\mu\nu|\lambda\sigma) = \int \phi_\mu^*(1) \, \phi_\nu(1) \cdot r_{12}^{-1} \cdot \phi_\lambda^*(2) \, \phi_\sigma(2) \cdot d\tau_1 \, d\tau_2 \tag{B.19}$$

It is seen directly that the different terms mean in order: kinetic energy of the electrons, attraction between nuclei and electrons, repulsions between electrons and repulsions between nuclei.

By analogy to eq. (B.5) we can write eqs. (B.15-17) as

$$E = \sum_{\mu\nu} P_{\mu\nu} H_{\mu\nu} + \frac{1}{2} \sum_{\mu\nu\lambda\sigma} P_{\mu\nu} P_{\lambda\sigma} \left[(\mu\nu|\lambda\sigma) - \frac{1}{2} (\mu\sigma|\lambda\nu) \right] \tag{B.20}$$

B.5. Self-repulsion

The electron-electron repulsion term (the first term in eq. (B.17)) is interesting. The summation over μ, ν, λ and σ means that also repulsion between $\mu\mu$ and $\mu\mu$ is included. Since there are two electrons in each orbital, half of this repulsion is "self-repulsion". But when $\mu = \nu = \lambda = \sigma$ the exchange term $-\frac{1}{2}(\mu\sigma|\lambda\nu)$, which is usually small, becomes $-\frac{1}{2}(\mu\mu|\mu\mu)$ and compensates the self-repulsion.

Slater [2] has studied the interpretation of the exchange term in Hartree-Fock theory.

Table B.5.

Atom	Total energy (eV)	Exchange term (eV)	Compensation of self-repulsion (eV)	Interelectronic exchange (eV)
He	77.8	27.9	27.9	0.0
Ne	3496.4	329.4	269.2	60.2
Ar	14329.4	821.0	617.6	203.4
Kr	74855.9	2560.9	1653.0	907.9

We observe that most of the exchange term is present to take care of the self-repulsion.

Compensation of self-repulsion

The compensation of self-repulsion takes place only for occupied orbitals. To prove this we have to make use of some results, deduced below in Sec.G.2. and Sec.G.5.

The Fock matrix element in Hartree-Fock is obtained from eq.(G.14) as

$$F_{\mu\nu} = H_{\mu\nu} + \sum_{\lambda\sigma} P_{\lambda\sigma} \left[(\mu\nu|\lambda\sigma) - \tfrac{1}{2} (\mu\sigma|\lambda\nu) \right] \qquad (B.21)$$

and the orbital energy ε_i is (see Sec. G.6.)

$$\varepsilon_i = \sum_{\mu\nu} c^*_{\mu i} F_{\mu\nu} c_{\nu i} \quad = H_{ii} + \qquad (B.22)$$

$$+ 2\sum_{j\mu\nu\lambda\sigma} c^*_{\mu i} c_{\nu i} c^*_{\lambda j} c_{\sigma j} (\mu\nu|\lambda\sigma) - \sum_{j\mu\nu\lambda\sigma} c^*_{\mu i} c_{\nu i} c^*_{\lambda j} c_{\sigma j} (\mu\sigma|\lambda\nu) \qquad (B.23)$$

The first term in (B.23) means repulsions between electron i and all electrons j . Here, j denotes an occupied orbital. j can therefore take n different values and the number of repulsions is $2n$. This includes self-repulsion. The self-repulsion is, however, compensated by the second sum, since for $j = i$ the two sums are equal. This is best seen after some change

$$\varepsilon_i = H_{ii} + \qquad (B.24)$$

$$+ \sum_{\mu\nu\lambda\sigma} (\mu\nu|\lambda\sigma) \sum_{j}^{occ} \left(2 c^*_{\mu i} c_{\nu i} c^*_{\lambda j} c_{\sigma j} - c^*_{\mu i} c_{\sigma i} c^*_{\lambda j} c_{\nu j} \right) \qquad (B.25)$$

The last term can be described as a repulsion integral multiplied by a coefficient. The coefficient has n terms which gives $2n$ repulsions. However, for $j = i$ the products in the parenthesis are equal which reduces this number to $(2n-1)$ repulsions. The self-repulsion has thus been compensated.

This is, however, true only if i is an occupied orbital. For an unoccupied orbital with $i \equiv a$ there is no j for which the two products are identical. We have therefore $2n$ repulsions between electron a and the other electrons. An electron in a moves under influence from the charges in the neutral molecule and forms with the molecule a negative ion. In Hartree-Fock theory the occupied and the unoccupied orbitals are not similar.

The unoccupied orbitals are therefore denoted as "virtual" in Hartree-Fock theory.

The complicated action of the exchange integral in its compensation of self-repulsion means a difficulty in the Hartree-Fock method. This was recognized especially by Slater [3] and resulted in the $X\alpha$ method.

In the $X\alpha$ method the exchange integral is replaced by an expression which depends on the electron density in the local point which is studied. The $X\alpha$ term eliminates therefore directly the self-repulsion, and since this is done in the same way for all orbitals, occupied and unoccupied (excited) orbitals are similar in the $X\alpha$ method.

It will be shown below that in HAM the self-repulsion is compensated by a term "-1" and that therefore occupied and unoccupied (excited) orbitals are similar also in the HAM model.

References:

1. J.A. Pople and D.L. Beveridge, Approximate Molecular Orbital Theory, McGraw-Hill, New York 1970.
2. J.C. Slater, Int. J. Quantum Chem. 4, 3 (1971).
3. J.C. Slater, The Self-consistent Field for Molecules and Solids, McGraw-Hill, New York 1974.

C. Density functional theory

C.1. Correlation

The orbital approximation means that an electron is described by a "cloud", called orbital. In this "cloud" the probability that the electron is just there, depends upon the coordinates x, y and z.

Another electron is described by a similar cloud.

It may then happen that the two electrons have the same coordinates in their clouds, which would result in a large repulsion energy. Since such a situation is comparatively rare, its contribution to the average energy is not large.

However, in nature such a situation never occurs, since the two electrons avoid each other due to their charge. If they approach each other with a certain velocity, they are slowed down by the repulsion and they will never come nearer each other than a certain distance, which depends upon the charges and the relative velocity.

If there are many electrons in the molecule, these repulsions will occur between every pair of electrons. The movements of the electrons is thus extremely complicated, and it is a hopeless task to get a picture of the complete situation.

In Sec.B.2. we discussed the total wavefunction Ψ in Hartree-Fock theory. Here the electrons are supposed to move independently of each other.

The corresponding total wavefunction, when we take the correlation into account, will be denoted Ψ_{exact}. It is very complicated.

C.2. Correlation energy

We showed in Chap.B. how the total energy E can be calculated if the correlation is neglected. We understand now, that a large number of electron-electron repulsions at small distances were incorrectly included in that calculation.

In nature such repulsions are absent, and therefore the true energy E_{exact} will be slightly lower that the calculated energy.

The energy difference is called "correlation energy" E_c.

The correlation energy depends according to this description only upon the electron-electron repulsion. We must add, that there is a contribution also from the kinetic energy.

C.3. Exact energy expression [1]

Even if we do not know Ψ_{exact} we can easily write down the corresponding exact total energy expression.

Consider in the molecule the volume element $d\tau_1$ with coordinates x_1, y_1 and z_1 which below are abbreviated as r_1. The probability that the first electron is in $d\tau_1$ is given by $d\tau_1$ multiplied by

$$\int |\Psi_{exact}|^2 \, d\tau_2 \, d\tau_3 \, d\tau_4 \, \cdots \, d\tau_N \tag{C.1}$$

This integral will be denoted as the probability density.

The probability density that any of the electrons is in $d\tau_1$ is then

$$\varrho(r_1) = N \cdot \int |\Psi_{exact}|^2 \, d\tau_2 \, d\tau_3 \, d\tau_4 \, \cdots \, d\tau_N \tag{C.2}$$

Evidently $\varrho \, d\tau_1$ measures the electric charge within $d\tau_1$ and we find then

$$\text{charge} = \int \varrho(r_1) \, d\tau_1 = N \tag{C.3}$$

since Ψ_{exact} is normalized. In these expressions $N = 2n$ measures the number of electrons in the molecule.

Consider now in the molecule the volume elements $d\tau_1$ and $d\tau_2$. The probability density that the first electron is in $d\tau_1$ and the second in $d\tau_2$ is

$$\int |\Psi_{exact}|^2 \, d\tau_3 \, d\tau_4 \, \cdots \, d\tau_N \tag{C.4}$$

and the probability density that any of the electrons in in $d\tau_1$ and any of the other electrons in $d\tau_2$ is

$$\Gamma(r_1 r_2) = \frac{1}{2} N(N-1) \int |\Psi_{exact}|^2 \, d\tau_3 \, d\tau_4 \, \cdots \, d\tau_N \tag{C.5}$$

Since all movements of the electrons is governed by the electronic hamiltonian operator, eq.(B.2), we find from eq.(B.3)

$$E_{exact} =$$

$$= \text{kinetic} + \text{attraction to nuclei} + \text{electron-electron repulsion} \tag{C.6}$$

$$= \text{kinetic} - \sum_A Z_A \int \frac{\varrho}{r_A} \, d\tau + \int \frac{\Gamma(r_1 r_2)}{r_{12}} \, d\tau_1 \, d\tau_2 \tag{C.7}$$

C.4. Exchange-correlation energy [1,2]

In eq. (C.7) $\Gamma(r_1\, r_2)$ is dependent upon the correlation and is therefore difficult to handle. We replace it therefore with the corresponding uncorrelated expression and add a correction term E_{xc}^{coul}, called exchange-correlation energy. The superscript coul means that only the coulombic part of the energy is considered, not the kinetic part. This gives

$$E_{exact} =$$

$$= \text{kinetic} - \sum_A Z_A \int \frac{\varrho}{r_A}\, d\tau + \frac{1}{2} \iint \frac{\varrho(r_1)\,\varrho(r_2)}{r_{12}}\, d\tau_1\, d\tau_2 + E_{xc}^{coul} \tag{C.8}$$

with

$$E_{xc}^{coul} = \int \frac{\Gamma(r_1, r_2)}{r_{12}}\, d\tau_1\, d\tau_2 - \frac{1}{2} \iint \frac{\varrho(r_1)\,\varrho(r_2)}{r_{12}}\, d\tau_1\, d\tau_2 \tag{C.9}$$

which can be simplified to

$$E_{xc}^{coul} = \frac{1}{2} \int \frac{\varrho(r_1)\,\varrho(r_2)}{r_{12}} \cdot \left[\frac{2\Gamma(r_1, r_2)}{\varrho(r_1)\,\varrho(r_2)} - 1 \right] d\tau_1\, d\tau_2 \tag{C.10}$$

We introduce abbreviations

$$\left[\frac{2\Gamma(r_1, r_2)}{\varrho(r_1)\,\varrho(r_2)} - 1 \right] = \left[g_{xc}^{coul} - 1 \right] = h_{xc}^{coul} \tag{C.11}$$

and obtain the following expression, first given in the general case by Gunnarsson and Lundqvist [3,4]

$$E_{xc}^{coul} = \frac{1}{2} \int \varrho(r_1)\, \frac{h_{xc}^{coul}}{r_{12}}\, \varrho(r_2)\, d\tau_1\, d\tau_2 \tag{C.12}$$

We assume now that g_{xc}^{coul} is a function of $r_{12} = |r_1 - r_2|$, which takes the following values. When r_{12} is large, the correlation can be neglected and we have $g_{xc}^{coul} \approx 1$. When r_{12} is small, the electrons avoid each other and $\Gamma \approx 0$ and $g_{xc}^{coul} \approx 0$.

This shows that around the electron at r_2 we have a hole in the cloud which was originally described as $\varrho(r_1)$.

The function h_{xc}^{coul} is shown schematically below:

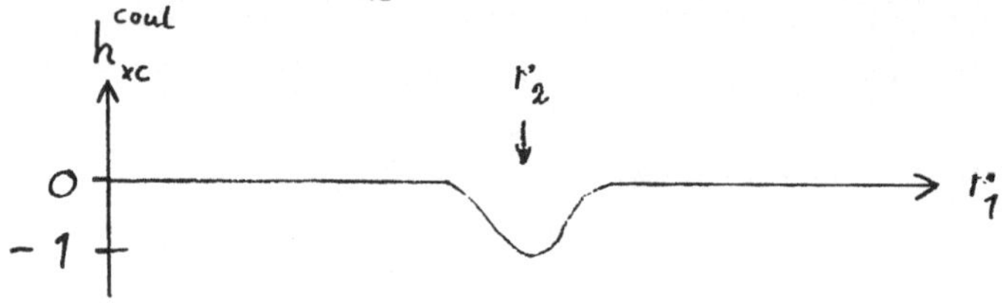

The hole itself has a density

$$\text{hole density} = \varrho(r_1)\cdot\left[g_{xc}^{coul} - 1\right] \tag{C.13}$$

and the total electric charge of the hole is

$$\text{hole charge} = \int \varrho(r_1)\left[g_{xc}^{coul} - 1\right] d\tau_1 = \int\left[\frac{2\Gamma(r_1 r_2)}{\varrho(r_2)} - \varrho(r_1)\right] d\tau_1 =$$

$$= \frac{2}{\varrho(r_2)}\int \Gamma(r_1 r_2)\, d\tau_1 - \int \varrho(r_1)\, d\tau_1 =$$

$$= N-1 - N = -1 \tag{C.14}$$

if eqs.(C.5) and (C.2) are used.

The result, -1, implies the following. In the uncorrelated expression

$$\frac{1}{2}\iint \frac{\varrho(r_1)\,\varrho(r_2)}{r_{12}}\, d\tau_1\, d\tau_2 \tag{C.15}$$

all electrons repel all electrons. Therefore an electron also repels itself (self-repulsion). In eqs.(C.4) and (C.5) the electrons in $d\tau_1$ and $d\tau_2$ are always different (no self-repulsion). The result, -1, means thus elimination of the self-repulsion.

A term, -1, is thus the natural way in quantum chemistry to describe elimination of self-repulsion. Such a term is obtained in the deduction of the HAM method (see eq.(D.23) and ref. [5]).

It is possible to change E_{xc}^{coul} in eq.(C.12) by varying Ψ in eqs.(C.2) and (C.5). Such a variation does not influence the compensation of self-repulsion in eq.(C.14) and changes therefore only the correlation energy E_c. It is evident that the correlation energy E_c has to be described by a formula like eq.(C.12) but with h_c^{coul} instead:

$$E_c^{coul} = \frac{1}{2} \int \varrho(r_1) \frac{h_c^{coul}}{r_{12}} \varrho(r_2) \, d\tau_1 \, d\tau_2 \tag{C.16}$$

We can therefore write E_{xc}^{coul} as a sum of two terms

$$E_{xc}^{coul} = E_x^{coul} + E_c^{coul} \tag{C.17}$$

The exchange energy, E_x^{coul}, takes care of the self-repulsion.

The function h_c^{coul} is shown below:

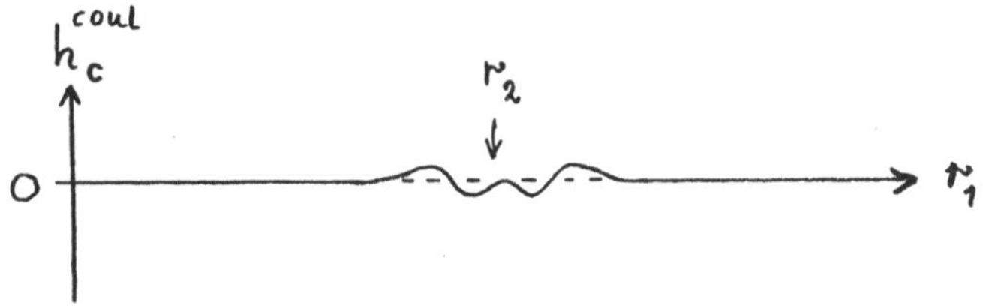

Corresponding to eq.(C.14) we have

$$\text{charge of correlation hole} = \int \varrho(r_1) \, h_c^{coul} \, d\tau_1 = 0 \tag{C.18}$$

The treatment up till here is valid both in Hartree-Fock theory and in density functional theory.

It has been demonstrated by Gunnarsson and Lundqvist [3,4] how this treatment can be extended to include the kinetic energy. Their result is that the total exchange-correlation energy E_{xc} has the same form as eq.(C.12) if the superscript ^{coul} is removed everywhere. For a simplified proof, see Sec.C.11.

C.5. Density functional theory: Kohn-Sham orbitals.

The energy expression, eq.(C.8), is seen to depend on the density ϱ. In a famous paper by Hohenberg and Kohn [6] this was extended, when they showed that the exact total energy of the ground state may be expressed as a functional of the density ϱ (for reviews see [7-10]). Furthermore, they showed that the energy expression is a minimum for correct ϱ.

In Sec.B.2. the Hartree-Fock wavefunction Ψ was described by use of molecular orbitals ψ_j. Such a molecular orbital describes the motion of the electrons in this orbital.

In Sec.C.1. the correlated wavefunction Ψ_{exact} was described as extremely complicated, and the orbital picture had to be abandoned.

In spite of this, Kohn and Sham [11] introduced orbitals ψ_j, which are called "Kohn-Sham" orbitals and which are purely mathematical constructions and do not describe the motion of the electrons in the orbital. Instead, they are defined by

$$\varrho = 2 \cdot \sum_{j}^{n} \psi_j^* \psi_j \tag{C.19}$$

The "Kohn-Sham" orbitals are orthonormal

$$\int \psi_j^* \psi_k \, d\tau = \delta_{jk} \tag{C.20}$$

and the "Kohn-Sham density matrix" is idempotent.

There are two electrons in each orbital as indicated in eq.(C.19).

The orbitals are used to obtain an approximate expression for the kinetic energy in eq.(C.8) above:

$$\text{kinetic energy} = -\sum_j \int \psi_j^* \nabla^2 \psi_j \, d\tau \tag{C.21}$$

The correction, which is necessary, can be included in E_{xc} as shown by Gunnarsson and Lundqvist (see Sec.C.4.).

The orbitals are finally used to study the minimum of E_{exact} at correct ϱ. The derivative of E is taken $= 0$, which gives (for details see page 91 in [8] or page 156 in [9])

$$\left[-\frac{1}{2}\nabla^2 - \sum_A \frac{Z_A}{r_A} + \int \frac{\varrho(r_2)}{r_{12}} d\tau_2 + \frac{\delta E_{xc}}{\delta \varrho} \right] \psi_j = \varepsilon_j \psi_j \tag{C.22}$$

In this equation, the "Kohn-Sham one-particle equation", the eigenvalue ε_j is introduced as a Lagrange multiplier. It is thus in density functional theory a purely mathematical concept without any direct physical meaning.

The Schrödinger-like equation (C.22) cannot be used for direct calculations, since h_{xc} in E_{xc} is generally not known. Instead, LCAO expansion of the problem can be performed, and h_{xc} can be determined indirectly from comparison with experiment.

Density functional theory is thus the natural starting point for the construction of semiempirical theories in quantum chemistry.

C.6. Introducing "Kohn-Sham orbitals" [12]

Using the Kohn-Sham orbitals ψ_j the total energy expression eqs.(C.8), (C.21), (C.17) and (C.16), takes the following form

$$E_{exact} =$$

$$= -\sum_j \int \psi_j^* \nabla^2 \psi_j \, d\tau - 2 \sum_j \sum_A \int \psi_j^* \frac{Z_A}{r_A} \psi_j \, d\tau + \qquad (C.23)$$

$$+ 2 \sum_{jk} \iint \psi_j^*(1) \, \psi_j(1) \cdot \frac{1}{r_{12}} \cdot \psi_k^*(2) \, \psi_k(2) \cdot d\tau_1 \, d\tau_2 + E_x$$

$$+ 2 \sum_{jk} \iint \psi_j^*(1) \, \psi_j(1) \cdot \frac{h_c}{r_{12}} \cdot \psi_k^*(2) \, \psi_k(2) \cdot d\tau_1 \, d\tau_2 \qquad (C.24)$$

If we neglect eq.(C.24) we can write eq.(C.23) as

$$E = 2 \sum_j H_{jj} + 2 \sum_{jk} J_{jk} + E_x \qquad (C.25)$$

which is analogous to the conventional Hartree-Fock formula in eq.(B.4).

The exchange energies E_x are, however, slightly different in eq.(C.23) and in Hartree-Fock [13] . The main part of E_x is in both cases devoted to the compensation of self-repulsion, which in both cases results in a complete compensation. In addition, however, the Hartree-Fock formula includes interelectronic exchange terms (see Table B.5.), which are present although small in the closed-shell molecule. In eq.(C.23) they are absent.

To have a complete analogy between eq. (C.23) and the Hartree-Fock formula it is necessary to introduce correction terms of interelectronic type.

If we denote the integral in eq. (C.24) as $\frac{1}{2}\varepsilon_{jk}$ we obtain the following expression for the correlation energy

$$E_c = \sum_{jk} \varepsilon_{jk} \tag{C.26}$$

This is a very well established formula from CI studies of correlation (see Sec.C.10.). The agreement proves therefore that the density functional treatment is correct.

We will now discuss the properties of the operator h_c in eq. (C.24). It is according to eq. (C.11) a function of r_1 and r_2 but may also be a function of ϱ . We will now try to see to what extent it depends on ϱ .

According to eq. (C.19) ϱ is a sum of terms with indices $j, k, l, m, ..$ If h_c depends on ϱ , the integral in eq. (C.16) will give one pair term depending only on j and k , but also triple terms depending on j , k and l and so on. Instead of eq. (C.26) we obtain

$$E_c = \sum_{jk} \varepsilon_{jk} + \sum_{jkl} \varepsilon_{jkl} + \sum_{jklm} \varepsilon_{jklm} + \cdots \tag{C.27}$$

It is, however, known or at least generally assumed [14 - 16] that the correlation energy of a molecule is very well described by a sum of pair energies only. This indicates that h_c has only a weak dependence of ϱ , which may be neglected in first order.

We will finally mention a possibility to simplify eqs. (C.23) and (C.24).

We observe that the repulsion integral and the correlation energy are very similar. We can therefore add them, obtaining

$$\iint \psi_j^*(1)\, \psi_j(1)\, \frac{1 + h_c}{r_{12}}\, \psi_k^*(r_2)\, \psi_k(r_2)\, d\tau_1\, d\tau_2 \tag{C.28}$$

This simplification was pointed out by Sinanoglu [17] but has never been used in calculations since h_c is generally not known. We will, however, show below that if comparison with experiment is used instead of calculation, this property is of great importance.

C.7. Introducing LCAO [12]

We assume now that the Kohn-Sham orbitals ψ_j can be LCAO expanded, using

$$\psi_j = \sum_{\mu} c_{\mu j} \phi_{\mu} \qquad (C.29)$$

where ϕ_{μ} are atomic orbitals. We obtain then from eq.(C.24)

$$E_c = \frac{1}{2} \sum_{\mu\nu\lambda\sigma} P_{\mu\nu} P_{\lambda\sigma} \iint \phi_{\mu}^*(1)\, \phi_{\nu}(1) \cdot \frac{h_c}{r_{12}} \cdot \phi_{\lambda}^*(2)\, \phi_{\sigma}(2) \cdot d\tau_1\, d\tau_2 \qquad (C.30)$$

which can be written

$$E_c = \frac{1}{2} \sum_{\mu\nu\lambda\sigma} P_{\mu\nu} P_{\lambda\sigma}\, \varepsilon_{\mu\nu\lambda\sigma}^{corr} \qquad (C.31)$$

with

$$P_{\mu\nu} = 2 \cdot \sum_{j} c_{\mu j}^* c_{\nu j} \qquad (C.32)$$

The integral in eq.(C.30) shows how the density $\phi_{\mu}^*(1)\, \phi_{\nu}(1)$ around r_2 is displaced from the region with very small r_{12} towards regions with larger r_{12} (in accordance with the picture of h_c in Sec.C.4.). It depends only upon the atomic orbitals, which have been used in the LCAO expansions.

We obtain now from eqs.(C.23) and (C.24) together with the nuclear repulsion

$$E_{exact} =$$

$$= -\frac{1}{2} \sum_{\mu\nu} P_{\mu\nu} \int \phi_{\mu}^* \nabla^2 \phi_{\nu}\, d\tau - \qquad (C.33)$$

$$- \sum_{\mu\nu} P_{\mu\nu} \int \phi_{\mu}^* \phi_{\nu} \sum_{A} \frac{Z_A}{r_A}\, d\tau + \qquad (C.34)$$

$$+ \frac{1}{2} \sum_{\mu\nu\lambda\sigma} P_{\mu\nu} P_{\lambda\sigma} \left[(\mu\nu|\lambda\sigma) - \frac{1}{2} (\mu\sigma|\lambda\nu) + \varepsilon_{\mu\nu\lambda\sigma}^{corr} \right] + \qquad (C.35)$$

$$+ \sum_{A>B} Z_A Z_B R_{AB}^{-1} \qquad (C.36)$$

Using Sinanoglu's simplification we combine in eq.(C.35) the integrals with the pair-correlation energies, writing

$$(\mu\nu|\lambda\sigma)' = (\mu\nu|\lambda\sigma) + \varepsilon_{\mu\nu\lambda\sigma}^{corr} \qquad (C.37)$$

We modify the exchange terms in the same way to take care of the correction terms, mentioned above, and obtain

$$E = \sum_{\mu\nu} P_{\mu\nu} H_{\mu\nu} + \frac{1}{2} \sum_{\mu\nu\lambda\sigma} P_{\mu\nu} P_{\lambda\sigma} \left[(\mu\nu|\lambda\sigma)' - \frac{1}{2} (\mu\sigma|\lambda\nu)' \right] \quad (C.38)$$

Eq.(C.38) shows the total energy expression according to Kohn and Sham after LCAO expansion. Carrying out the variation gives the Kohn-Sham one-particle Schrödinger equation, and solving this gives the Kohn-Sham orbitals.

This description means a description of all semiempirical MO methods: PPP, MINDO, MNDO, CNDO/S, SPINDO, HAM. The main difference between these methods concerns then the more or less practical ways in which the parameters have been determined.

The orbitals, which are obtained from such calculations, give, in principle, exact total energy of the ground state and exact density of this state, but nothing more.

The Kohn-Sham orbitals do not, for instance, give the wavefunction, but it is in many cases, in spite of this limitation, valuable to get information concerning the energies and the densities for the molecule studied. --- In practical applications the limitation is of little importance. Since $\varepsilon^{corr}_{\mu\nu\lambda\sigma}$ in eq.(C.35) is small, the Hartree-Fock orbital is a good approximation to the Kohn-Sham orbital (Sinanoglu [17]). Further, several properties of the molecule, e.g. the dipole moment, are functions of the density which is correctly obtained from the Kohn-Sham procedure.

Another limitation is that the Hohenberg-Kohn theorem applies only to the ground state of the molecule. With this limitation our description would be valid only for MINDO and MNDO which are used mainly to study the total energy of a molecule in its ground state. It is, however, also valid for the ionization process with the lowest ionization energy and for the addition of an electron with the lowest electron affinity, since ions in their ground states may be handled.

However, Gunnarsson and Lundqvist [3] have shown that the Hohenberg-Kohn-Sham scheme can be extended to the lowest excited state of each symmetry of the molecule (see also [9]). This is not unexpected if we remember, that the main part of the exchange-correlation energy, E^{coul}_{xc} , was handled in Sec.C.4. without use of density functional theory, and considering just this part the validity of our method is general.

This extension increases the range of validity a good deal. For e.g. ethylene we can now handle the $\pi\pi^*$ excitation and five ionization processes out of six in the valence region and also the electron affinity. This means that when considering a semiempirical MO method only the ionization process at 23.6 eV lies outside the validity of eq.(C.38).

We find it in this situation reasonable to assume that for those excited states, for which the validity has not been strictly proven, eq.(C.38) is approximately valid. This means that in eq.(C.31) the pair-correlation energy $\varepsilon^{corr}_{\mu\nu\lambda\sigma}$ is assumed to be independent of the state of the molecule in agreement with the preliminary discussion above.

Eq.(C.38) can then be considered as the basis also for the other semiempirical methods: PPP, CNDO/S, SPINDO and HAM.

The addition of the pair-correlation energy in eq.(C.37) gives rise to the Trees' correction in atomic spectroscopy [18], which is necessary to bring Slater's famous expression for the splitting in multiplets [19] into agreement with experiment (see Sec.E.8.). The Trees' correction means probably the only direct proof that the addition in eq.(C.37) is reasonable.

C.8. Pair-correlation energies $\varepsilon_{\mu\nu}$

We have now shown that eq.(C.31) is the basis for the creation of semiempirical methods in quantum chemistry. It is then remarkable that no proof for this expression has been published previously.

In his fundamental work on correlation Sinanoglu [17] proved the expression $E_c = \sum \varepsilon_{jk}$ (eq.(C.26)), and in later papers [20,21] he proposed an approximate form of eq.(C.31), namely

$$E_c = \frac{1}{2} \sum_{\mu\nu} P_{\mu\mu} P_{\nu\nu} \varepsilon_{\mu\nu} \qquad (C.39)$$

In one of these papers [20] he indicated that the proof should appear in a forthcoming paper, but this paper seems not to have been published.

However, in a series of papers [20-28] he and his coworkers tested the use of eq.(C.39) in studies of small molecules. Later, Verhaegen [29] simply postulated that an "atoms-in-molecule" expansion is possible and tested the expression together with his coworkers [29-37].

Only few tables of $\varepsilon_{\mu\nu}$ have been published. Brown and Ruby [21] give the following table.

Table C.8.a. Pair-correlation energies ($-\varepsilon_{\mu\nu}$), eV.

	Be	B	C	N	O	F
1s-1s	1.205	1.219	1.227	1.233	1.238	1.241
1s-2s	0.115	0.115	0.115	0.115	0.115	0.115
2s-2s	1.132	0.811	0.457	0	0	0
2s-2p		1.14	1.19	1.22	1.0	1.0

Pamuk [22] and Snyder and Basch [25] have given similar tables.

These tables have been obtained from ab-initio calculations of a few values, followed by extrapolation, and also from comparison with experimental atomic energies. The numbers in Table C.8.a. are therefore instructive but probably mainly of qualitative value.

Asbrink [18,38] has studied a large number of atoms (see Chap.E.) and has determined the $(\mu\nu|\lambda\sigma)'$ in eq.(C.37) with a probably high accuracy. However, since $(\mu\nu|\lambda\sigma)$ is more uncertain, it is questionable whether his work can give better $\varepsilon_{\mu\nu}$.

Ab-initio calculations have been performed by Jankowski and Polasik [39] . Their results correspond to only a certain part of our $\varepsilon_{\mu\nu}$. Table C.8.b. below has been deduced from numbers in their work.

Table C.8.b. Pair-correlation energies of "all-external" type, obtained from data by Jankowski and Polasik [39] ($-\varepsilon_{\mu\nu}$), eV.

	Be	B	C	N	O	F
1s-1s	1.133	1.126	1.119	1.113	1.106	1.099
1s-2s	0.083	0.087	0.092	0.096	0.101	0.105
2s-2s	0.315	0.316	0.318	0.319	0.320	0.322
2s-2p	0.576	0.569	0.563	0.557	0.551	0.544

These pair energies cannot be used directly in eq.(C.39). Other types of correlation energies ("internal" and "semi-internal") have first to be calculated and added.

C.9. Semiempirical methods

Eq.(C.38) is in two respects a suitable starting point for a semiempirical method:

a) The Kohn-Sham orbitals are one-particle orbitals but are still able to take care of the correlation energy,

b) If $(\mu\nu|\lambda\sigma)'$ is replaced by a parameter, this parameter will simultaneously take care of both the repulsion integral and the pair-correlation energy.

Different semiempirical methods are obtained depending on the tricks which are used during the parametrization. This depends mainly upon the way in which the exchange term, i.e. the last term in eq.(C.38), is handled. We can distinguish three groups of methods.

a) In the first group the exchange term is used without any changes also in the parametrized method. Small terms in eq.(C.38) are neglected (ZDO), and parametrization is then performed by comparison with different experimental data. In this way PPP, MINDO, MNDO, CNDO/S and SPINDO have been created.

The PPP method by Pariser and Parr [40] and Pople [41] has been successful in calculations of excitation energies and electron affinities of planar organic molecules with π-electrons.

MINDO and MNDO by Dewar and coworkers [42,43] has been successful for calculation of heats of formation, which is the property which governs the geometry of the molecule and also chemical reactions.

CNDO/S [44] has been successful for calculation of excitation energies and SPINDO [45] for calculation of ionization energies of hydrocarbons.

Most of these methods are useful for only one or a few properties of a molecule. The reason for this is probably that there is an error or an approximation in the theoretical model (the mathematical framework), which has been compensated by the parameters for the property in question.

b) In the second group the exchange term is approximated by use of a local density approximation, introducing a $\varrho^{1/3}$ dependence (LCAO - Xα [46]). Since this is difficult to handle in the computer, other approximations must be introduced simultaneously.

c) In the third group the property of the Kohn-Sham orbitals as being orthonormal is exploited. An idempotency relation can be added to the total energy expression, eq.(C.38), which is changed into

$$E =$$

$$= -\frac{1}{2}\sum_A \sum_\mu N_\mu \zeta_\mu^2 - \sum_{\mu_A \nu_B} P_{\mu\nu}\beta_{\mu\nu} + \text{electrostatic energy between atoms} + \qquad (C.40)$$

$$+ \text{ correction}$$

Here ζ_μ is the Slater-type orbital exponent, which depends on the atomic $P_{\mu\nu}$ and in which the exchange term now appears as a term -1 in the expression for the orbital exponent ζ_μ. The correction is composed of two small terms.

This is the basis for the HAM method [38,18] and will be discussed in detail in Chaps. D, E, F and G.

C.10. A general comment on semiempirical theories

It is known that in semiempirical theories different parameters are required to handle different properties, e.g. bond energies, spectra or ionization potentials [47]. The variation of the parameters with the properties is serious and has reduced the confidence in such theories.

Since the reason for this variation seems not to be generally understood, we will present an explanation, starting from eq.(C.38).

When in the future a perfect parametrization has been performed, eq.(C.38) gives the correct total energy of a molecule including all correlation energies. MINDO and HAM can be considered as approximations to this perfect semiempirical method.

In studies of ionization energies we have always two difficulties, the correlation energy change and the reorganization energy, both caused by the reduced number of electrons in the ion. Using the perfect semiempirical method, the correlation energies are always taken care of, but we must calculate the reorganization energy by use of the ΔE_{SCF} method. If we instead use Koopmans' theorem here, we neglect the reorganization energy. This means that we introduce a theoretical error.

If, however, we should find it necessary to make this theoretical error, we can, of course, manipulate the parameters to take care of the reorganization during the ionization. SPINDO is such a method. It is then self-evident that the manipulated method cannot be used to calculate electron affinities, since in the attachment process the number of electrons is increased and the reorganization goes in the reverse direction. This method can also not be used to handle excitation energies since now the number of electrons is unchanged and the reorganization small. These studies require therefore separate parametrizations.

In conclusion: The main explanation for the variation of the parameters with the properties in semiempirical theories is that there is an error in the theoretical model (e.g. incorrect use of Koopmans' theorem) which one has tried to compensate by change of the parameters.

C.11. The conventional CI method to handle correlation [14,15,16]

In the conventional methods, the molecular orbitals ψ_j, ψ_k, ψ_l ... are first obtained from a Hartree-Fock calculation. Then the total wavefunction Ψ is constructed.

The true wavefunction Ψ_{exact} is now written

$$\Psi_{exact} = \Psi + \chi \tag{C.41}$$

from which the exact total energy (including correlation energy) is obtained as

$$E_{exact} = \int \Psi^*_{exact} \mathcal{H} \Psi_{exact} \, d\tau \tag{C.42}$$

which is often written

$$E_{exact} = E + E_c \tag{C.43}$$

According to mathematics χ can always be written as a sum of a large number of functions. We have therefore

$$\Psi_{exact} = \Psi + \sum_j \sum_a \Psi_j^a C_j^a + \sum_{jk} \sum_{ab} \Psi_{jk}^{ab} C_{jk}^{ab} + \sum_{jkl} \sum_{abc} \Psi_{jkl}^{abc} C_{jkl}^{abc} + \cdots \tag{C.44}$$

where j, k and l mean occupied orbitals and a, b and c mean unoccupied orbitals. Ψ_j^a means the wavefunction for an excited state, when one electron has been excited from j to a, and Ψ_{jk}^{ab} means the same when two electrons have been excited. C are coefficients which can be determined by CI (configuration interaction).

The total energy is then obtained from eq.(C.42). It can be shown that many terms here are zero and that the remaining terms give

$$E_{exact} = E + \sum_{jk} \varepsilon_{jk} \qquad (C.45)$$

with

$$\varepsilon_{jk} = \sum_{ab} \int \Psi \mathcal{H} \Psi_{jk}^{ab} C_{jk}^{ab} \, d\tau \qquad (C.46)$$

ε_{jk} are called "pair-correlation energies". They reflect our idea that the correlation is due to pair-wise interaction of the electrons.

This method, the ab-initio CI method, to handle correlation is the prevailing method nowadays.It has given good results for atoms and small molecules, although the large number of terms in eq.(C.46) has made the calculations complicated and expensive.

For medium-size molecules, however, the wavefunctions Ψ_{exact} seem to be more difficult to describe. In order to increase the number of terms it is common to include more atomic orbitals, and in this way diffuse Rydberg orbitals are used.

It can be questioned whether this is practical.

According to eq.(C.24) the main contribution to the pair-correlation energy comes from those regions in space, where simultaneously ρ_j ($= \psi_j^* \psi_j$) and ρ_k are large. The correlation of the two electrons is thus largest here in agreement with our intuition.

The diffuse orbitals, which may enter the calculation of eq.(C.46), have, however, their main density outside the region where the densities ρ_j and ρ_k of the valence orbitals are large. It is therefore necessary to have positive and negative contributions in the outer region from these orbitals in order to influence only the inner region, where Rydberg-type orbitals have a small density but where the main correlation takes place. It is obvious that this will require a large number of terms in the CI expansions.

It is not quite certain that the number of terms in recent CI studies is sufficient.

C.12. Proof for the Gunnarsson-Lundqvist expression for E_{xc}.

It can be shown that the exchange-correlation energy is given by the formula

$$E_{xc} = \frac{1}{2} \int \varrho(r_1) \frac{h_{xc}}{r_{12}} \varrho(r_2) \, d\tau_1 \, d\tau_2 \tag{C.47}$$

This result was first derived by Gunnarsson and Lundqvist [3]. Since their proof is not elementary (see also refs. [7, 8, 9, 13]), we give here a simplified argument.

Since the correlation represents a small effect, we can in the first approximation neglect the correlation term, eq. (C.24). We find then that the remainder, eq. (C.23), represents the Hartree-Fock energy. Addition of the Hartree-Fock energy and the correlation term gives then the approximate total energy E.

In the formula for the correlation energy

$$E_c = E_c^{coul} + E_c^{kin} \tag{C.48}$$

the coulombic part is given by eq. (C.16).

It was pointed out by Brown and Ruby [21] that the virial theorem can be used here. This gives the relation

$$E_c^{kin} = -\frac{1}{2} E_c^{coul} \tag{C.49}$$

which holds in equilibrium, i.e. in the ground state.

We can use this to express E_{xc} in terms of E_{xc}^{coul}, which shows that E_{xc} has the same structure as the Gunnarsson-Lundqvist formula.

We observe finally the minus sign in eq. (C.49), which means, that when the electrons avoid each other due to the correlation, their motion is more complex with increase of their kinetic energy.

References

1. P.O. Löwdin, Adv.Phys. 5, 1 (1956).
2. S. Lundqvist, in Quantum Science, Methods and Structure, A Tribute to Per-Olov Löwdin (edited by J.L. Calais, O. Goscinski, J. Linderberg and Y. Öhrn), Plenum Press, New York (1976).
3. O. Gunnarsson and B.I. Lundqvist, Phys.Rev. B13, 4274 (1976).
4. O. Gunnarsson, M. Jonsson and B.I. Lundqvist, Phys.Rev. B20, 3136 (1979).
5. L. Åsbrink, C. Fridh, E. Lindholm and S. de Bruijn, Chem.Phys. Letters 66, 411 (1979).
6. P. Hohenberg and W. Kohn, Phys.Rev. 136B, 864 (1964).
7. A.K. Rajagopal, Adv.Chem.Phys. 41, 59 (1980).
8. W. Kohn and P. Vashishta, in Theory of the Inhomogeneous Electron Gas (edited by S. Lundqvist and N.H. March), p.79, Plenum Press, New York (1983).
9. J. Callaway and N.H. March, Solid State Physics 38, 135 (1984).
10. R.G. Parr, Ann.Rev.Phys.Chem. 34, 631 (1983).
11. W. Kohn and L.J. Sham, Phys.Rev. 140A, 1133 (1965).
12. E. Lindholm and S. Lundqvist, Phys. Scripta (1985).
13. J. Harris and R.O. Jones, J.Phys.F: Metal Phys. 4, 1170 (1974).
14. A.C. Hurley, Electron Correlation in Small Molecules, Academic Press, London (1976).
15. W. Kutzelnigg, in Methods of Electronic Structure Theory (edited by H.F. Schaefer III), p.129, Plenum Press, New York (1977).
16. E. Lindholm, L. Åsbrink and R. Manne, Phys.Scripta 28, 377 (1983).
17. O. Sinanoglu, J.Chem.Phys. 36, 706 (1962).
18. L. Åsbrink, Phys.Scripta 28, 394 (1983).
19. J.C. Slater, Quantum Theory of Atomic Structure, McGraw-Hill, New York (1960), Table 15-3.
20. O. Sinanoglu and H.Ö. Pamuk, Theoret.Chim.Acta 27, 289 (1972).
21. R.D. Brown and K.R. Ruby, Theoret.Chim.Acta 16, 291 (1970).
22. H.Ö. Pamuk, Theoret.Chim.Acta 29, 85 (1972).
23. O. Sinanoglu and H.Ö. Pamuk, J.Am.Chem.Soc. 95, 5435 (1973).
24. C. Hollister and O. Sinanoglu, J.Am.Chem.Soc. 88, 13 (1966).
25. L.C. Snyder and H. Basch, J.Am.Chem.Soc. 91, 2189 (1969).
26. H.Ö. Pamuk, J.Am.Chem.Soc. 98, 7948 (1976).

References (cont.)

27. A.J. Duben, L. Goodman, H.Ö. Pamuk and O. Sinanoglu, Theoret. Chim.Acta 30, 177 (1973).
28. H.Ö. Pamuk and C. Trindle, IntlJ.Quantum Chem. S 12, 271 (1978).
29. H.P.D. Liu and G. Verhaegen, J.Chem.Phys. 53, 735 (1970).
30. H.P.D. Liu and G. Verhaegen, Int.J.Quantum Chem. 5, 103 (1971).
31. R. Colin, D. de Greef, P. Goethals and G. Verhaegen, Chem.Phys. Letters 25, 70 (1974).
32. J. Lievin and G. Verhaegen, Theoret.Chim.Acta 42, 47 (1976).
33. D. Gervy and G. Verhaegen, Int.J.Quantum Chem. 12, 115 (1977).
34. J. Lievin, J. Breulet and G. Verhaegen, Theoret.Chim.Acta 60, 339 (1981).
35. J. Lievin, J. Breulet, P. Clerq and J.Y. Metz, Theoret.Chim. Acta 61, 513 (1982).
36. J. Breulet and J. Lievin, Theoret.Chim.Acta 61, 59 (1982).
37. J.Y. Metz and J. Lievin, Theoret.Chim.Acta 62, 195 (1983).
38. L. Åsbrink, C. Fridh and E. Lindholm, Chem.Phys.Letters 52, 63 (1977).
39. K. Jankowski and M. Polasik, J.Phys.B: At.Mol.Phys. 17, 2393 (1984).
40. R. Pariser and R.G. Parr, J.Chem.Phys. 21, 466, 767 (1953).
41. J.A. Pople, Trans.Faraday Soc. 49, 1375 (1953).
42. N.C. Baird and M.J.S. Dewar, J.Chem.Phys. 50, 1262 (1969).
43. M.J.S. Dewar and W. Thiel, J.Am.Chem.Soc. 99, 4899 (1977).
44. J. Del Bene and H.H. Jaffé, J.Chem.Phys. 48, 1807 (1968).
45. C. Fridh, L, Åsbrink and E. Lindholm, Chem.Phys.Letters 15, 282 (1972).
46. H. Sambe and R.H. Felton, J.Chem.Phys. 62, 1122 (1975).
47. K.F. Freed, in Modern Theoretical Chemistry (edited by G.A. Segal), Vol.7, p.201, Plenum Press, New York (1977).

D. Total energy of molecules and atoms

Up till here we have neglected the electron spin in order to simplify the introduction. Instead, we have assumed that each orbital contains two electrons.

We will now distinguish btween α spin and β spin and will assume that in each "spinorbital" 1 or 0 electrons can be accomodated. Many relations in the preceding sections will therefore be changed.

The LCAO expansion of the molecular orbital ψ_j^α with α spin is

$$\psi_j^\alpha = \sum_\mu c_{\mu j}^\alpha \phi_\mu \tag{D.1}$$

where ϕ_μ is an atomic orbital which is assumed to be hydrogenic. (ϕ_μ may be different for α and β spin. For simplicity this difference is omitted from equations (D.1) to (D.30)).

The density matrix elements $P_{\mu\nu}^\alpha$ are defined

$$P_{\mu\nu}^\alpha = \sum_j^{occ} c_{\mu j}^{\alpha *} c_{\nu j}^\alpha \tag{D.2}$$

We write

$$P_{\mu\nu} = P_{\mu\nu}^\alpha + P_{\mu\nu}^\beta \tag{D.3}$$

where the α and β terms are not necessarily equal.

We introduce this into the total energy expression, eqs. (C.33 – C.37), from density functional theory. All molecular orbitals here are thus "Kohn-Sham" orbitals. This gives

$$E = -\frac{1}{2}\sum_{\alpha,\beta}\sum_{\mu\nu} P_{\mu\nu}^\alpha \int \phi_\mu^* \nabla^2 \phi_\nu \, d\tau \; - \tag{D.4}$$

$$-\sum_{\alpha,\beta}\sum_{\mu\nu} P_{\mu\nu}^\alpha \int \phi_\mu^* \phi_\nu \sum_B Z_B r_B^{-1} \, d\tau \; + \tag{D.5}$$

$$+\sum_{\mu\nu\lambda\sigma}\left\{P_{\mu\nu}P_{\lambda\sigma}\cdot\frac{1}{2}(\mu\nu|\lambda\sigma)' - \frac{1}{2}\left[P_{\mu\nu}^\alpha P_{\lambda\sigma}^\alpha + P_{\mu\nu}^\beta P_{\lambda\sigma}^\beta\right](\mu\sigma|\lambda\nu)'\right\} \tag{D.6}$$

$$+\sum_{A>B} Z_A Z_B R_{AB}^{-1} \tag{D.7}$$

where $(\mu\nu|\lambda\sigma)'$ describes the repulsion energy together with the corresponding pair-correlation energy.

The total number of electrons in the molecule is given by

$$N = \sum_{\mu\nu} P_{\mu\nu} S_{\mu\nu} = \sum_{\mu} N_{\mu} = \sum_{\mu} \left(N_{\mu}^{\alpha} + N_{\nu}^{\beta} \right) \tag{D.8}$$

where

$$N_{\mu}^{\alpha} = \sum_{\nu} \frac{1}{2} \left(P_{\mu\nu}^{\alpha} S_{\mu\nu} + P_{\nu\mu}^{\alpha} S_{\nu\mu} \right) \tag{D.9}$$

N_{μ}^{α} is interpreted as the number of electrons with α spin in orbital ϕ_{μ} and is in an atom an integer but in a molecule usually a fraction number $0 \leqslant N_{\mu}^{\alpha} \leqslant 1$.

D.1. Rearrangement of the total energy expression

We will now rearrange the total energy expression (D.4-7) in order to present it in a mathematical form in which the different parts can be given a physical interpretation.

One important aspect is then that the new expression must have different terms for one-center and two-center energies. The one-center energies in a molecule are often understood as atomic energy contributions and the two-center energies are called bond energies.

Another important aspect is that it is desirable to compensate the self-repulsion in a simple way, or better, not to have it at all included in the expression.

The mathematical rearrangement of (D.4-7) will be performed by adding a number of "zeroes".

The first zero (D.12) is due to the idempotency of density matrices. We have pointed out in Sec.C.5. that the Kohn-Sham orbitals have an idempotent density matrix. Starting from eq.(A.14) we get

$$P_{\mu\nu}^{\alpha} = \sum_{\lambda\sigma} P_{\mu\lambda}^{\alpha} S_{\lambda\sigma} P_{\sigma\nu}^{\alpha} \tag{D.10}$$

which gives

$$N_{\mu}^{\alpha} = \frac{1}{2} \sum_{\nu\lambda\sigma} P_{\mu\lambda}^{\alpha} S_{\lambda\sigma} P_{\sigma\nu}^{\alpha} S_{\nu\mu} + \frac{1}{2} \sum_{\nu\lambda\sigma} P_{\nu\lambda}^{\alpha} S_{\lambda\sigma} P_{\sigma\mu}^{\alpha} S_{\mu\nu} \tag{D.11}$$

and then

$$0 = -\frac{1}{2}\sum_{\mu} N_{\mu}(\mu\mu|\mu\mu)' + \frac{1}{2}\sum_{\mu\nu\lambda\sigma}\left[P^{\alpha}_{\mu\nu}P^{\alpha}_{\sigma\lambda} + P^{\beta}_{\mu\nu}P^{\beta}_{\sigma\lambda}\right] S_{\lambda\mu} S_{\nu\sigma}(\mu\mu|\mu\mu)' \qquad (D.12)$$

The use of the idempotency relation has been discussed in ref. [1] without help of the density functional theory. Related work by Davidson [2] and Ponec [3,4] is also discussed there.

The second zero (D.13) is obtained from eq. (D.9)

$$0 = \frac{1}{2}\sum_{\mu\nu} N_{\mu} N_{\nu}(\mu\mu|\nu\nu)' - \frac{1}{2}\sum_{\mu\nu\lambda\sigma} P_{\mu\nu} P_{\lambda\sigma} S_{\mu\nu} S_{\lambda\sigma}(\mu\mu|\lambda\lambda)' \qquad (D.13)$$

The third zero (D.14) is an identity

$$0 = \frac{1}{2}\sum_{A}\sum_{\substack{\mu\nu \\ \mu\neq\nu}}^{A}\left[N^{\alpha}_{\mu}N^{\alpha}_{\nu} + N^{\beta}_{\mu}N^{\beta}_{\nu}\right]\cdot(\mu\nu|\mu\nu)' - \text{the same} \qquad (D.14)$$

The fourth zero (D.16) is obtained from a discussion of the kinetic energy term eq. (D.4). When μ is on atom A and ν on atom B we can write the integral in (D.4) as

$$\int \phi^*_{\mu}\nabla^2\phi_{\nu}\, d\tau = \frac{1}{2} S_{\mu\nu}\left[\int \phi^*_{\mu}\nabla^2\phi_{\mu}\, d\tau + \int \phi^*_{\nu}\nabla^2\phi_{\nu}\, d\tau\right] + \beta_{kin_{\mu\nu}}(\mu,\nu,R_{AB}) \qquad (D.15)$$

where β_{kin} is a function of ϕ_{μ}, ϕ_{ν} and the internuclear distance R_{AB}. The idea behind this relation is that the two-center integral is replaced by two one-center integrals. Since there is no mathematics involved in this relation, eq. (D.15) is acceptable only after addition of the correction term β_{kin}.

Since $S_{\mu_A\mu_A} = 1$ and $S_{\mu_A\nu_A} = 0$ for $\mu_A \neq \nu_A$, the relation (D.15) is valid also when μ and ν are on the same atom (but then β is zero.)

From (D.15) we obtain the "zero"

$$0 = -(D.4) - \frac{1}{2} \sum_{\alpha\beta} \sum_{\mu} N_{\mu}^{\alpha} \int \phi_{\mu}^{*} \nabla^{2} \phi_{\mu} \, d\tau - \frac{1}{2} \sum_{\substack{\mu_A \nu_B \\ A \neq B}} P_{\mu\nu} \beta_{kin_{\mu\nu}} \qquad (D.16)$$

These relations have been discussed previously in different contexts. Eq. (D.15) with $\beta = 0$ is called the Mulliken approximation [5] , but it is known [6] that it is very inaccurate. It was later suggested [7-10] that if the correction term β_{kin} is added, the relation is exact.

The fifth zero (D.18) is obtained from a discussion of the electron-nucleus attraction term eq. (D.5). We have

$$\int \phi_{\mu_A}^{*} \phi_{\nu_B} \frac{Z_A}{r_A} d\tau = \frac{1}{2} S_{\mu_A \nu_B} \left[\int \phi_{\mu_A}^{2} \frac{Z_A}{r_A} d\tau + \int \phi_{\nu_B}^{2} \frac{Z_A}{r_A} d\tau \right] + \beta_{en_{\mu\nu}} (\mu, \nu, R_{AB}) \qquad (D.17)$$

and

$$\left. \begin{aligned} 0 = -(D.5) &- \sum_{\alpha\beta} \sum_{\mu} N_{\mu_A}^{\alpha} Z_A \int \phi_{\mu_A}^{2} r_A^{-1} d\tau - \\ &- \sum_{\mu} N_{\mu_A} \sum_{B \neq A} Z_B V_{AB} - \\ &- \sum_{\substack{\mu_A \nu_B \\ A \neq B}} P_{\mu\nu} \beta_{en_{\mu\nu}} \end{aligned} \right\} \quad (D.18)$$

where V_{AB} describes the attraction between one positive charge in the nucleus of atom B and one electron around atom A.

Our treatment of the fifth zero is somewhat incomplete. In eq. (D.17) we consider the interaction of $\mu_A \nu_B$ with the nucleus on A (or on B), but we have neglected the other nuclei, C. However, in principle the same results are obtained if also these nuclei are included in the treatment (for proof see ref. [17]).

On the other hand, if μ and ν are both on atom A and different, we have a situation which may not be neglected. The interaction with Z_c will then result in new terms in eq. (D.18), but since these terms are complicated, we will postpone this discussion until Sec. F.2.

If we now add eqs. (D.6), (D.12) and (D.13) we obtain

$$(D.6) = \frac{1}{2}\sum_{\mu\nu} N_\mu N_\nu (\mu\mu|\nu\nu)' - \frac{1}{2}\sum_{\mu} N_\mu (\mu\mu|\mu\mu)' \qquad (D.19)$$

$$+ \frac{1}{2}\sum_{\mu\nu\lambda\sigma} P_{\mu\nu} P_{\lambda\sigma}\left[(\mu\nu|\lambda\sigma)' - S_{\mu\nu} S_{\lambda\sigma} (\mu\mu|\lambda\lambda)'\right] \qquad (D.20)$$

$$- \frac{1}{2}\sum_{\mu\nu\lambda\sigma}\left[P^{\alpha}_{\mu\nu} P^{\alpha}_{\lambda\sigma} + P^{\beta}_{\mu\nu} P^{\beta}_{\lambda\sigma}\right]\left[(\mu\sigma|\lambda\nu)' - S_{\sigma\mu} S_{\nu\lambda} (\mu\mu|\mu\mu)'\right] \qquad (D.21)$$

Addition of eqs. (D.4), (D.5), (D.7'), (D.14), (D.16), (D.18), (D.19), (D.20) and (D.21) gives the expression for the molecular energy

$E =$

$$= -\sum_{A}\sum_{\mu}^{A}\sum_{\alpha,\beta} N_{\mu}^{\alpha}\cdot\Big[+\tfrac{1}{2}\int \phi_{\mu}^{*}\nabla^{2}\phi_{\mu}\,d\tau + \tag{D.22}$$

$$+ Z_A\cdot\int \phi_{\mu}^{2} r_A^{-1} d\tau - \tfrac{1}{2}(N_{\mu}^{\alpha}-1)\gamma_{\mu\mu}^{\alpha\alpha} - \tfrac{1}{2}N_{\mu}^{\beta}\gamma_{\mu\mu}^{\beta\alpha} - \tag{D.23}$$

$$-\tfrac{1}{2}\sum_{\nu\neq\mu}^{A}\Big\{N_{\nu}^{\alpha}\big[\gamma_{\nu\mu}^{\alpha\alpha} - (\mu\nu|\mu\nu)'\big] + N_{\nu}^{\beta}\cdot\gamma_{\nu\mu}^{\beta\alpha}\Big\}\Big] + \tag{D.24}$$

$$-\sum_{\mu_A\nu_B} P_{\mu\nu}\big[\beta_{kin} + \beta_{en}\big] + \tag{D.25}$$

$$+\sum_{A>B}\Big[Z_A Z_B R_{AB}^{-1} - Z_A\sum_{\nu}^{B} N_{\nu}V_{BA} - Z_B\sum_{\mu}^{A} N_{\mu}V_{AB} + \tag{D.26}$$

$$+\sum_{\mu}^{A}\sum_{\nu}^{B}\Big\{\big(N_{\mu}^{\alpha}N_{\nu}^{\alpha} + N_{\mu}^{\beta}N_{\nu}^{\beta}\big)\gamma_{\mu\nu}^{\alpha\alpha} + \big(N_{\mu}^{\alpha}N_{\nu}^{\beta} + N_{\mu}^{\beta}N_{\nu}^{\alpha}\big)\gamma_{\mu\nu}^{\alpha\beta}\Big\}\Big] +$$

$$+\tfrac{1}{2}\sum_{\mu\nu\lambda\sigma} P_{\mu\nu}P_{\lambda\sigma}\big[(\mu\nu|\lambda\sigma)' - S_{\mu\nu}S_{\lambda\sigma}(\mu\mu|\lambda\lambda)'\big] - \tag{D.27}$$

$$-\tfrac{1}{2}\sum_{\mu\nu\lambda\sigma}\big[P_{\mu\nu}^{\alpha}P_{\lambda\sigma}^{\alpha} + P_{\mu\nu}^{\beta}P_{\lambda\sigma}^{\beta}\big]\big[(\mu\sigma|\lambda\nu)' - S_{\sigma\mu}S_{\nu\lambda}(\mu\mu|\mu\mu)'\big] + \tag{D.28}$$

$$+\tfrac{1}{2}\sum_{A}\sum_{\substack{\mu\nu\\ \mu\neq\nu}}^{A}\big[N_{\mu}^{\alpha}N_{\nu}^{\alpha} + N_{\mu}^{\beta}N_{\nu}^{\beta}\big]\cdot(\mu\nu|\mu\nu)' \quad . \tag{D.29}$$

We have here used an abbreviation

$$\gamma_{\nu\mu}^{\alpha\alpha} = (\nu\nu|\mu\mu)'^{\,\alpha\alpha} = (\nu\nu|\mu\mu)^{\alpha\alpha} + \varepsilon_{\nu\mu}^{\alpha\alpha} \tag{D.30}$$

and similarly for different spins.

Eqs.(D.22-29) give the Kohn-Sham total energy expression for a molecule.

In our deduction no small terms have been omitted (in contrast to the common procedure in ZDO methods). According to density functional theory there is further no error due to the orbital approximation, but, of course, the use of LCAO means an approximation, which gives an unknown error.

The expression, eqs.(D.22-29), may look complicated, but since it seems to correspond to Nature, it can be simplified and its final form in Sec.F. is very brief and easy to understand.

We can already interpret the different parts of the expression.

The first part,, eqs. (D.22-24),mean the one-center energies in the molecule, which can also be called the atomic energies in the molecule.The parenthesis shows the energy of one electron in orbital μ on atom A, and the different terms there represent in order: the kinetic energy of electron μ, the attraction between nucleus A and electron μ, the repulsion between the two electrons in μ, and the repulsions between the electron in μ and all electrons in other orbitals on A.

We observe a term "-1" in $-\frac{1}{2}\left(N_{\mu}^{\alpha}-1\right)\gamma_{\mu\mu}^{\alpha\alpha}$. The effect of this term is that no self-repulsion is present. No elimination of self-repulsion is therefore necessary. It follows from the deduction, that "-1" is obtained from the exchange integral by use of the idempotency relation.

Next part, eq. (D.25), is a two-center term and represents part of the energy in the bond.

Next part, eq. (D.26), is also a two-center-term and represents the repulsions between the atoms in the molecule.

The last part, eqs.(D.27-29), means correction terms, which will be discussed later (in Sec.F.1.).

D2. Shielding efficiencies $\sigma_{\nu\mu}$ in the one-center terms

It is possible to simplify eqs. (D.23) and (D.24) by introduction of shielding efficiencies, which were called "shielding constants" by Slater [11, 12]

$$\sigma_{\mu\mu}^{\alpha\alpha} = \frac{(\mu\mu|\mu\mu)^{\alpha\alpha} + \varepsilon_{\mu\mu}^{\alpha\alpha}}{2\int \phi_{\mu}^{\alpha\,2} r^{-1}\, d\tau} \tag{D.31}$$

$$\sigma_{\mu\mu}^{\beta\alpha} = \frac{(\mu\mu|\mu\mu)^{\beta\alpha} + \varepsilon_{\mu\mu}^{\beta\alpha}}{2\int \phi_{\mu}^{\alpha\,2} r^{-1}\, d\tau} \tag{D.32}$$

$$\sigma_{\nu\mu}^{\alpha\alpha} = \frac{(\nu\nu|\mu\mu)^{\alpha\alpha} + \varepsilon_{\nu\mu}^{\alpha\alpha} - (\mu\nu|\mu\nu)'}{2\int \phi_{\mu}^{\alpha\,2} r^{-1}\, d\tau} \tag{D.33}$$

$$\sigma_{\nu\mu}^{\beta\alpha} = \frac{(\nu\nu|\mu\mu)^{\beta\alpha} + \varepsilon_{\nu\mu}^{\beta\alpha}}{2\int \phi_{\mu}^{\alpha\,2} r^{-1}\, d\tau} \tag{D.34}$$

This gives directly the sum of the terms (D.22-24) as

$$(D.22-24) = -\sum_{A}\sum_{\mu}^{A}\sum_{\alpha,\beta} N_{\mu}^{\alpha} \cdot \tag{D.35}$$

$$\cdot \int \phi_{\mu}^{\alpha *} \left[\frac{1}{2}\nabla^2 + \frac{Z - (N_{\mu}^{\alpha}-1)\sigma_{\mu\mu}^{\alpha\alpha} - N_{\mu}^{\beta}\sigma_{\mu\mu}^{\beta\alpha} - \sum_{\nu}(N_{\nu}^{\alpha}\sigma_{\nu\mu}^{\alpha\alpha} + N_{\nu}^{\beta}\sigma_{\nu\mu}^{\beta\alpha})}{r} \right] \phi_{\mu}^{\alpha}\, d\tau =$$

$$= -\sum_{A}\sum_{\mu}^{A}\sum_{\alpha,\beta} N_{\mu}^{\alpha} \int \phi_{\mu}^{\alpha *} \left[\frac{1}{2}\nabla^2 + \frac{Z - s_{\mu}^{\alpha}}{r} \right] \phi_{\mu}^{\alpha}\, d\tau \tag{D.36}$$

with the shielding .

$$s_{\mu}^{\alpha} = (N_{\mu}^{\alpha}-1)\sigma_{\mu\mu}^{\alpha\alpha} + N_{\mu}^{\beta}\sigma_{\mu\mu}^{\beta\alpha} + \sum_{\nu\neq\mu}^{A}(N_{\nu}^{\alpha}\sigma_{\nu\mu}^{\alpha\alpha} + N_{\nu}^{\beta}\sigma_{\nu\mu}^{\beta\alpha}) =$$

$$= -\sigma_{\mu\mu}^{\alpha\alpha} + \sum_{\nu}^{A}(N_{\nu}^{\alpha}\sigma_{\nu\mu}^{\alpha\alpha} + N_{\nu}^{\beta}\sigma_{\nu\mu}^{\beta\alpha}) \tag{D.37}$$

which is just the expression for the shielding s_μ which was given by Slater [11, 12]. The term "-1" guarantees that only "the other electrons" shield, as pointed out by Slater.

D.3. The one-center energies in a molecule.

The operator in eq. (D.36) is known to have eigenfunctions which are hydrogenic orbitals. Introduction of such orbitals gives then a very simple form of the complicated one-center energy, eqs. (D.22-24), namely

$$(D.22-24) = -\frac{1}{2}\sum_A \sum_\mu^A \sum_{\alpha,\beta} N_\mu^\alpha \zeta_\mu^{\alpha\,2} \tag{D.38}$$

where the orbital exponent is

$$\zeta_\mu^\alpha = \frac{1}{n_\mu}\left[Z + \sigma_{\mu\mu}^{\alpha\alpha} - \sum_\nu^A \left(N_\nu^\alpha \sigma_{\nu\mu}^{\alpha\alpha} + N_\nu^\beta \sigma_{\nu\mu}^{\beta\alpha}\right)\right] \tag{D.39}$$

This means that the "Kohn-Sham" molecular orbital ψ_i^α is written $\sum_\mu c_{\mu j}\phi_\mu^\alpha$ where ϕ_μ^α are hydrogenic.

The use of hydrogenic orbitals simplifies also the shielding efficiencies, since the denominator in the shielding efficiencies

$$2\int \phi_\mu^{\alpha\,2} r^{-1}\, d\tau$$

is then changed into

$$\frac{2}{n_\mu}\cdot \zeta_\mu^\alpha \tag{D.40}$$

D.4. Further study of the shielding efficiencies

We will now show that the complicated expression for the shielding efficiencies, eqs. (D.31 - 34), can be written in a very simple way

$$\sigma_{\nu\mu} = \alpha_{\nu\mu} - \frac{\beta_{\nu\mu}}{\zeta_\mu} \tag{D.41}$$

in which $\alpha_{\nu\mu}$ and $\beta_{\nu\mu}$ are positive numbers and ζ_μ is the orbital exponent of the shielded electron.

We have already shown in eq. (D.40) that the denominator in the shielding efficiencies is proportional to ζ_μ.

The parameter α in eq. (D.41) is a number if the first term in the numerator of (D.31-34) is proportional to ζ_μ. This is, in fact, the case for hydrogenic orbitals.

Crossley and Coulson [13] have calculated $(\nu\nu|\mu\mu)$ for such orbitals and found it to be proportional to the effective nuclear charge. The first part of $\sigma_{\nu\mu}$ is then obtained for 1s-1s as 0.312 and for 2s-2s as 0.301. Recently, Teruya and Anno [14] and Kregar [15] have given the same numbers. Bessis and Bessis [16] have obtained shielding constants directly from an approximate solution of the Schrödinger equation with for instance the following results for $\alpha_{\nu\mu}$:

for ns-ns: 0.353, for 1s-2s: 0.913, for 2s-1s: 0.014.

The extra term in eq. (D.33), $(\mu\nu|\mu\nu)$, is proportional to the effective nuclear charge in the same way as $(\nu\nu|\mu\mu)$ and causes only a small change of $\alpha_{\nu\mu}$.

The parameter β in eq. (D.41) corresponds to the second term, $\varepsilon_{\nu\mu}$, in the numerator of eq. (D.31-34).

The pair-correlation energies have been tabulated in Sec. C.7., and it can be seen that they are negative and slightly dependent upon the atom. It is therefore appropriate to write $\beta_{\nu\mu}$ as $(b_{\nu\mu} + c_{\nu\mu} \cdot Z_A)$ where b and c are constants. The same Z-dependence has recently been obtained from ab-initio calculations by Jankowski and Polasik [17].

The shielding efficiency can therefore be written

$$\sigma_{\nu\mu} = a_{\nu\mu} - \frac{b_{\nu\mu} + c_{\nu\mu} \cdot Z_A}{\zeta_\mu} \qquad (D.42)$$

A final problem concerns the fact that e.g. 1s and 2s hydrogenic orbitals, satisfying eq. (D.36), are not orthogonal due to different shielding.

This problem has been solved by Slater ([12] p. 348), however using STO´s, and has been discussed in a previous paper [18]. It appears that the interaction between 1s and 2s can be taken care of by calculating the energies using slightly changed shielding efficiencies. If they are determined from comparison with atomic spectroscopy, this problem will therefore be taken care of without any difficulties.

References

1. E. Lindholm, L. Åsbrink and R. Manne, Phys.Scripta 28, 377 (1983).
2. E.R. Davidson, J.Chem.Phys. 57, 1999 (1972).
3. R. Ponec, Theoret.Chim.Acta 59, 629 (1980).
4. R. Ponec, Coll.Czechoslovak Chem.Comm. 47, 1479 (1982).
5. R.S. Mulliken, J.Chim.Phys. 46, 497, 675 (1949).
6. M.D. Newton, F.P. Boer and W.N. Lipscomb, J.Am.Chem.Soc. 88, 2353 (1966).
7. G. Berthier, I. Baudet and M. Suard, Tetrahedron 19, Suppl. 2, 1 (1963).
8. G. Berthier, P. Millié and A. Veillard, J.Chim.Phys.62, 8 (1965).
9. G. Del Re, Gazzetta Chim.It. 102, 929 (1972).
10. G. Del Re, Adv. Quantum Chem. 8, 95 (1974).
11. J.C. Slater, Phys.Rev. 36, 57 (1930).
12. J.C. Slater, Quantum Theory of Atomic Structure, Vol.1, McGraw-Hill, New York (1960).
13. R.J.S. Crossley and C.A. Coulson, Proc.Phys. Soc. 81, 211 (1963).
14. H. Teruya and T. Anno, J.Chem.Phys. 75, 4997 (1981).
15. M. Kregar, Phys.Scripta 31, 246 (1985).
16. N. Bessis and G. Bessis, J.Chem.Phys. 74, 3628 (1981).
17. K. Jankowski and M. Polasik, J.Phys.B: At.Mol.Phys. 17, 2393 (1984).
18. L. Åsbrink, C. Fridh, E. Lindholm, S. de Bruijn and D.P. Chong, Phys. Scripta 22, 475 (1980).

E. Atoms

E.1. The simple atom.

We will now study the total energies of atoms, using the results from Chap.D. It is then important that we study only such atoms for which this deduction is valid. Most atomic states which are observed in atomic spectroscopy, e.g. ^1D, can therefore not be used directly in our work.

Only atoms, which are described by a one-determinant wavefunction, can be used, since with a many-determinant wavefunction the idempotency lacks meaning. The wavefunction must therefore be a "pure spin-configuration" such as e.g. $x^\alpha y^\beta$ in which one electron in $2px$ has α spin and one electron in $2py$ has β spin. Such an atom was discussed in Sec. A.5. and it was shown there that $P_{\mu\nu} = 0$ when $\mu \neq \nu$.

The total energy expression in eqs. (D.22-29) becomes now very simple since some terms from (D.27-29) which are proportional to $P_{\mu\nu}^2$ then vanish and the total energy of the atom becomes

$$E = -\frac{1}{2} \sum_{\mu} \sum_{\alpha,\beta} N_\mu^\alpha \, \zeta_\mu^{\alpha\,2} \qquad (E.1)$$

This simple formula is very convenient for the calculation of the total energy of an atom, supposed that the parameters $a_{\nu\mu}$, $b_{\nu\mu}$ and $c_{\nu\mu}$ in the shielding efficiencies are known. It is obvious that they can be determined empirically from the total energies of a sufficiently large number of atoms.

The results from such a determination are presented in Table E.1.

Table E.1. Empirical shielding efficiencies $\sigma_{\nu\mu}$ in HAM/3

	$a_{\nu\mu}$	$b_{\nu\mu}$	$c_{\nu\mu}$
$\sigma^{\alpha\beta}$			
σ_{1s1s}	0.42320018	0.00535688	0.10850046
σ_{2s1s}	0.0	0.0	0.0
σ_{x1s}	0.0	0.0	0.0
σ_{1s2s}	0.81147656	-0.01045298	-0.01468092
σ_{ss}	0.26099092	0.04027565	-0.01683921
σ_{xs}	0.09380453	-0.13861674	0.00639068
σ_{1sx}	0.89614688	0.05252657	-0.03895827
σ_{sx}	0.48411734	0.25746323	-0.05210169
σ_{xx}	0.45657165	0.02202149	0.02620898
σ_{Yx}	0.37960710	0.02913969	0.01217109
$\sigma^{\alpha\alpha}$			
σ_{1s1s}	—	—	—
σ_{2s1s}	0.0	0.0	0.0
σ_{x1s}	0.0	0.0	0.0
σ_{1s2s}	0.72573164	-0.03315235	-0.01407734
σ_{ss}	—	—	—
σ_{xs}	0.25728828	-0.12933014	0.03192253
σ_{1sx}	0.82960674	0.02391131	-0.04136912
σ_{sx}	0.25996183	0.17935061	-0.04316377
σ_{xx}	—	—	—
σ_{Yx}	0.36334920	0.01268583	0.01364199

The relativistic corrections r_μ :

$\mu = 1s$ $r_\mu = 0.00000616$

$2s$ $r_\mu = 0.00000807$

$2p$ $r_\mu = 0.00000389$

E.2. The energies of the spin-configurations

The atoms, which are studied in atomic spectroscopy, have always defined angular momenta. (The orbital angular momentum is denoted S, P, D,... and the spin angular momentum is denoted singlet, doublet, triplet,...). The energies of pure spin-configurations are therefore usually not observed in atomic spectroscopy.

It is therefore necessary to know the relation between the energies of the configurations and the energies of the states.

Let us as an example study the excited helium atom 1s2p, for which the pure spin-configurations can be written

$s^{\alpha} x^{\alpha}$ and $s^{\alpha} x^{\beta}$

The observed states are ^1P and ^3P.

First, we write the spinfunctions for a singlet and a triplet:

$$\text{singlet} = \alpha\beta - \beta\alpha \tag{E.2}$$

$$\text{triplet} = \begin{cases} \alpha\alpha \\ \alpha\beta + \beta\alpha \\ \beta\beta \end{cases} \tag{E.3}$$

Then, we see directly that $s^{\alpha} x^{\alpha}$ has the spinfunction $\alpha\alpha$ and therefore must be a triplet. Since according to the Pauli principle the total wavefunction must be antisymmetric, we understand, that the space orbital must be

$$s(1)\,x(2) - s(2)\,x(1)$$

and therefore the energy of the configuration must be that of ^3P with experimental energy 58.0432 eV.

Finally, we see that $s^{\alpha} x^{\beta}$ has the spinfunction $\alpha\beta$, which is neither symmetric nor antisymmetric. We write therefore

$$\alpha\beta = \tfrac{1}{2}(\alpha\beta - \beta\alpha) + \tfrac{1}{2}(\alpha\beta + \beta\alpha) \tag{E.4}$$

and find that this configuration corresponds to $\frac{1}{2}$ (^1P + ^3P). Its energy is therefore the average of ^1P and ^3P.

Our results can therefore be written

$$^3P = E\,(s^{\alpha} x^{\alpha}) \tag{E.5}$$

$$^1P = 2 \cdot E\,(s^{\alpha} x^{\beta}) - E\,(s^{\alpha} x^{\alpha}) \tag{E.6}$$

In Table E.2. similar results are shown for atoms with

Table E.2. Comparison of energies of configurations and observable energies of states.

spin-configuration		states
ss'	$s^\alpha s'^\alpha$	^3S
	$s^\alpha s'^\beta$	$(^3S + {}^1S)/2$
sp	$s^\alpha x^\alpha$	^3P
	$s^\alpha x^\beta$	$(^3P + {}^1P)/2$
p^2	$x^\alpha x^\beta$	$(2\cdot{}^1D + {}^1S)/3$
	$x^\alpha y^\alpha$	^3P
	$x^\alpha y^\beta$	$(^3P + {}^1D)/2$
sp^2	$s^\alpha x^\alpha y^\alpha$	^4P
	$s^\beta x^\alpha y^\alpha$	$(2\cdot{}^2P + {}^4P)/3$
	$s^\alpha x^\alpha x^\beta$	$(2\cdot{}^2D + {}^2S)/3$
	$s^\alpha x^\alpha y^\beta$	$(3\cdot{}^2D + 2\cdot{}^4P + {}^2P)/6$
p^3	$x^\alpha y^\alpha z^\alpha$	^4S
	$x^\alpha y^\alpha x^\beta$	$(^2P + {}^2D)/2$
	$x^\alpha y^\alpha z^\beta$	$(2\cdot{}^2D + {}^4S)/3$
sp^3	$s^\alpha x^\alpha y^\alpha z^\alpha$	^5S
	$s^\alpha x^\alpha y^\alpha x^\beta$	$(^3P + {}^3D)/2$
	$s^\beta x^\alpha y^\alpha z^\alpha$	$(^5S + 3\cdot{}^3S)/4$
	$s^\beta x^\alpha y^\alpha x^\beta$	$(^3D + {}^3P + {}^1D + {}^1P)/4$
	$s^\beta x^\alpha y^\alpha z^\beta$	$(2\cdot{}^3D + 2\cdot{}^1D + {}^3S + {}^5S)/6$
	$s^\alpha x^\alpha y^\alpha z^\beta$	$(8\cdot{}^3D + 3\cdot{}^5S + {}^3S)/12$
pp'	$x^\alpha x'^\alpha$	$(2\cdot{}^3D + {}^3S)/3$
	$x^\alpha x'^\beta$	$(2\cdot{}^3D + {}^3S + 2\cdot{}^1D + {}^1S)/6$
	$x^\alpha y'^\alpha$	$(^3P + {}^3D)/2$
	$x^\alpha y'^\beta$	$(^3P + {}^1D + {}^3D + {}^1P)/4$

s and p electrons as deduced by Åsbrink and Fridh (see ref. [1,2]).

No similar complete table seems to have been published previously. Moffitt [3] has published a similar table in which he, however, did not distinguish between the spins of the electrons. Klopman [4] has published a less complete table in which he, however, did not distinguish between x, y and z, which were all denoted p. On the other hand the results were classified with respect to the spin. McGlynn et al. [5] have on p.103 a complete and simple treatment of the p^2 configuration. Turner [6] discusses the wavefunctions of the p^3 configuration. The results in Moffitt's paper were used by Pritchard and Skinner [7] without any extensions of Moffitt's results. Hinze and Jaffe´ [8] reviewed previous work.

It is clear that the energies of the states, e.g. ^3P, ^1D and ^1S, can be calculated in a simple way as soon as the energies of all corresponding spin-configurations are known. This will be discussed in detail in Sec.E.8.

E.3. Comments on the shielding efficiencies

In Table E.1.,which gives the shielding efficiencies, there are no values for

$$\sigma^{\alpha\alpha}_{1s1s}, \quad \sigma^{\alpha\alpha}_{2s2s} \quad \text{and} \quad \sigma^{\alpha\alpha}_{xx} .$$

This is expected, since in an atom there are never two electrons in the 1s orbital with α spin owing to the Pauli principle.

In the molecule, on the other hand, these shielding efficiencies may play some role. In eq. (D.37) we have a term $(N^{\alpha}_{\mu} - 1)\,\sigma^{\alpha\alpha}_{\mu\mu}$ and if here N^{α}_{μ} is less than 1, this term will influence the energy of the molecule. These shielding efficiencies can therefore be determined from molecular studies.

An interesting aspect is that it can easily be proved that in some cases the total energy is dependent not upon the shielding efficiencies separately but upon their sum;

$$\left(\sigma_{2s1s} + \tfrac{1}{4}\sigma_{1s2s}\right), \quad \left(\sigma_{x1s} + \tfrac{1}{4}\sigma_{1sx}\right) \quad \text{and} \quad \left(\sigma_{x2s} + \sigma_{2sx}\right).$$

The shielding efficiencies in these parentheses can therefore in the first approximation not be determined unequivocally.

The relativistic corrections r_μ are used in the expression (D.39) to correct the nuclear charge in the following way

$$\zeta_\mu^\alpha = \frac{1}{n_\mu}\left[Z + r_\mu \cdot Z^3 - S_\mu^\alpha\right] \tag{E.7}$$

Since in the energy formula ζ_μ is squared, the relativistic correction of the energy will be proportional to Z^4 as is well known.

Since the relativity correction is due to the high velocity of the electron near the nucleus, we can use the same r_μ independently of the number of electrons in the atom [2].

In molecules, finally, we do not distinguish between spins. We use therefore the average of the two shielding efficiencies

$$\sigma_{\nu\mu} = \frac{1}{2}\left[\sigma_{\nu\mu}^{\alpha\alpha} + \sigma_{\nu\mu}^{\alpha\beta}\right] \tag{E.8}$$

E.4. Previous work on shielding efficiencies

In 1930 Slater [9,10] proposed the following shielding constants to be used in an atom with 1s, 2s and 2p electrons;

σ_{1s1s} = 0.30

σ_{1s2s} and σ_{1s2p} = 0.85

σ_{2s1s} and σ_{2p1s} = 0

σ_{2s2s} and σ_{2s2p} and σ_{2p2s} and σ_{2p2p} = 0.35

These results, which were extended to higher n, are really good. Later efforts have not resulted in any improvements until it now is understood that the shielding efficiencies must be functions of the nature of the atom.

Shielding constants have been discussed in only six papers [11-16].

The earlier difficulties are demonstrated by a statement by Burden and Wilson [15] who write on p. 902, that "they have tried many approaches to obtain the effective nuclear charge for the valence state of the atom, but these have failed. Vladimiroff (private communication 1971) has come to the same conclusion, and, as yet, very little progress has been made in this field."

E.5. Total energies of atoms and atomic ions in HAM/3

In the table the atomic species is described by the nucleus and the atomic charge, followed by the number of electrons in the different orbitals in the following order:

Atom Charge $1s^\alpha$ $2s^\alpha$ $2px^\alpha$ $2py^\alpha$ $2pz^\alpha$ $1s^\beta$ $2s^\beta$ $2px^\beta$ $2py^\beta$ $2pz^\beta$

Then the theoretical total energy, the experimental total energy and their difference are given in eV. The energy scale should of course start at the bare nucleus, but in several cases the experimental data are insufficient. The zero of the energy scale is then given for both theoretical and experimental energies at the top of this part of the table. Finally, the atomic states are given, which participate in the "average state" of the atomic species in question. Here 2*1D means $2 \cdot {}^1D$.
When no experimental data are known this has been marked with —— in the table.

Atomic species			E_{HAM}	E_{exp}	$E_{exp}-E_{HAM}$	Average state
H 1	00000	00000	0.0	0.0	0.0	
H-1	10000	10000	14.5923	14.3512	-0.2411	1S
H-1	01000	01000	4.3297	4.0460	-0.2837	1S
H-1	00110	00000	3.4365	3.4083	-0.0282	3P
H-1	00001	00001	3.1864	3.4083	0.2220	2*1D+1S
H-1	00001	00010	3.6562	3.4083	-0.2479	3P+1D
H-2	00111	00000	2.4378	——	——	4S
HE 2	00000	00000	0.0	0.0	0.0	
HE 0	10000	10000	79.3013	79.0100	-0.2912	1S
HE 0	11000	00000	59.0789	59.1888	0.1099	3S
HE 0	10000	01000	58.6731	58.7918	0.1187	3S+1S
HE 0	10001	00000	58.2440	58.0432	-0.2008	3P
HE 0	00001	10000	58.2183	57.9194	-0.2989	3P+1P
HE 0	01001	00000	20.9494	20.6976	-0.2518	3P
HE 0	00001	01000	20.3687	19.7909	-0.5779	3P+1P
HE 0	01000	01000	20.7540	21.1396	0.3857	1S
HE 0	00110	00000	19.2957	19.3348	0.0391	3P
HE 0	00001	00001	18.1758	18.3725	0.1967	2*1D+1S
HE 0	00001	00010	19.2910	19.2291	-0.0620	3P+1D
HE-1	11000	01000	59.5778	59.6895	0.1117	2S
HE-1	01001	01000	22.5210	21.7995	-0.7215	2P

Atomic species	E_{HAM}	E_{exp}	$E_{exp}-E_{HAM}$	Average state
LI 3 00000 00000	0.0	0.0	0.0	
LI 1 11000 00000	139.0229	139.0914	0.0685	3S
LI 1 10000 01000	138.0125	138.1316	0.1192	3S+1S
LI 1 10001 00000	137.0972	136.8124	-0.2848	3P
LI 1 00001 10000	136.6357	136.3827	-0.2530	3P+1P
LI 1 01001 00000	51.3468	51.1275	-0.2192	3P
LI 1 00110 00000	48.7457	48.8891	0.1434	3P
LI 1 10000 10000	198.3061	198.1167	-0.1894	1S
LI 0 10001 10000	201.1507	201.6609	0.5102	2P
LI 0 11000 10000	203.6741	203.5090	-0.1650	2S
LI-1 11000 11000	203.6411	204.1290	0.4879	1S
BE 4 00000 00000	0.0	0.0	0.0	
BE 2 10000 10000	371.7405	371.6444	-0.0962	1S
BE 2 11000 00000	252.9896	253.0376	0.0480	3S
BE 2 10000 01000	251.3843	251.5080	0.1237	3S+1S
BE 2 10001 00000	249.9792	249.7202	-0.2590	3P
BE 2 00001 10000	249.0925	248.8456	-0.2470	3P+1P
BE 2 00110 00000	91.8045	92.0487	0.2441	3P
BE 1 11000 10000	390.1588	389.8569	-0.3018	2S
BE 1 10001 10000	385.8086	385.8975	0.0889	2P
BE 0 11000 11000	399.4522	399.1802	-0.2720	1S
BE 0 11001 10000	396.3392	396.4548	0.1155	3P
BE 0 10001 11000	394.9332	395.1786	0.2455	3P+1P
BE 0 10110 10000	391.4588	391.7782	0.3194	3P
BE 0 10001 10001	390.8772	391.5603	0.6830	2*1D+1S
BE 0 10001 10010	391.3583	391.9807	0.6223	3P+1D
BE-1 11001 11000	397.6356	~398.7302	1.0946	2P
B 5 00000 00000	0.0	0.0	0.0	
B 3 10000 10000	599.6502	599.6393	-0.0109	1S
B 3 11000 00000	401.0104	401.0729	0.0625	3S
B 3 10001 00000	396.9117	396.6844	-0.2273	3P
B 3 00001 10000	395.6025	395.3750	-0.2275	3P+1P
B 2 11000 10000	637.8689	637.5739	-0.2950	2S
B 2 10001 10000	631.5494	631.5738	0.0244	2P
B 1 11000 11000	662.9843	662.7303	-0.2540	1S
B 1 11001 10000	658.2108	658.0994	-0.1115	3P
B 1 10001 11000	655.9569	655.8644	-0.0925	3P+1P
B 1 10110 10000	650.4480	650.4641	0.0161	3P
B 1 10001 10001	648.8844	648.9925	0.1081	2*1D+1S
B 1 10001 10010	650.0678	650.2511	0.1834	3P+1D
B 0 11001 11000	670.8947	671.0278	0.1331	2P
B 0 11110 10000	667.3023	667.4566	0.1544	4P
B 0 10110 11000	663.4709	663.8418	0.3709	2*2P+4P
B 0 11001 10001	664.2839	664.4462	0.1623	2*2D+2S
B 0 11001 10010	665.0287	665.3722	0.3435	3*2D+2*4P+2P
B 0 10111 10000	659.0455	659.1333	0.0878	4S
B 0 10110 10010	657.6275	—	—	2P+2D
B 0 10110 10001	658.4439	—	—	2*2D+4S
B-1 11110 11000	669.9462	671.3078	1.3616	3P

Atomic species			E_{HAM}	E_{exp}	$E_{exp}-E_{HAM}$	Average state
C 6	00000	00000	0.0	0.0	0.0	
C 5	10000	00000	490.0332	490.0266	-0.0066	2S
C 4	10000	10000	882.0760	882.1324	0.0563	1S
C 4	11000	00000	583.1112	583.1658	0.0546	3S
C 4	10001	00000	577.9166	577.7150	-0.2016	3P
C 4	00001	10000	576.1858	575.9786	-0.2072	3P+1P
C 3	10110	00000	643.7382	—	—	4P
C 3	11000	10000	946.9006	946.6306	-0.2700	2S
C 3	10001	10000	938.5908	938.6262	0.0354	2P
C 2	11000	11000	994.6221	994.5117	-0.1104	1S
C 2	11001	10000	988.1806	988.0185	-0.1621	3P
C 2	10001	11000	985.0578	984.9197	-0.1381	3P+1P
C 2	10110	10000	977.5232	977.4718	-0.0514	3P
C 2	10001	10001	975.0122	974.9096	-0.1026	2*1D+1S
C 2	10001	10010	976.8752	976.9481	0.0729	3P+1D
C 1	11001	11000	1018.9441	1018.8933	-0.0508	2P
C 1	11110	10000	1013.6284	1013.5602	-0.0682	4P
C 1	10110	11000	1008.0880	1007.9710	-0.1170	2*2P+4P
C 1	11001	10001	1008.7979	1008.7137	-0.0841	2*2D+2S
C 1	11001	10010	1010.2249	1010.0037	-0.2212	3*2D+2*4P+2P
C 1	10111	10000	1001.5631	1001.2859	-0.2771	4S
C 1	10110	10010	998.9352	999.1061	0.1709	2P+2D
C 1	10110	10001	1000.4273	1000.5882	0.1609	2*2D+4S
C 0	11110	11000	1029.8749	1030.1605	0.2855	3P
C 0	11001	11001	1028.3013	1028.4268	0.1255	2*1D+1S
C 0	11001	11010	1029.2631	1029.5304	0.2672	3P+1D
C 0	11111	10000	1025.9458	1025.9812	0.0354	5S
C 0	10111	11000	1018.6905	1019.2796	0.5891	5S+3*3S
C 0	11110	10010	1021.3742	1021.5251	0.1510	3P+3D
C 0	10110	11010	1018.9540	—	—	3D+3P+1D+1P
C 0	11110	10001	1022.4271	1022.7274	0.3003	8*3D+3*5S+3S
C 0	10110	11001	1020.0066	—	—	2*3D+2*1D+3S+5S
C 0	10001	10111	1010.7841	—	—	3P
C 0	10110	10110	1009.1628	—	—	2*1D+1S
C 0	10110	10101	1010.2955	—	—	3P+1D
C-1	11111	11000	1030.1307	1031.4325	1.3018	4S
C 3	10110	00000	643.7382	—	—	4P
C 3	11001	00000	652.8091	9.2115	0.1406	4P

Atomic species			E_{HAM}	E_{exp}	$E_{exp}-E_{HAM}$	Average state
N 7	00000	00000	0.0	0.0	0.0	
N 6	10000	00000	667.0964	667.0873	-0.0091	2S
N 6	01000	00000	166.8054	166.8041	-0.0013	2S
N 6	00001	00000	166.7370	166.7370	0.0000	2P
N 5	10000	10000	1219.0635	1219.1561	0.0926	1S
N 5	11000	00000	799.3217	799.3362	0.0144	3S
N 5	10000	01000	795.9380	796.0245	0.0864	3S+1S
N 5	10001	00000	793.0196	792.8155	-0.2042	3P
N 5	00001	10000	790.8677	790.6274	-0.2403	3P+1P
N 4	11000	10000	1317.3138	1317.0604	-0.2534	2S
N 4	10001	10000	1306.9985	1307.0620	0.0636	2P
N 3	11000	11000	1394.4542	1394.5369	0.0827	1S
N 3	11001	10000	1386.3356	1386.1956	-0.1400	3P
N 3	10001	11000	1382.3379	1382.2637	-0.0741	3P+1P
N 3	10110	10000	1372.7710	1372.7642	-0.0068	3P
N 3	10001	10001	1369.3223	1369.1954	-0.1268	2*1D+1S
N 3	10001	10010	1371.8583	1371.9405	0.0822	3P+1D
N 2	11001	11000	1441.9739	1441.9650	-0.0089	2P
N 2	11110	10000	1434.9441	1434.8768	-0.0673	4P
N 2	10110	11000	1427.6671	1427.5465	-0.1205	2*2P+4P
N 2	11001	10001	1428.3015	1428.2139	-0.0876	2*2D+2S
N 2	11001	10010	1430.4036	1430.3322	-0.0714	3*2D+2*4P+2P
N 2	10111	10000	1419.0669	1418.8175	-0.2494	4S
N 2	10110	10010	1415.2397	1415.1051	-0.1346	2P+2D
N 2	10110	10001	1417.4028	1417.4720	0.0691	2*2D+4S
N 1	11110	11000	1471.5393	1471.5835	0.0441	3P
N 1	11001	11001	1469.0035	1468.9775	-0.0260	2*1D+1S
N 1	11001	11010	1470.6515	1470.6395	-0.0120	3P+1D
N 1	11111	10000	1465.8337	1465.7461	-0.0876	5S
N 1	10111	11000	1455.9970	1455.7063	-0.2907	5S+3*3S
N 1	11110	10010	1459.1905	1459.1050	-0.0855	3P+3D
N 1	10110	11010	1455.9102	1455.7107	-0.1995	3D+3P+1D+1P
N 1	11110	10001	1460.9163	1460.9049	-0.0114	8*3D+3*5S+3S
N 1	10110	11001	1457.6360	1457.6422	0.0062	2*3D+2*1D+3S+5S
N 1	10001	10111	1444.5062	—	—	3P
N 1	10110	10110	1441.9567	—	—	2*1D+1S
N 1	10110	10101	1443.7544	—	—	3P+1D
N 0	11111	11000	1485.8579	1486.1443	0.2865	4S
N 0	11110	11010	1482.9883	1483.1645	0.1762	2P+2D
N 0	11110	11001	1484.2518	1484.5550	0.3032	2*2D+4S
N 0	10001	11111	1475.2019	1475.2163	0.0143	4P
N 0	11001	10111	1465.3591	—	—	2*2P+4P
N 0	10110	11110	1470.1854	—	—	2*2D+2S
N 0	10110	11101	1471.5430	—	—	3*2D+2*4P+2P
N 0	10110	10111	1456.9808	—	—	2P
N-1	11001	11111	1484.5908	~1486.0743	~1.4835	3P

Atomic species			E_{HAM}	E_{exp}	$E_{exp}-E_{HAM}$	Average state
O 8	00000	00000	0.0	0.0	0.0	
O 7	10000	00000	871.4706	871.4597	-0.0109	2S
O 7	01000	00000	217.9210	217.9194	-0.0016	2S
O 7	00001	00000	217.8044	217.8042	-0.0002	2P
O 6	10000	10000	1610.6653	1610.7834	0.1181	1S
O 6	11000	00000	1049.6766	1049.6712	-0.0055	3S
O 6	10000	01000	1045.7003	1045.7570	0.0567	3S+1S
O 6	10001	00000	1042.2511	1042.0653	-0.1858	3P
O 6	00001	10000	1039.6783	1039.4244	-0.2539	3P+1P
O 5	11000	10000	1749.1713	1748.9097	-0.2616	2S
O 5	10001	10000	1736.8333	1736.9160	0.0827	2P
O 4	11000	11000	1862.5569	1862.8207	0.2638	1S
O 4	11001	10000	1852.7475	1852.6022	-0.1453	3P
O 4	10001	11000	1847.8725	1847.8804	0.0079	3P+1P
O 4	10110	10000	1836.2596	1836.2953	0.0357	3P
O 4	10001	10001	1831.8771	1831.7669	-0.1102	2*1D+1S
O 4	10001	10010	1835.0835	1835.1922	0.1087	3P+1D
O 3	11001	11000	1940.0805	1940.2085	0.1280	2P
O 3	11110	10000	1931.3301	1931.3902	0.0601	4P
O 3	10110	11000	1922.3060	1922.3583	0.0523	2*2P+4P
O 3	11001	10001	1922.8763	1922.9541	0.0778	2*2D+2S
O 3	11001	10010	1925.6502	1925.6875	0.0374	3*2D+2*4P+2P
O 3	10111	10000	1911.6279	1911.5643	-0.0636	4S
O 3	10110	10010	1906.6054	1906.5031	-0.1023	2P+2D
O 3	10110	10001	1909.4377	1909.5893	0.1516	2*2D+4S
O 2	11110	11000	1995.0785	1995.1668	0.0883	3P
O 2	11001	11001	1991.5945	1991.7323	0.1378	2*1D+1S
O 2	11001	11010	1993.9207	1993.9231	0.0024	3P+1D
O 2	11111	10000	1987.6095	1987.7145	0.1050	5S
O 2	10111	11000	1975.1644	1974.9954	-0.1691	5S+3*3S
O 2	11110	10010	1978.8937	1978.9236	0.0298	3P+3D
O 2	10110	11010	1974.7443	1974.7362	-0.0081	3D+3P+1D+1P
O 2	11110	10001	1981.2905	1981.3641	0.0736	8*3D+3*5S+3S
O 2	10110	11001	1977.1410	1977.1813	0.0403	2*3D+2*1D+3S+5S
O 2	10001	10111	1960.1055	1959.9957	-0.1098	3P
O 2	10110	10110	1956.6276	1956.3470	-0.2806	2*1D+1S
O 2	10110	10101	1959.0911	1959.1015	0.0104	3P+1D
O 1	11111	11000	2030.2523	2030.3504	0.0982	4S
O 1	11110	11001	2028.1046	2028.1338	0.0291	2*2D+4S
O 1	11110	11010	2026.1616	2026.1791	0.0175	2P+2D
O 1	10001	11111	2015.4503	2015.4798	0.0294	4P
O 1	11001	10111	2007.8790	2007.8158	-0.0632	2*2P+4P
O 1	10110	11110	2008.6361	2008.5407	-0.0954	2*2D+2S
O 1	10110	11101	2010.6612	2010.7084	0.0472	3*2D+2*4P+2P
O 0	11001	11111	2043.7642	2043.9596	0.1954	3P
O 0	11110	11110	2041.2093	2041.2610	0.0517	2*1D+1S
O 0	11110	11101	2042.7745	2042.9808	0.2063	3P+1D
O 0	10110	11111	2028.2869	2028.3086	0.0217	3P
O 0	11110	10111	2024.8641	2024.3699	-0.4942	3P+1P
O-1	11110	11111	2044.0800	2045.4246	1.3446	2P

Atomic species			E_{HAM}	E_{exp}	$E_{exp}-E_{HAM}$	Average state
F 9	00000	00000	0.0	0.0	0.0	
F 8	10000	00000	1103.1858	1103.1739	-0.0119	2S
F 8	01000	00000	275.8820	275.8803	-0.0017	2S
F 8	00001	00000	275.6951	275.6948	-0.0003	2P
F 7	10000	10000	2056.9414	0.0	0.0	1S
F 6	11000	10000	2242.5433	185.1994	-0.4025	2S
F 6	10001	10000	2228.1616	171.1795	-0.0407	2P
F 5	11001	10000	2387.4928	330.3221	-0.2293	3P
F 5	10001	11000	2381.7393	324.7616	-0.0362	3P+1P
F 5	11000	11000	2399.0120	342.3678	0.2973	1S
F 5	10110	10000	2368.0608	311.1990	0.0796	3P
F 5	10001	10001	2362.7465	305.6558	-0.1492	2*1D+1S
F 5	10001	10010	2366.6220	309.8018	0.1212	3P+1D
F 4	11001	11000	2513.3548	456.5574	0.1440	2P
F 4	11110	10000	2502.8685	445.9032	-0.0239	4P
F 4	10111	10000	2479.3217	422.3170	-0.0632	4S
F 4	10110	10010	2473.1057	416.0332	-0.1310	2P+2D
F 4	10110	10001	2476.6064	419.7897	0.1247	2*2D+4S
F 3	11110	11000	2600.5972	543.8281	0.1724	3P
F 3	11001	11001	2596.1719	539.5804	0.3498	2*1D+1S
F 3	11001	11010	2599.1724	542.2892	0.0582	3P+1D
F 3	11111	10000	2591.3583	534.6418	0.2249	5S
F 3	10111	11000	2576.2925	519.4108	0.0597	5S+3*3S
F 3	11110	10010	2580.5701	523.8487	0.2200	3P+3D
F 3	10110	11010	2575.5472	518.7904	0.1846	3D+3P+1D+1P
F 3	11110	10001	2583.6365	526.8854	0.1904	8*3D+3*5S+3S
F 3	10110	11001	2578.6136	521.8437	0.1715	2*3D+2*1D+3S+5S
F 3	10001	10111	2557.6590	500.6649	-0.0527	3P
F 3	10110	10110	2553.2526	496.0646	-0.2466	2*1D+1S
F 3	10110	10101	2556.3824	499.4927	0.0517	3P+1D
F 2	11111	11000	2663.4364	606.5913	0.0962	4S
F 2	11110	11010	2658.1357	601.2812	0.0869	2P+2D
F 2	11110	11001	2660.7538	603.7728	-0.0396	2*2D+4S
F 2	10001	11111	2644.4931	587.7333	0.1817	4P
F 2	11001	10111	2635.1806	578.2614	0.0222	2*2P+4P
F 2	10110	11110	2635.8760	578.9523	0.0177	2*2D+2S
F 2	10110	11101	2638.5689	581.7604	0.1330	3*2D+2*4P+2P
F 2	10110	10111	2613.9117	556.8257	-0.1446	2P
F 1	11001	11111	2698.5527	641.5434	-0.0678	3P
F 1	11110	11110	2695.0553	637.9821	-0.1318	2*1D+1S
F 1	11110	11101	2697.2950	640.2597	-0.0938	3P+1D
F 1	10110	11111	2678.0572	621.1116	-0.0042	3P
F 1	11110	10111	2673.7673	616.4838	-0.3421	3P+1P
F 0	11110	11111	2715.8919	658.9775	0.0270	2P
F 0	10111	11111	2694.6845	638.0889	0.3458	2S
F-1	11111	11111	2718.1034	662.3935	1.2315	1S

Atomic species			E_{HAM}	E_{exp}	$E_{exp}-E_{HAM}$	Average state
NE10	00000	00000	0.0	0.0	0.0	
NE 9	10000	00000	1362.2764	1362.2670	-0.0094	2S
NE 9	01000	00000	340.6995	340.6979	-0.0016	2S
NE 9	00001	00000	340.4147	340.4145	-0.0002	2P
NE 6	11000	11000	3003.9098	0.0	0.0	1S
NE 5	11001	11000	3161.8941	157.8512	-0.1331	2P
NE 5	11110	10000	3149.6492	145.6480	-0.0915	4P
NE 5	10110	11000	3137.1180	133.2048	-0.0034	2*2P+4P
NE 5	11001	10001	3137.5742	133.5511	-0.1133	2*2D+2S
NE 5	11001	10010	3141.6875	137.5960	-0.1817	3*2D+2*4P+2P
NE 4	11110	11000	3288.2010	284.3306	0.0393	3P
NE 4	11001	11001	3282.8382	279.2793	0.3509	2*1D+1S
NE 4	11001	11010	3286.5107	282.4995	-0.1015	3P+1D
NE 4	11111	10000	3277.1742	273.4090	0.1446	5S
NE 4	10111	11000	3259.4811	255.6923	0.1209	5S+3*3S
NE 4	11110	10010	3264.3137	260.6167	0.2128	3P+3D
NE 4	10110	11010	3258.4152	254.7167	0.2113	3D+3P+1D+1P
NE 4	11110	10001	3268.0491	264.2465	0.1072	8*3D+3*5S+3S
NE 4	10110	11001	3262.1506	258.3646	0.1238	2*3D+2*1D+3S+5S
NE 4	10001	10111	3237.2541	233.2134	-0.1309	3P
NE 3	11111	11000	3385.5249	381.6181	0.0030	4S
NE 3	11110	11010	3379.0200	375.2255	0.1152	2P+2D
NE 3	11110	11001	3382.3108	378.2328	-0.1682	2*2D+4S
NE 3	10001	11111	3362.4282	358.7769	0.2585	4P
NE 3	11001	10111	3351.3680	347.5551	0.0968	2*2P+4P
NE 3	10110	11110	3352.0057	348.2665	0.1706	2*2D+2S
NE 3	10110	11101	3355.3664	351.6573	0.2007	3*2D+2*4P+2P
NE 3	10110	10111	3325.6211	321.4911	-0.2202	2P
NE 2	11001	11111	3449.0728	445.3230	0.1600	3P
NE 2	11110	11110	3444.6370	440.9277	0.2005	2*1D+1S
NE 2	11110	11101	3447.5491	443.7434	0.1040	3P+1D
NE 2	10110	11111	3423.5391	420.0012	0.3719	3P
NE 2	11110	10111	3418.3774	414.7386	0.2710	3P+1P
NE 1	11110	11111	3490.2616	486.3005	-0.0514	2P
NE 1	10111	11111	3463.1532	459.4214	0.1780	2S
NE 0	11111	11111	3511.7496	507.8982	0.0585	1S
NA 9	10000	10000	3113.7964	0.0	0.0	1S
NA 8	11000	10000	3414.1566	299.8741	-0.4861	2S
NA 8	10001	10000	3395.6025	281.7960	-0.0100	2P
NA 7	11000	11000	3677.3509	564.1155	0.5610	1S
NA 7	11001	10000	3662.3319	548.2955	-0.2399	3P
NA 7	10001	11000	3654.8199	541.1274	0.1039	3P+1P
NA 7	10110	10000	3636.9275	523.2904	0.1593	3P
NA 7	10001	10001	3629.7517	515.8354	-0.1199	2*1D+1S
NA 7	10001	10010	3634.9641	521.3206	0.1529	3P+1D
NA 6	11001	11000	3885.8053	772.4514	0.4425	2P
NA 6	11110	10000	3871.7720	758.2046	0.2290	4P
NA 6	10110	11000	3857.4843	744.2836	0.5957	2*2P+4P
NA 6	11001	10001	3857.8867	744.7206	0.6303	2*2D+2S
NA 6	11001	10010	3862.6688	749.2011	0.3287	3*2D+2*4P+2P
NA 6	10110	10010	3831.8524	718.2335	0.1776	2P+2D
NA 6	10110	10001	3836.6888	723.3569	0.4645	2*2D+4S
NA 6	10111	10000	3840.4528	727.0652	0.4088	4S

Atomic species			E_{HAM}	E_{exp}	$E_{exp}-E_{HAM}$	Average state
NA 3	11001	11111	4295.4485	0.0	0.0	3P
NA 3	11110	11101	4293.6599	-1.8949	-0.1063	3P+1D
NA 3	11110	11110	4290.0767	-5.2640	0.1078	2*1D+1S
NA 3	10110	11111	4264.8484	-30.2090	0.3911	3P
NA 3	11110	10111	4258.8125	-36.3947	0.2413	3P+1P
NA 5	10110	10110	3992.7250	-303.2413	-0.5178	2*1D+1S
NA 4	11111	11000	4196.6382	-98.8450	-0.0347	4S
NA 4	11110	11010	4188.9327	-106.2879	0.2279	2P+2D
NA 4	11110	11001	4192.8947	-102.7778	-0.2241	2*2D+4S
NA 4	10001	11111	4169.3645	-125.6845	0.3995	4P
NA 4	11001	10111	4156.5530	-138.5706	0.3249	2*2P+4P
NA 4	10110	11110	4157.1353	-137.8696	0.4436	2*2D+2S
NA 4	10110	11101	4161.1639	-133.9062	0.3785	3*2D+2*4P+2P
NA 4	10110	10111	4126.2962	-169.2854	-0.1331	2P
NA 5	11110	11000	4058.0037	-237.6446	-0.1998	3P
NA 5	11001	11010	4056.0489	-239.7581	-0.3585	3P+1D
NA 5	11001	11001	4051.7055	-243.4800	0.2629	2*1D+1S
NA 5	11111	10000	4045.1618	-250.3214	-0.0348	5S
NA 5	11110	10010	4030.2292	-265.1075	0.1119	3P+3D
NA 5	10111	11000	4024.8377	-270.5077	0.1031	5S+3*3S
NA 5	10110	11010	4023.4538	-271.8284	0.1663	3D+3P+1D+1P
NA 5	10110	11001	4027.8578	-267.5922	-0.0015	2*3D+2*1D+3S+5S
NA 5	11110	10001	4034.6331	-260.8862	-0.0708	8*3D+3*5S+3S
NA 2	11110	11111	4367.3170	71.6798	-0.1887	2P
NA 2	10111	11111	4334.2672	38.9480	0.1293	2S
NA 1	11111	11111	4414.9068	119.0422	-0.4162	1S
MG12	00000	00000	0.0	0.0	0.0	
MG11	10000	00000	1962.7407	1962.7524	0.0117	2S
MG11	01000	00000	490.9556	490.9571	0.0015	2S
MG11	00001	00000	490.3649	490.3650	0.0001	2P
MG10	10000	10000	3724.5351	3724.5489	0.0138	1S
MG 9	11000	10000	4092.5833	4092.0312	-0.5522	2S
MG 9	10001	10000	4071.8877	4071.8671	-0.0206	2P
MG 8	11000	11000	4419.4461	4420.0394	0.5933	1S
AL13	00000	00000	0.0	0.0	0.0	
AL12	10000	00000	2304.2031	2304.2384	0.0353	2S
AL12	01000	00000	576.4233	576.4283	0.0049	2S
AL12	00001	00000	575.6095	575.6100	0.0005	2P
AL11	10000	10000	4390.2677	4390.3198	0.0521	1S
AL10	11000	10000	4832.8960	4832.3211	-0.5749	2S
AL10	10001	10000	4810.0120	4810.0224	0.0103	2P

Atomic species	E_{HAM}	E_{exp}	$E_{exp}-E_{HAM}$	Average state
LI 0 11000 10000	203.6741	203.5090		2S
LI 1 11000 00000	139.0229	139.0914	0.0685	3S
LI 1 10000 01000	138.0125	138.1316	0.1192	3S+1S
BE 0 11000 11000	399.4522	399.1802		1S
BE 1 11000 01000	274.8650	275.85	0.9850	2S
B 0 11001 11000	670.8947	671.0278		2P
B 1 11001 01000	470.2884	470.09	−0.1984	3P
B 1 01001 11000	469.4174	—	—	3P+1P
C 0 11110 11000	1029.8749	1030.1605		3P
C 1 11110 01000	734.7123	733.67	−1.0423	4P
C 1 01110 11000	732.4563	—	—	2*2P+4P
C 0 10001 10111	1010.7841	—	—	3P
C 1 00001 10111	714.4591	—	—	4P
C 1 10001 00111	711.9552	—	—	2*2P+4P
N 0 11111 11000	1485.8579	—	—	4S
N 1 11111 01000	1077.5969	—	—	5S
N 1 01111 11000	1073.4378	—	—	5S+3*3S
O 0 11001 11111	2043.7642	—	—	3P
O 1 01001 11111	1500.5948	—	—	4P
O 1 11001 01111	1497.3044	—	—	2*2P+4P
F 0 11110 11111	2715.8919	—	—	2P
F 1 01110 11111	2018.7810	—	—	3P
F 1 11110 01111	2016.8767	—	—	3P+1P
NE 0 11111 11111	3511.7496	0.0	0.0	1S
NE 1 01111 11111	2641.6278	−869.9999	0.1219	2S
NE 2 01110 11111	2594.4330	−917.2498	0.0667	3P
NE 2 11110 01111	2592.1091	−919.5748	0.0656	3P+1P

E.6. The multiplet split in atomic spectroscopy.

It is well known that e.g. the carbon atom, C $1s^2 2s^2 2p^2$, exists in three states: ^3P, ^1D and ^1S, with energies, which are well known from spectroscopy.

The relative energies of these three states were studied by Slater [10] :

$$\left.\begin{aligned} ^3P &= E_{av} - \frac{3}{25} F^2 \\ ^1D &= E_{av} + \frac{3}{25} F^2 \\ ^1S &= E_{av} + \frac{12}{25} F^2 \end{aligned}\right\} \quad (E.9)$$

where F^2 is a constant which can be calculated.

The two energy intervals are then $\frac{6}{25} F^2$ and $\frac{9}{25} F^2$ and the ratio between the intervals is therefore 1.50.

Experimental studies of many isoelectronic atoms have, however, shown that the experimental ratio is about 1.14 [17-21].

The difficulty was solved by Trees [22,23], who proposed that a correction term $\alpha L(L+1)$ has to be added to the energies in eq. (E.9). Here L means the azimuthal quantum number. Slater's formulas are then changed into

$$\left.\begin{aligned} ^3P &= E_{av} - \frac{3}{25} F^2 + 2\alpha \\ ^1D &= E_{av} + \frac{3}{25} F^2 + 6\alpha \\ ^1S &= E_{av} + \frac{12}{25} F^2 \end{aligned}\right\} \quad (E.10)$$

Later, Rajnak and Wybourne [24] and Lindgren [25] presented second order perturbation proofs for the Trees' correction. Recently, Freed [26] has studied the problem using his "effective valence shell hamiltonian".

All these theoretical treatments are rather complicated, and the important Trees' correction seems to have been more or less neglected in recent comprehensive works on atomic spectroscopy [17-20].

Below we present a proof, recently given by Åsbrink [2], which is simpler than the perturbation theory proofs.

It is obvious that a satisfactory description of the atomic states is to be expected from our general methods, since our description of the atoms starts from eqs. (D.22-29) which are equivalent to the Hartree-Fock method with correlation energies included.

One may, however, wonder, how states such as ^3P and ^1D may be studied using our methods which are valid only for one-determinant wavefunctions, i.e. for pure spin-configurations, in which the angular momenta are undefined. The answer is given by Table E.2. Our proof can therefore be performed by dealing only with spin-configurations, and the results can afterwards be translated into the language of states by use of this table.

E.7. The average state

In our study of the multiplet energies, e.g. those in eq. (E.10), we must first of all have expressions for the average. It is well known that for the states in eq. (E.10) the "average energy" is

$$E_{av} = \frac{9E(^3P) + 5E(^1D) + E(^1S)}{15} \tag{E.11}$$

since the statistical weight of e.g. ^3P is 3x3 = 9.

We propose now a new theoretical expression for the average energy

$$E_{av} = -\frac{1}{2}\sum_{\mu} N_{\mu}\, \overline{\zeta}_{\mu}^{\,2} \tag{E.12}$$

where $\overline{\zeta}_{\mu}$ is the "average orbital exponent", which depends upon the "average shielding parameters"

$$\overline{\zeta}_{\mu} = \frac{1}{n_{\mu}}\left[Z + \sigma_{\mu\mu} - \sum_{\nu} N_{\nu}\, \sigma_{\nu\mu} \right] \tag{E.13}$$

Here, N_{ν} is the number of electrons in ν where ν means 1s, 2s, 2p, 3s, 3p, 3d, 4s or 4p. We distinguish thus neither between different spins, nor between the different orbitals in e.g. 2p, and this averaging concerns both numbers of electrons and shielding efficiencies.

E.8. Energies of terms in relation to energies of average states

We start our discussion from eq. (E.1)

$$E = -\frac{1}{2}\sum_{\mu}\left[N_{\mu}^{\alpha}\, \zeta_{\mu}^{\alpha\,2} + N_{\mu}^{\beta}\, \zeta_{\mu}^{\beta\,2} \right] \tag{E.14}$$

and introduce average orbital exponents $\overline{\zeta}$ and average shielding efficiencies σ_{pp}, σ_{sp}, σ_{ps} and σ_{ss}. We have further

$$\zeta_{\mu}^{\alpha} = \overline{\zeta}_{\mu} + \Delta\zeta_{\mu}^{\alpha} \tag{E.15}$$

and correspondingly for β spin. This gives

$$\zeta_{\mu}^{\alpha\,2} = \overline{\zeta}_{\mu}^{\,2} + 2\,\overline{\zeta}_{\mu}\cdot\Delta\zeta_{\mu}^{\alpha} \tag{E.16}$$

if we neglect the square of $\Delta\zeta_{\mu}^{\alpha}$.

We restrict now the treatment to 2s and 2p electrons only, and introduce the abbreviations

$$\begin{aligned}
a &= \sigma_{pp} - \sigma_{xx}^{\alpha\alpha} & \bar{a} &= \sigma_{pp} - \sigma_{xx}^{\alpha\beta} \\
b &= \sigma_{pp} - \sigma_{xy}^{\alpha\alpha} & \bar{b} &= \sigma_{pp} - \sigma_{xy}^{\alpha\beta} \\
c &= \sigma_{sp} - \sigma_{sx}^{\alpha\alpha} & \bar{c} &= \sigma_{sp} - \sigma_{sx}^{\alpha\beta} \\
d &= \sigma_{ps} - \sigma_{xs}^{\alpha\alpha} & \bar{d} &= \sigma_{ps} - \sigma_{xs}^{\alpha\beta} \\
e &= \sigma_{ss} - \sigma_{ss}^{\alpha\alpha} & \bar{e} &= \sigma_{ss} - \sigma_{ss}^{\alpha\beta}
\end{aligned} \tag{E.17}$$

We find then

$$E =$$

$$= -\tfrac{1}{2} N_s \overline{\zeta}_s^{\,2} - \tfrac{1}{2} \sum_{\mu = x,y,z} N_\mu \overline{\zeta}_p^{\,2} - \tag{E.18}$$

$$\begin{aligned}
&- \overline{\zeta}_s \cdot \frac{1}{n_s} \cdot N_s^{\alpha} \cdot \Big[N_s^{\alpha}\cdot e + N_s^{\beta}\cdot \bar{e} + \sum_{\mu = x,y,z} N_\mu^{\alpha}\cdot d + \sum_{\mu = x,y,z} N_\mu^{\beta}\cdot \bar{d} \Big] - \\
&- \overline{\zeta}_s \cdot \frac{1}{n_s} \cdot N_s^{\beta} \cdot \Big[N_s^{\beta}\cdot e + N_s^{\alpha}\cdot \bar{e} + \sum_{\mu = x,y,z} N_\mu^{\beta}\cdot d + \sum_{\mu = x,y,z} N_\mu^{\alpha}\cdot \bar{d} \Big] -
\end{aligned} \tag{E.19}$$

$$\begin{aligned}
&- \overline{\zeta}_p \cdot \frac{1}{n_p} \cdot \sum_{\mu = x,y,z} N_\mu^{\alpha} \cdot \Big[N_s^{\alpha}\cdot c + N_s^{\beta}\cdot \bar{c} + N_\mu^{\alpha}\cdot a + N_\mu^{\beta}\cdot \bar{a} + \sum_{\nu \neq \mu} N_\nu^{\alpha}\cdot b + \sum_{\nu \neq \mu} N_\nu^{\beta}\cdot \bar{b} \Big] \\
&- \overline{\zeta}_p \cdot \frac{1}{n_p} \cdot \sum_{\mu = x,y,z} N_\mu^{\beta} \cdot \Big[N_s^{\beta}\cdot c + N_s^{\alpha}\cdot \bar{c} + N_\mu^{\beta}\cdot a + N_\mu^{\alpha}\cdot \bar{a} + \sum_{\nu \neq \mu} N_\nu^{\beta}\cdot b + \sum_{\nu \neq \mu} N_\nu^{\alpha}\cdot \bar{b} \Big]
\end{aligned} \tag{E.20}$$

If eq.(E.18) represents the average E_{av}, then eq.(E.19) and eq.(E.20) must represent the deviations from the energy average (energy splitting). This means that if the configuration consists of only one term, then this term is identical to the average, and the sum of eqs.(E.19) and (E.20) must be zero. This gives us a possibility to reduce the number of parameters in eqs.(E.19) and (E.20).

First, consider an atom with only one s electron. Then $N_s^{\alpha} = 1$ and all other $N = 0$. There is only one state: ^2S. This gives $e = 0$.

Second, consider an atom with only two s electrons. Then $N_s^{\alpha} = 1$ and $N_s^{\beta} = 1$ and all other $N = 0$. There is only one state: ^1S. This gives $\bar{e} = 0$.

Third, consider an atom with only one p electron. Then $N_x^{\alpha} = 1$ and all other $N = 0$. The state is ^2P. This gives

$$a = 0 \qquad (E.21)$$

Fourth, consider an atom with six p electrons and no s electrons. All $N_{\mu} = 1$ and all $N_s = 0$. The state is ^1S. This gives directly from eq.(E.19)

$$a + \bar{a} + 2b + 2\bar{b} = 0 \qquad (E.22)$$

or, since $a = 0$

$$\bar{a} + 2b + 2\bar{b} = 0 \qquad (E.23)$$

Fifth, consider an atom with five p electrons and no s electrons. Take $N_x^{\alpha} = -1$ in the third case and superimpose on the fourth case. This gives again eq.(E.23).

Sixth, consider an atom with six p electrons and one s electron ($N_s^{\alpha} = 1$). This gives

$$\bar{\zeta}_s \cdot \frac{1}{n_s} \cdot \left[d + \bar{d} \right] + \bar{\zeta}_p \cdot \frac{1}{n_p} \cdot \left[c + \bar{c} \right] = 0 \qquad (E.24)$$

Using these results we can simplify eqs.(E.19) and (E.20). We write also E_{av} for eq.(E.18) and obtain

$$E = E_{av} -$$

$$-\left[\bar{\zeta}_s \cdot \frac{1}{n_s} \cdot \bar{d} + \bar{\zeta}_p \cdot \frac{1}{n_p} \cdot \bar{c}\right]\left[N_s^{\alpha} - N_s^{\beta}\right] \cdot \sum_{\substack{\mu = \\ x,y,z}} \left(N_\mu^{\alpha} - N_\mu^{\beta}\right) - \qquad (E.25)$$

$$-\bar{\zeta}_p \cdot \frac{1}{n_p} \cdot \bar{a} \cdot \left[2 \cdot \sum_{\substack{\mu = \\ x,y,z}} N_\mu^{\alpha} N_\mu^{\beta} - \frac{1}{2} \sum_{\substack{\mu,\nu = \\ x,y,z \\ \mu \neq \nu}} \left(N_\mu^{\alpha} N_\nu^{\beta} + N_\mu^{\beta} N_\nu^{\alpha}\right)\right] - \qquad (E.26)$$

$$-\bar{\zeta}_p \cdot \frac{1}{n_p} \cdot b \cdot \sum_{\substack{\mu,\nu = \\ x,y,z \\ \mu \neq \nu}} \left(N_\mu^{\alpha} - N_\mu^{\beta}\right)\left(N_\nu^{\alpha} - N_\nu^{\beta}\right) \qquad (E.27)$$

It is interesting to observe, that the parenthesis in eq. (E.26) can be written in a condensed form:

$$[\quad] = \sum_{\mu > \nu} \left(N_\mu^{\alpha} - N_\nu^{\alpha}\right)\left(N_\mu^{\beta} - N_\nu^{\beta}\right)$$

We introduce abbreviations:

$$A = \bar{\zeta}_p \cdot \frac{1}{n_p} \cdot \bar{a} \qquad (E.28)$$

$$B = \bar{\zeta}_p \cdot \frac{1}{n_p} \cdot b \qquad (E.29)$$

$$C = \left[\bar{\zeta}_s \cdot \frac{1}{n_s} \cdot \bar{d} + \bar{\zeta}_p \cdot \frac{1}{n_p} \cdot \bar{c}\right] \qquad (E.30)$$

which gives with summations over p orbitals

$$E = E_{av} - C \cdot \left[N_s^{\alpha} - N_s^{\beta}\right] \cdot \sum_{\mu} \left(N_\mu^{\alpha} - N_\mu^{\beta}\right) - \qquad (E.31)$$

$$- A \cdot \left[2 \cdot \sum_{\mu} N_\mu^{\alpha} N_\mu^{\beta} - \frac{1}{2} \sum_{\mu \neq \nu} \left(N_\mu^{\alpha} N_\nu^{\beta} + N_\mu^{\beta} N_\nu^{\alpha}\right)\right] \qquad (E.32)$$

$$- B \cdot \sum_{\mu \neq \nu} \left(N_\mu^{\alpha} - N_\mu^{\beta}\right)\left(N_\nu^{\alpha} - N_\nu^{\beta}\right) \qquad (E.33)$$

Study of the p^2 configuration

The expressions (E.32) and (E.33) can be used directly to obtain the energies of the terms of the p^2 configuration.

If we take $N_x^\alpha = N_Y^\alpha = 1$ we obtain

$$E = E_{av} - 2B$$

since in eq.(E.33) μ is x or Y . According to Table E.2. this is ^3P.

If we take $N_x^\alpha = N_x^\beta = 1$ we obtain

$$E = E_{av} - 2A$$

which according to Table E.2. is $(2 \cdot {}^1D + {}^1S)/3$.

If we take $N_x^\alpha = N_Y^\beta = 1$ we obtain

$$E = E_{av} + A + 2B$$

which according to Table E.2. is $({}^3P + {}^1D)/2$.

We have thus

$$\begin{cases} {}^3P = E_{av} - 2B \\ 2 \cdot {}^1D + {}^1S = 3E_{av} - 6A \\ {}^3P + {}^1D = 2E_{av} + 2A + 4B \end{cases}$$

Subtraction gives

$$\begin{cases} {}^3P = E_{av} - 2B \\ {}^1D = E_{av} + 2A + 6B \\ {}^1S = E_{av} - 10A - 12B \end{cases}$$

We observe that if we form the average of ^3P , ^1D and ^1S , giving them the weights 9, 5 and 1, respectively, we obtain just E_{av}.

We observe further, that if we assume that $A = -2B$ and $B = 3/50\, F^2$, then we obtain (as already shown in eqs.(E.9))

$$\begin{cases} {}^3P = E_{av} - 3/25\, F^2 \\ {}^1D = E_{av} + 3/25\, F^2 \\ {}^1S = E_{av} + 12/25\, F^2 \end{cases}$$

which is the result of Slater's work (see ref. [10] p.343, Table 15-3, or ref. [17] , p.575, App.3, case 3).

Study of the p^3 configuration

If we take $N_x^\alpha = N_y^\alpha = N_z^\alpha = 1$ we obtain

$$E = E_{av} - 6B$$

which is ^4S.

If we take $N_x^\alpha = N_y^\alpha = N_x^\beta = 1$ we obtain

$$E = E_{av} - A\cdot\left[2 - \tfrac{1}{2}\cdot(1+1)\right] - B\cdot\left[2-2\right] = E_{av} - A$$

which is $(^2P + {}^2D)/2$.

If we take $N_x^\alpha = N_y^\alpha = N_z^\beta = 1$ we obtain

$$E = E_{av} - A\cdot\left[-\tfrac{1}{2}\cdot(1+1+1+1)\right] - B\cdot\left[2-2-2\right] =$$

$$= E_{av} + 2A + 2B$$

which is $(2\cdot {}^2D + {}^4S)/3$

This gives

$$\begin{cases} ^4S = E_{av} \qquad\quad - 6B \\ ^2D = E_{av} + 3A + 6B \\ ^2P = E_{av} - 5A - 6B \end{cases}$$

We observe that if we form the average, using the weights 4, 10 and 6, respectively, we obtain just E_{av} , and if we assume $A = -2B$ with $B = 3/50\ F^2$ we obtain

$$\begin{cases} ^4S = E_{av} - 9/25\ F^2 \\ ^2D = E_{av} \\ ^2P = E_{av} + 6/25\ F^2 \end{cases}$$

which is again the result of Slater's work.

Study of the sp configuration

If we take $N_s^\alpha = N_x^\alpha = 1$ we obtain

$$E = E_{av} - C\cdot 1\cdot 1 = E_{av} - C$$

which is ^3P.

If we take $N_s^\alpha = N_x^\beta = 1$ we obtain

$$E = E_{av} - C\cdot 1\cdot(-1) = E_{av} + C$$

which is $(^3P + {}^1P)/2$.

This gives

$$\begin{cases} {}^3P = E_{av} - C \\ {}^1P = E_{av} + 3C \end{cases}$$

If we here take $C = 1/6\ G^1$ we obtain

$$\begin{cases} {}^3P = E_{av} - 1/6\ G^1 \\ {}^1P = E_{av} + 3/6\ G^1 \end{cases}$$

which is again Slater's result.

Energies of terms arising from configurations of s and p electrons

If our studies above are extended to other configurations, we obtain the results presented in Table E.8.a.

If we make the following substitutions in Table E.8.a.

$$A = -2B \tag{E.34}$$

$$B = 3/50\ F^2(pp) \tag{E.35}$$

$$C = 1/6\ G^1(sp) \tag{E.36}$$

then Table E.8.a. will be transformed into the similar table presented by Slater [10] and reproduced as Table E.8.b.

To achieve the introduction of the Trees correction we make the following substitutions in Table E.8.a.

$$A = -3/25\ F^2 - 0.4\ \alpha \tag{E.37}$$

$$B = +3/50\ F^2 + 0.6\ \alpha \tag{E.38}$$

$$C = +1/6\ G^1 \tag{E.39}$$

In for instance the p^2 case this gives

$${}^3P = E_{av} - 3/25F^2 - 1.2\alpha = (E_{av} - 3.2\alpha) - 3/25F^2 + 2\alpha \tag{E.40}$$

$${}^1D = E_{av} + 3/25F^2 + 2.8\alpha = (E_{av} - 3.2\alpha) + 3/25F^2 + 6\alpha \tag{E.41}$$

$${}^1S = E_{av} + 12/25F^2 - 3.2\alpha = (E_{av} - 3.2\alpha) + 12/25F^2 \tag{E.42}$$

It is satisfactory that the same substitution can be used in all cases in Table E.8.a. and still in all cases gives the correct correction proportional to $L(L+1)$.

Table E.8.a. Energies of terms, arising from configurations of s and p electrons.

sp or sp^5	$^3P^o$:	E_{av} $-$ C
	$^1P^o$:	E_{av} + 3C
p^2 or s^2p^2 or p^4 or s^2p^4	^3P:	E_{av} $-$ 2B
	^1D:	E_{av} + 2A + 6B
	^1S:	E_{av} $-$ 10A $-$ 12B
sp^2 or sp^4	^4P:	E_{av} $-$ 2B $-$ 2C
	^2P:	E_{av} $-$ 2B + 4C
	^2D:	E_{av} + 2A + 6B
	^2S:	E_{av} $-$ 10A $-$ 12B
p^3 or s^2p^3	$^4S^o$:	E_{av} $-$ 6B
	$^2D^o$:	E_{av} + 3A + 6B
	$^2P^o$:	E_{av} $-$ 5A $-$ 6B
sp^3	$^5S^o$:	E_{av} $-$ 6B $-$ 3C
	$^3S^o$:	E_{av} $-$ 6B + 5C
	$^3D^o$:	E_{av} + 3A + 6B $-$ C
	$^1D^o$:	E_{av} + 3A + 6B + 3C
	$^3P^o$:	E_{av} $-$ 5A $-$ 6B $-$ C
	$^1P^o$:	E_{av} $-$ 5A $-$ 6B + 3C

Table E.8.b. Energies of terms according to Slater [10].

sp or sp^5	$^3P^o$:	$E_{av} - \frac{1}{6}G^1(sp)$
	$^1P^o$:	$E_{av} + \frac{3}{6}G^1(sp)$
p^2 or s^2p^2 or p^4 or s^2p^4	^3P:	$E_{av} - \frac{3}{25}F^2(pp)$
	^1D:	$E_{av} + \frac{3}{25}F^2(pp)$
	^1S:	$E_{av} + \frac{12}{25}F^2(pp)$
sp^2 or sp^4	^4P:	$E_{av} - \frac{3}{25}F^2(pp) - \frac{1}{3}G^1(sp)$
	^2P:	$E_{av} - \frac{3}{25}F^2(pp) + \frac{2}{3}G^1(sp)$
	^2D:	$E_{av} + \frac{3}{25}F^2(pp)$
	^2S:	$E_{av} + \frac{12}{25}F^2(pp)$
p^3 or s^2p^3	$^4S^o$:	$E_{av} - \frac{9}{25}F^2(pp)$
	$^2D^o$:	E_{av}
	$^2P^o$:	$E_{av} + \frac{6}{25}F^2(pp)$
sp^3	$^5S^o$:	$E_{av} - \frac{9}{25}F^2(pp) - \frac{1}{2}G^1(sp)$
	$^3S^o$:	$E_{av} - \frac{9}{25}F^2(pp) + \frac{5}{6}G^1(sp)$
	$^3D^o$:	$E_{av} - \frac{1}{6}G^1(sp)$
	$^1D^o$:	$E_{av} + \frac{1}{2}G^1(sp)$
	$^3P^o$:	$E_{av} + \frac{6}{25}F^2(pp) - \frac{1}{6}G^1(sp)$
	$^1P^o$:	$E_{av} + \frac{6}{25}F^2(pp) + \frac{1}{2}G^1(sp)$

E.9. The physical meaning of the parameters

The meaning of Åsbrink's parameter A is found from its definition in eq. (E.28) together with the expressions (D.31) and (D.32) for the shielding efficiencies

$$A = \bar{\zeta}_p \cdot \frac{1}{n_p} \cdot \left[\sigma_{xx}^{\alpha\alpha} - \sigma_{xx}^{\alpha\beta} \right] =$$

$$= \bar{\zeta}_p \cdot \frac{1}{n_p} \cdot \left[\frac{(xx|xx)^{\alpha\alpha} + \varepsilon_{xx}^{\alpha\alpha}}{\frac{2}{n_x} \cdot \zeta_x^{\alpha}} - \frac{(xx|xx)^{\alpha\beta} + \varepsilon_{xx}^{\alpha\beta}}{\frac{2}{n_x} \cdot \zeta_x^{\alpha}} \right] =$$

$$= \frac{1}{2} \left(\varepsilon_{xx}^{\alpha\alpha} - \varepsilon_{xx}^{\alpha\beta} \right) \tag{E.43}$$

if we assume that $(xx|xx)$ is independent of the spins.

In the same way we find B by use of the relations (see p.81 in ref. [27])

$$(xx|xx) = F^0 + \frac{4}{25} F^2 \tag{E.44}$$

$$(xx|yy) = F^0 - \frac{2}{25} F^2 \tag{E.45}$$

$$(xy|xy) = \frac{3}{25} F^2 \tag{E.46}$$

together with eqs. (D.31) and (D.33)

$$B = \bar{\zeta}_p \cdot \frac{1}{n_p} \cdot \left[\sigma_{xx}^{\alpha\alpha} - \sigma_{yx}^{\alpha\alpha} \right] =$$

$$= \bar{\zeta}_p \cdot \frac{1}{n_p} \cdot \left[\frac{(xx|xx)^{\alpha\alpha} + \varepsilon_{xx}^{\alpha\alpha}}{\frac{2}{n_x} \cdot \zeta_x^{\alpha}} - \frac{(xx|yy)^{\alpha\alpha} + \varepsilon_{yx}^{\alpha\alpha} - (xy|xy)}{\frac{2}{n_x} \cdot \zeta_x^{\alpha}} \right] =$$

$$= \frac{9}{50} F^2 + \frac{1}{2} \left(\varepsilon_{xx}^{\alpha\alpha} - \varepsilon_{yx}^{\alpha\alpha} \right) \tag{E.47}$$

Comparing eq. (E.43) with (E.37) and eq. (E.47) with (E.38) we obtain

$$A = -\frac{3}{25} F^2 - 0.4\,\alpha = \frac{1}{2} \left(\varepsilon_{xx}^{\alpha\alpha} - \varepsilon_{xx}^{\alpha\beta} \right) \tag{E.48}$$

$$B = +\frac{3}{50} F^2 + 0.6\,\alpha = \frac{1}{2} \left(\varepsilon_{xx}^{\alpha\alpha} - \varepsilon_{yx}^{\alpha\alpha} \right) + \frac{9}{50} F^2 \tag{E.49}$$

Subtraction gives

$$\alpha = \frac{1}{2}\left(\varepsilon_{xx}^{\alpha\beta} - \varepsilon_{yx}^{\alpha\alpha}\right) \tag{E.50}$$

Trees´s parameter can therefore be interpreted as a difference of two pair-correlation energies. Such a description is evidently identical to the results from the conventional deduction of the Trees correction using second-order perturbation theory [24, 25].

The Slater approximation, which gives Table E.8.b.from Table E.8.a. is obtained by taking $A = -2B$, (eq. (E.34)).This implies therefore that $\varepsilon_{xx}^{\alpha\beta}$ is assumed to be equal to $\varepsilon_{yx}^{\alpha\alpha}$.

It can finally be mentioned that α is small compared both with ε and with F^2. For carbon (see page E.9. lines 25, 26 and 27) $\alpha = 0.039$ eV, $F^2 = 4.597$ eV and ε of the order of 1 eV (according to Sec.C.3.).

E.10. The semiempirical methods HAM/3 and HAM/4

The main ideas in the semiempirical method HAM/3 were proposed already in 1972 by Åsbrink but the method was not published until 1977 [28].

The HAM/3 method was constructed by analogy with earlier methods for molecular calculations and was not deduced from the Hartree-Fock expressions. The parametrized expression

$$\sigma_{\nu\mu} = \alpha_{\nu\mu} - \frac{\beta_{\nu\mu}}{\zeta_{\mu}} \qquad (E.51)$$

was obtained empirically during the study of the atomic energies, tabulated in Sec.E.5. In this table the calculated energies of 311 configurations from hydrogen to neon are compared with experiment. The average error is only 0.16 eV, which indicates that nearly all correlation energies have been taken care of by Åsbrink's empirical formula for $\sigma_{\nu\mu}$. The mechanism behind this result (see Sec.D.4.) was not understood at that time.

The HAM/3 method was extended to molecules by determination of molecular parameters [28] and the computer program was published by Quantum Chemistry Program Exchange, QCPE, [29, 31].

HAM/4, on the other hand, was deduced from quantum mechanics (see Chaps.D. and E.) when Åsbrink in 1979 was able to deduce eqs.(D.22-29) and Åsbrink's formula (E.51) has now got an interpretation. At the same time Åsbrink has extended the table in Sec.E.5. to 2200 atomic configurations from H up to Kr (Z = 36) and improved the accuracy of the calculated energies considerably [2] . The average error in the new table is only 0.11 eV.

This result has been achieved partly by the introduction of some extra terms in the first part of eq.(E.51)

$$\alpha_{\nu\mu} = a_{\nu\mu} + d_{\nu\mu} \cdot Z - e_{\nu\mu} \cdot N_{\nu} \qquad (E.52)$$

The interpretation of these small terms is not clear, although it is possible that the last term describes the influence of the correlation between the two electrons in ν upon the correlation between μ and ν .

The HAM/4 computer program to calculate atomic energies is available [30] but the extension to molecules has not yet been completed.

References

1. L. Åsbrink and C. Fridh, unpublished results.
2. L. Åsbrink, Phys.Scripta 28, 394 (1983).
3. W. Moffitt, Ann.Rep.Progr.Phys. 17, 173 (1954).
4. G. Klopman, J.Am.Chem.Soc. 86, 1463 (1964).
5. S.P. McGlynn, L.G. Vanquickenborne, M. Kinoshita and D.G. Carroll, Introduction to Applied Quantum Chemistry, Holt, Rinehart and Winston, New York (1972).
6. A.G. Turner, Methods in Molecular Orbital Theory, Prencice-Hall, Engelwood Cliffs, N.J. (1974).
7. H.O. Pritchard and H.A. Skinner, Chem.Rev. 55, 745 (1955).
8. J. Hinze and H.H. Jaffé, J.Am.Chem.Soc. 84, 540 (1962).
9. J.C. Slater, Phys.Rev. 36, 57 (1930).
10. J.C. Slater, Quantum Theory of Atomic Structure, Vol.1, McGraw-Hill, New York (1960).
11. D. Layzer, Annals of Physics 8, 271 (1959).
12. E. Clementi and D.L. Raimondi, J.Chem.Phys. 38, 2686 (1963).
13. G. Burns, J.Chem.Phys. 41, 1521 (1964).
14. T. Anno, Theoret.Chim.Acta 18, 223 (1970).
15. F.R. Burden and R.M. Wilson, Adv.Phys. 21, 825 (1972).
16. Y. Sakai and T. Anno, J.Chem.Phys. 60, 620 (1974).
17. E.U. Condon and H. Odabasi, Atomic Structure, Cambridge University Press, Cambridge (1980).
18. I.I. Sobel'man, Introduction to the Theory of Atomic Spectra, Pergamon, Oxford (1972).
19. B.W. Shore and D.H. Menzel, Principles of Atomic Spectra, Wiley, New York (1968).
20. R.D. Cowan, The Theory of Atomic Structure and Spectra, University of California Press, Berkeley (1981), p. 343 and 480.
21. B. Edlén, Atomic Spectra, in Handbuch der Physik (Edited by S. Flügge), p.80. Band XXVII, Spektroskopie I (1964).
22. R.E. Trees, Phys.Rev. 83, 756 (1951).
23. R.E. Trees and C.K. Jørgensen, Phys.Rev. 123, 1278 (1961).
24. K. Rajnak and B.G. Wybourne, Phys.Rev. 132, 280 (1963).

References (cont.)

25. I. Lindgren and J. Morrison, Atomic Many-Body Theory, Springer Verlag, Berlin (1982).

26. J.J. Oleksik and K.F. Freed, J.Chem.Phys. 79, 1396 (1983).

27. J.A. Pople and D.L. Beveridge, Approximate Molecultar orbital Theory, McGraw-Hill, New York (1970).

28. L. Åsbrink, C. Fridh and E. Lindholm, Chem.Phys.Letters 52, 63, 69, 72 (1977).

29. L. Åsbrink, C. Fridh and E. Lindholm, QCPE No. 393 (1980).

30. L. Åsbrink, TRITA-FYS-1001, Physics Department, The Royal Institute of Technology, Stockholm (1980).

31. D.P. Chong, QCPE QCMP005 (1985).

F. Molecules

The most important calculations on molecules are for their ground states, which are usually "closed shells" with 2 electrons in each orbital.

We will therefore no longer distinguish between α spin and β spin. The relations in Chap.D. will be simplified and the notations in Chap.A. and Chap.B. will again be valid.

F.1. Interpretation of the energy expression for a molecule.

For a molecule the total energy is given by eqs. (D.22-29).

The first part of this expression, given by (D.22-24), can be treated in the same way as in atoms, although s_μ in the molecule is calculated with the actual charges there. The first part can therefore be written

$$-\frac{1}{2}\sum_\mu N_\mu \zeta_\mu^2 \qquad (F.1)$$

The characteristics of this expression will now be discussed.

The expression (F.1) takes care of the main part of the total energy of a molecule. It is important to observe that both one-center electron correlation and self-repulsion are handled in a reasonable way by this expression.

The one-center pair-correlation energies are taken care of by the shielding efficiencies $\sigma_{\nu\mu}$. This is the main part of the correlation energy in a molecule as shown by numerical studies by for instance Sinanoglu and Pamuk [1]. They find that the one-center pair-correlation energy in ethylene is about 88 % of the total correlation energy in this molecule, using an expression corresponding to eq. (C.39). For benzene and naphthalene they obtain 81 %. This result is, according to Dewar [2], very satisfactory. He points out that it is not possible by use of CI, even with the largest basis sets in current use for molecules, to take care of more than 80 % of the total correlation energy.

The self-repulsion is compensated with the term "-1" in the shielding. This was introduced by Slater and in HAM/3 in an intuitive way. The compensation is complete if no other terms can be found in (D.22-29), which can be attributed to self-repulsion. It will be shown below that such terms are small. The term "-1" in the atomic part of the energy expression therefore takes care of a large part of the exchange effects and the compensation of self-repulsion also in the bonds. This was questioned previously [3] but was later shown to be true [4] .

It may finally be pointed out that the shielding also in a molecule can be treated in a very simple way. The shielding is caused on one hand by the electrons in the atomic orbital ϕ_ν whose number is $P_{\nu\nu}$ and on the other hand by the electrons in the bonds whose number is $\sum_{\lambda \neq \nu} P_{\nu\lambda} S_{\nu\lambda}$. Eq. (D.37) shows that it is possible to use the same shielding efficiencies $\sigma_{\nu\mu}$ for both types of shielding electrons (cf. ref. [5] , where it was feared that these shielding efficiencies have to be different.

The second part, eq. (D.25), will now be discussed.

We observe that the terms in (D.25) are proportional to $P_{\mu_A \nu_B}$ and therefore responsible for the chemical bonding. Each term is further proportional to a sum of several β's which can be replaced by a single function $\beta_{\mu\nu}$, depending upon μ , ν and R_{AB} . The second part should therefore be written

$$-\sum P_{\mu\nu} \beta_{\mu\nu} \qquad (F.2)$$

The function $\beta_{\mu\nu}$ is unknown. It can be determined in two ways, either from ab-initio studies of the defects of the Mulliken approximation, performed for different bonds in different molecules, or empirically from comparison of calculated and experimental properties of molecules. Let us discuss the theoretical approach first.

An ab-initio study of the Mulliken approximation has been performed by Jesaitis [6] . He points out that the Mulliken approximation is very unsatisfactory for kinetic energy but reasonably successful for potential energy. This means that in eq. (D.25) β_{kin} is the predominant term.

Jesaitis has given the true kinetic energy and the Mulliken kinetic energy for a 2s-2s interaction in a carbon-carbon bond, from which our β_{kin} can be obtained by subtraction. His results together with the difference are shown in Table F.1.

Table F.1. Kinetic energy in a carbon-carbon bond.

Distance (Å)	True T (a.u.)	Mulliken T (a.u.)	$\frac{1}{2}\cdot\beta_{kin}$
0	0.440	0.440	0
1.30	0.076	0.201	0.125
1.95	-0.008	0.082	0.090
2.60	-0.011	0.029	0.040

It follows from the definition of β in eqs. (D.15) and (D.17) that β is a complicated function of $S_{\mu\nu}$. It must, however, behave like $S_{\mu\nu}$ even if it is not proportional to $S_{\mu\nu}$.

The following properties of β are therefore obvious.

a) β must decrease exponentially with R like $S_{\mu\nu}$. This follows from Jesaitis' calculations.

b) β must behave like $S_{\mu\nu}$ with respect to symmetry. This means that for the interaction between two carbon atoms we have four cases: $2s-2s$, $2s-2p\sigma$, $2p\sigma-2p\sigma$ and $2p\pi-2p\pi$ but the interactions $2s-2p\pi$ and $2p\sigma-2p\pi$ are zero.

c) β must be different in these four cases.

d) β must probably depend upon the nature of the interacting atoms and must be different for C-C, C-N and N-N . . .

e) If $S_{\mu\nu}$ is very small, then also $\beta_{\mu\nu}$ is very small. This is important for the interaction between 1s on one atom and a valence orbital on another.

The inaccuracies of the Mulliken approximation have recently been discussed in two papers [7, 8]. The problem was avoided by exact calculation of the kinetic energy. The problem has further been discussed by Nanda and Jug [9] in their SINDO/1 method.

Previously, these problems have only been handled in an intuitive way in several semiempirical theories. In CNDO [10] the bond term was assumed to be $\beta\cdot S_{\mu\nu}$ in order to obtain in a simple way the required distance dependence and rotational invariance. In the Extended Hückel Method [11] it was also assumed, that the overlap integral $S_{\mu\nu}$ must be a factor in the bond term.

Therefore, in the HAM/3 method it was believed that it is necessary to have $S_{\mu\nu}$ as a factor [12]. This unjustified assumption is probably one explanation why the determination of the parameter $\beta_{\mu\nu}$ was complicated and difficult in HAM/3. The use of eq. (F.2) in a future parametrization will probably be much easier.

The third part, eq. (D.26), will now be discussed.

Its first term, $\sum Z_A Z_B R_{AB}^{-1}$, describes the nucleus-nucleus repulsion.

Its last term describes the electron-electron repulsion and can be written

$$\sum_{A>B} \sum_{\mu}^{A} \sum_{\nu}^{B} N_\mu N_\nu \gamma_{\mu\nu} \qquad (F.3)$$

According to eq. (D.30)

$$\gamma_{\mu\nu} = (\mu\mu|\nu\nu) + \varepsilon_{\mu\nu} \qquad (F.4)$$

It is possible here to calculate $(\mu\mu|\nu\nu)$ from ϕ_{μ_A} and ϕ_{ν_B}, but it appears from comparison with experimental results that the two-center pair-correlation $\varepsilon_{\mu\nu}$ is not small compared to $(\mu\mu|\nu\nu)$, especially for small internuclear distances. One has therefore preferred to resort to empirical expressions for γ.

Empirical values were first used in the PPP method by Pariser and Parr [13] and later converted into a formula by Ohno and Klopman

$$\gamma_{AB} = -14.399 \left[R_{AB}^2 + (\rho_A + \rho_B)^2 \right]^{-\frac{1}{2}} \quad (eV) \qquad (F.5)$$

where R_{AB} is measured in Ångströms and ρ_A is given in MINDO/1 [14]. Since the formula is approximate, there is no reason to distinguish between different orbitals. The indices μ and ν have therefore been replaced by A and B.

The corresponding formula with exponent $n = 1$ instead of $n = 2$ is less used (the Mataga-Nishimoto formula).

The second and third terms in eq. (D.26) describe the nucleus-electron attraction. The functions V_{BA} and V_{AB} are smaller than R_{AB}^{-1} and larger than γ_{AB} but have not been discussed much. Finding suitable forms for these functions constitutes an important problem (the "penetration" problem). In the HAM/3 program the following function is used:

$$V_{AB} = \gamma_{AB} \left[1 + a \cdot \exp\left(-b \cdot R_{AB}\right) \right] \qquad (F.6)$$

where the parameter a is characteristic for atom A and the parameter b for atom B.

The last part, eqs. (D.27-29), can be written for molecules

$$\sum_{\mu\nu\lambda\sigma} P_{\mu\nu} P_{\lambda\sigma} \left[\underset{a)}{\tfrac{1}{2}(\mu\nu|\lambda\sigma)'} - \underset{b)}{\tfrac{1}{4}(\mu\sigma|\lambda\nu)'} + \underset{c)}{\tfrac{1}{4} S_{\sigma\mu} S_{\nu\lambda} (\mu\mu|\mu\mu)'} - \underset{d)}{\tfrac{1}{2} S_{\mu\nu} S_{\lambda\sigma} (\mu\mu|\lambda\lambda)'} \right] +$$

$$+ \sum_{A} \sum_{\substack{\mu\nu \\ \mu \neq \nu}}^{A} \underset{e)}{N_\mu N_\nu \cdot \tfrac{1}{4} (\mu\nu|\mu\nu)'} \qquad (F.7)$$

This is the "remaining term" which should be small. If not, the HAM/3 expression for the total energy is a bad approximation to the Hartree-Fock energy. This will be analyzed in detail below but can also be recognized by inspection.

Term a) in eq. (F.7) describes in the Hartree-Fock energy expression the electron-electron repulsion and is very large, but it is in eq. (F.7) compensated to a large extent by the term d). Term b) describes the exchange, which is also large but in eq. (F.7) compensated partly by c) and e).

We are thus able to show in a qualitative way that eq.(F.7) is small. The quantitative proof must, however, be performed for specified values of μ, ν, λ and σ. These discussions must be performed for no less than 22 different situations. We will here discuss only the ten most important situations and we denote the corresponding energy contributions as (F.7.1) - (F.7.10).

The one-center contributions to (F.7) can be expected to be the largest. We characterize the different situations by specifying in order μ, ν, λ and σ, all on atom A. We have;

For $\mu_A\ \mu_A\ \mu_A\ \mu_A$: (F.7.1) = 0

For $\mu_A\ \mu_A\ \mu_A\ \sigma_A$: (F.7.2) = 0

For $\mu_A\ \mu_A\ \lambda_A\ \sigma_A$: (F.7.3) = 0

For $\mu_A\ \nu_A\ \mu_A\ \sigma_A$: (F.7.4) = 0

For $\mu_A\ \nu_A\ \lambda_A\ \sigma_A$: (F.7.5) = 0

For $\mu_A\ \mu_A\ \nu_A\ \nu_A$: (F.7.6) = 0

For $\mu_A \nu_A \mu_A \nu_A$ we have

$$(F.7.7) = \sum_{A}^{} \sum_{\substack{\mu\nu \\ \mu>\nu}}^{A} P_{\mu\nu}^{2} \cdot \left[\frac{1}{4}(\mu\mu|\mu\mu)' + \frac{1}{4}(\nu\nu|\nu\nu)' - \frac{1}{2}(\mu\mu|\nu\nu)' + \frac{3}{2}(\mu\nu|\mu\nu)' \right]$$

This term is important in molecules. If it is neglected the rotational invariance of the SCF result is impaired. Approximately this expression has therefore been included in the HAM/3 computer program. Typical terms in (F. 7.7) are

$$P_{sx}^{2} \cdot \left[\frac{1}{25} F^{2} + \frac{1}{2} G^{1} \right] \quad \text{and} \quad P_{xy}^{2} \cdot \frac{3}{10} F^{2}$$

(In the study by Dewar and Lo [15] the last two terms in (F. 7. 7) can be identified in their eq.(4) and (5) but not the first two terms in (F. 7. 7). The explanation is that they originate from the idempotency relation.)

The two-center contributions to (F. 7) will now be discussed.

For $\mu_A \nu_A \lambda_A \sigma_B$ we have (F. 7. 8) = 0

For $\mu_A \nu_B \mu_A \nu_B$ we have

$$(F.7.9) = \sum_{\substack{A \\ A>B}} \sum_{B} \sum_{\mu}^{A} \sum_{\nu}^{B} P_{\mu\nu}^{2} \cdot \left\{ \frac{1}{4}(\mu\mu|\mu\mu)' + \frac{1}{4}(\nu\nu|\nu\nu)' - \frac{1}{2}(\mu\mu|\nu\nu)' + \right.$$

$$\left. + \frac{3}{2}(\mu\nu|\mu\nu)' - S_{\mu\nu}^{2} \cdot \left[\frac{1}{4}(\mu\mu|\mu\mu)' + \frac{1}{4}(\nu\nu|\nu\nu)' + (\mu\mu|\nu\nu)' \right] \right\}$$

It can be shown by use of Mulliken´s approximation [16,17] that the last four terms in (F. 7. 9) may be neglected.

In CNDO only the third term $-\frac{1}{2} P_{\mu\nu}^{2} \gamma_{AB}$ is present (see eq.(3.42) in ref. [10]). The term in CNDO is thus larger and compensates for the self-repulsion in the bond [3].

For $\mu_A \mu_A \nu_B \nu_B$ we have

$$(F.7.10) = \sum_{A} \sum_{B} \sum_{\mu}^{A} \sum_{\nu}^{B} P_{\mu\mu} P_{\nu\nu} \cdot \left\{ -\frac{1}{2}(\mu\nu|\mu\nu)' + \frac{1}{4} S_{\mu\nu}^{2} \left[(\mu\mu|\mu\mu)' + (\nu\nu|\nu\nu)' \right] \right\}$$

According to Mulliken´s approximation this is small and will be neglected.

F.2. Local dipoles

Many molecules contain strong local dipoles. This will be illustrated in Sec.N.2. for HCN, but also the lone-pair orbital in pyridine has a strong local dipole on N.

The local dipole moments can influence the energy of the molecule a good deal and they must therefore be added to our expression for the molecular energy.

The case $\mu_A \nu_A \lambda_B \sigma$ is important and will be discussed now together with some terms which originate from eq.(D.5) when both μ and ν are on atom A and $\mu \neq \nu$.

The two-center terms of this type in eq. (D.5) are

$$-\sum_A \sum_{\mu \neq \nu}^{A} P_{\mu\nu} \cdot \int \phi_\mu^* \phi_\nu \sum_{B \neq A} Z_B r_B^{-1} \, d\tau \tag{F.8}$$

where r_B means the distance between nucleus B and the electron in atom A. Taking nucleus B as origin we can describe the position of nucleus A with $\overline{R}_{AB}$ and the position of the electron with $\overline{r}_B$. With nucleus A as origin we describe the position on the electron in atom A with $\bar{\varrho}$. The projection of $\bar{\varrho}$ on R_{AB} is denoted t.

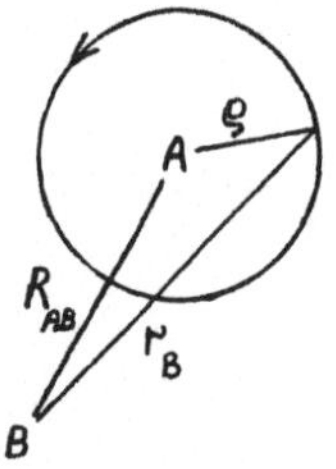

or

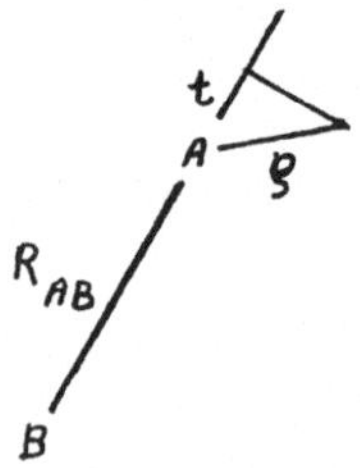

We have then

$$\frac{1}{r_B} \approx \frac{1}{R+t} = \frac{R-t}{R^2-t^2} \approx \frac{R-t}{R^2} \tag{F.9}$$

and

$$(F.8) = -\sum_A \sum_{B \neq A} Z_B \sum_{\mu \neq \nu}^{A} P_{\mu\nu} \cdot \int \phi_\mu^* \phi_\nu r_B^{-1} \, d\tau = \tag{F.10}$$

$$= \sum_A \sum_{B \neq A} Z_B R_{AB}^{-2} \left(P_{st} + P_{ts}\right) \cdot \int \phi_{2s}^* \, t \, \phi_{2pt} \, d\tau = \tag{F.11}$$

$$= \sum_A \sum_{B \neq A} Z_B R_{AB}^{-2} \mu_{t_A} \tag{F.12}$$

where μ_{t_A} is the hybridization dipole moment on atom A in the direction towards atom B. It is given by

$$\mu_{t_A} = \left(P_{st} + P_{ts}\right) \cdot \int \phi_{2s}^{*}\, t\, \phi_{2pt}\, d\tau \tag{F.13}$$

If we select the corresponding two-center terms from eq. (F.7) we obtain

$$\sum_{A} \sum_{B \neq A} \left(P_{st_A} + P_{ts_A}\right) \cdot \sum_{\lambda}^{B} \sum_{\sigma} P_{\lambda\sigma}\, (st|\lambda\sigma) = \tag{F.14}$$

$$= - \sum_{A} \sum_{B \neq A} \sum_{\lambda}^{B} N_{\lambda} \cdot R_{AB}^{-2} \cdot \mu_{t_A} \tag{F.15}$$

The integral in eq. (F.14) is given for Slater type orbitals (using the notation $Z = 2\zeta$) by McGlynn et al. [18]

$$\int \phi_{2s}^{*}\, t\, \phi_{2pt}\, d\tau = \left[\frac{\zeta_s^5\, \zeta_t^5}{3}\right]^{1/2} \cdot \frac{5 \cdot 2^5}{\left(\zeta_s + \zeta_t\right)^6} \tag{F.16}$$

which with atomic units for Z and t and assuming that $\zeta_s = \zeta_t$ gives

$$\mu_{t_A} = \left(P_{st} + P_{ts}\right) \cdot \frac{1.44}{\zeta} \tag{F.17}$$

Local dipole moments have recently been given a very general treatment by Dewar and Thiel [19] . It can be remarked that our treatment does not give any interaction between two local dipole moments. In a recent study Thiel [20] has pointed out, that inclusion of monopole-dipole interactions (as here) is very important, whereas inclusion of dipole-dipole- and higher interactions gives much smaller improvement.

The inclusion of local dipoles in the expression for the electrostatic interaction is of great importance for many molecules with lone-pair orbitals. If it is omitted, as is done in HAM/3 and most other semiempirical methods, it will be very difficult to find parameters, which give good results also for molecules with lone-pair orbitals.

F.3. The final expression for the total energy

The total energy of a molecule is therefore to the approximation, discussed above,

$$E =$$

$$= -\frac{1}{2}\sum_{\mu} N_{\mu} \zeta_{\mu}^{2} - \qquad (F.19)$$

$$-\sum_{\mu_A \nu_B} P_{\mu\nu} \beta_{\mu\nu} + \qquad (F.20)$$

$$+\sum_{A>B}\left[Z_A Z_B R_{AB}^{-1} - Z_A \sum_{\nu}^{B} N_{\nu} V_{BA} - Z_B \sum_{\mu}^{A} N_{\mu} V_{AB} + \sum_{\mu}^{A} N_{\mu} \sum_{\nu}^{B} N_{\nu} \gamma_{AB}\right] \qquad (F.21)$$

$$+\sum_{\substack{A,B \\ A \neq B}} \sum_{x,y,z} \left[Z_A - \sum_{\lambda}^{A} N_{\lambda}\right] \cdot R_{AB}^{-2} \cdot \mu_{t_B} \qquad (F.22)$$

$$+\frac{1}{2}\sum_{A}\sum_{\substack{\mu\nu \\ \mu \neq \nu}}^{A} P_{\mu\nu}^{2}\left[\frac{1}{4}(\mu\mu|\mu\mu) + \frac{1}{4}(\nu\nu|\nu\nu) - \frac{1}{2}(\mu\mu|\nu\nu) + \frac{3}{2}(\mu\nu|\mu\nu)\right] + \qquad (F.23)$$

$$+\frac{1}{2}\sum_{\substack{A,B \\ A \neq B}} \sum_{\mu}^{A} \sum_{\nu}^{B} P_{\mu\nu}^{2}\left[\frac{1}{4}(\mu\mu|\mu\mu) + \frac{1}{4}(\nu\nu|\nu\nu) - \frac{1}{2}(\mu\mu|\nu\nu)\right] \qquad (F.24)$$

The one-center pair-correlation energy is taken care of by eq.(F.19) and the two-center pair-correlation energy by the last term in (F.21), since the contributions from (F.23) and (F.24) are small.

If the pair-correlation energies are omitted, the energy above is approximately identical with the Hartree-Fock total energy.

Eq.(F.21) looks complicated, but assuming $\gamma_{AB} = V_{AB} - V_{BA} = R_{AB}^{-1}$ we obtain

$$(F.21) = \sum_{A>B} Q_A Q_B R_{AB}^{-1} \qquad (F.25)$$

which means the electrostatic interaction between the atoms in the molecule.

F.4. The parametrization of HAM/3

The one-center parameters, which are given in Table E.1., were obtained from the energies of 311 different atomic species, which are tabulated in Sec.E.5.

The two-center parameters were obtained from comparisons of calculated and experimental photoelectron spectra of the following molecules:

H_2, methane, acetylene, ethylene, ethane, cyclopropene, cyclopropane, propane, diacetylene, butatriene, butenyne, 1,3-butadiene, methylene cyclopropane, dimethylacetylene, cyclobutane, cis-butene, n-pentane, neopentane, triacetylene, benzene, Dewar benzene, norbornadiene, N_2, NH_3, hydrazine, hydrogen azide, cyanogen azide, hydrogen cyanide, diazirine, cyanamide, diazomethane, methylamine, cyanogen, acetonitrile, ethylenimine, acrylonitrile, s-triazine, cis-1,2-dicyano-ethylene, O_2, O_3, H_2O, H_2O_2, CO, CO_2, C_2O_3, formaldehyde, formic acid, methanol, ketene, glyoxal, oxalic acid, ethylene oxide, acetaldehyde, acetic acid, dimethyl ether, maleic anhydride, acrolein, acetone, furan, F_2, HF, CF_2, trifluoromethane, difluoromethane, methyl fluoride, tetrafluoroethylene, fluoroacetylene, 1,1-difluoroethylene, fluorobenzene, N_2O, HNO_3, hydroxylamine, formamide, nitromethane, urea, F_2O, HOF, NF_3, N_2F_2.

The parameters were changed until the sum of the squares of the errors was as small as possible.

A difficulty was that in a large molecule there are many bands in the PES. It is then necessary to compare each calculated IP with the right band in the PES. Interpretations of the bands in the PES were usually available from ab-initio calculations, chemical intuition and other sources. To check that the right band had been chosen, the overlap between the calculated MO and a standard MO was calculated every time the comparison of the ionization energies was performed for a certain molecule.

Since the two-center parameters determine the properties of both occupied and unoccupied MO's, it should be possible to determine also UV spectra and electron affinities from studies of PES. To check these results and to increase the number of data in the parameter determination, also UV spectra of 34 of these molecules were used together with electron affinities of 13 molecules.

References

1. O. Sinanoglu and H.Ö. Pamuk, J.Am.Chem.Soc. 95, 5435 (1973).
2. M.J.S. Dewar and G.P. Ford, J.Am.Chem.Soc. 101, 5558 (1979).
3. S. de Bruijn, Chem.Phys.Letters 52, 76 (1977).
4. L. Åsbrink, C. Fridh, E. Lindholm and S. de Bruijn, Chem.Phys.Letters 66, 411 (1979).
5. L. Åsbrink, C. Fridh and E. Lindholm in Chemical Spectroscopy and Photochemistry in the Vacuum-Ultraviolet (Edited by C. Sandorfy, P. Ausloos and M.B. Robin), Reidel, Dordrecht (1974).
6. R.G. Jesaitis, J.Am.Chem.Soc. 93, 3849 (1971).
7. H.A. Germer Jr., J.Chem.Soc. 58, 3524 (1973).
8. F.E. Harris, A. Trautwein and J. Delhalle, Chem.Phys.Letters 72, 315 (1980).
9. D.N. Nanda and K. Jug, Theoret.Chim.Acta 57, 95 (1980).
10. J.A. Pople and D.L. Beveridge, Approximate Molecular Orbital Theory, McGraw-Hill, New York (1970).
11. R. Hoffmann, J.Chem.Phys. 39, 1397 (1963).
12. L. Åsbrink, C. Fridh and E. Lindholm, Chem.Phys.Letters 52, 63 (1977).
13. R. Pariser and R.G. Parr, J.Chem.Phys. 21, 466, 767 (1953).
14. N.C. Baird and M.J.S. Dewar, J.Chem.Phys. 50, 1262 (1969).
15. M.J.S. Dewar and D.H. Lo, J.Am.Chem.Soc. 93, 7201 (1971).
16. R.S. Mulliken, J.Chim.Phys. 46, 497, 675 (1949).
17. M.D. Newton, F.P. Boer and W.N. Lipscomb, J.Am.Chem.Soc. 88, 2353 (1966).
18. S.P. McGlynn, L.G. Vanquickenborne, M. Kinoshita and D.G. Carroll, Introduction to Applied Quantum Chemistry, Holt, Rinehart and Winston, New York (1972) p.327.
19. M.J.S. Dewar and W. Thiel, J.Am.Chem.Soc. 99, 4899 (1977).
20. W. Thiel, J.C.S. Faraday II, 76, 302 (1980).

G. Solving the Schrödinger equation

G.1. Variational calculus

It is very difficult to solve the Schrödinger equation. The usual procedure is therefore to make a number of guesses concerning the total wavefunction Ψ and then to choose the best of them as the approximate solution.

To be able to discuss which of the solutions is the best, one calculates the average of the hamiltonian $\mathcal{H}$ using one of the guesses Φ

$$\text{average of } \mathcal{H} = \int \Phi^* \mathcal{H} \Phi \, d\tau \qquad (G.1)$$

It is assumed here that Φ is normalized.

We expand now Φ using the unknown eigenfuntions Ψ_n and obtain

$$\Phi = \sum_n a_n \Psi_n \qquad (G.2)$$

and find

$$\int \Phi^* \mathcal{H} \Phi \, d\tau = \int \sum_n a_n^* \Psi_n^* \cdot \mathcal{H} \cdot \sum_n a_n \Psi_n \, d\tau =$$

$$= \sum_n |a_n|^2 E_n \qquad (G.3)$$

since Ψ_n are normalized and orthogonal.

We can now obtain a useful inequality by replacing each E_n in this sum by the lowest eigenvalue E_0

$$\int \Phi^* \mathcal{H} \Phi \, d\tau \geqslant \sum_n |a_n|^2 E_0 = E_0 \cdot \sum_n |a_n|^2 = E_0 \qquad (G.4)$$

The reason for this simplification is that Φ is normalized which gives

$$1 = \int \Phi^* \Phi \, d\tau = \int \sum_n a_n^* \Psi_n^* \cdot \sum_n a_n \Psi \, d\tau = \sum_n |a_n|^2 \qquad (G.5)$$

The result is therefore that the true (experimental) energy E_0 is always below $\int \Phi^* \mathcal{H} \Phi \, d\tau$. The lowest possible value of this integral indicates therefore that the Φ here is the best of all guesses.

The variational calculus means that one takes as Φ a function containing some parameters. One varies these parameters so that $\int \Phi^* \mathcal{H} \Phi \, d\tau$ becomes a minimum.

The experience from the use of variational calculus is that the value, obtained for the energy, is often very good even if the guess concerning the wavefunction is less successful.

G.2. Deduction of Roothaan's equations

Our discussion up till now has concerned only the total energy of the molecule. First the Hartree-Fock total energy was deduced and afterwards it was transformed to give the HAM total energy.

Our goal is now to solve the Schrödinger equation using variational calculus. We will vary the total wavefunction until the total energy E is a minimum. The variation of the total wavefunction will be performed by varying the coefficients $c_{\mu i}$ in the LCAO expressions for the molecular orbitals ψ_i. The atomic orbitals ϕ_μ are not varied.

During all the variational procedures the molecular orbitals ψ_j must be kept normalized and orthogonal:

$$\int \psi_j^* \psi_k \, d\tau - \delta_{jk} = 0 \tag{G.6}$$

It is obvious that this normalization means that our molecular orbitals accomodate only one electron, i.e. they mean ψ_j^α or ψ_j^β with $q_j = 1$.

The variational calculus will be performed by use of Lagrange's method with undetermined multipliers. This means that we look for the minimum not of E but of G, defined as

$$G = E - \sum_{jk} \varepsilon_{jk} \left\{ \sum_{\mu\nu} c_{\mu j}^* c_{\nu k} S_{\mu\nu} - \delta_{jk} \right\} \tag{G.7}$$

where ε_{jk} are the undetermined multipliers, and the constraint is taken from eq. (A.5). Unfortunatoly, we have to use the same notation here as for the pair-correlation energies.

To find the minimum of E we change all parameters, i.e. $c_{\mu j}^*$ and $c_{\nu k}$ and ε_{jk}, by infinitesimal increments $\delta c_{\mu j}^*$, $\delta c_{\nu k}$ and $\delta \varepsilon_{jk}$, respectively. In the minimum the resulting increment δG of G must be = 0.

Since

$$E = \int \Psi^* \mathcal{H} \Psi \, d\tau \tag{G.8}$$

and since Ψ depends upon $c_{\mu j}$, it is reasonable to consider E as a function of the parameters $c_{\mu j}$.

However, we have shown above that in the Hartree-Fock total energy (B.15) --- (B.18) and in the HAM total energy (D.22) --- (D.29) the total energy is a function not of $c_{\mu j}$ but of $P_{\mu\nu}$ and N_ν. It is therefore reasonable by the derivation to consider E as a function of $P_{\mu\nu}$, which gives

$$\delta \mathcal{G} = \sum_{\mu\nu j} \frac{\partial E}{\partial P_{\mu\nu}} \frac{\partial P_{\mu\nu}}{\partial c^*_{\mu j}} \delta c^*_{\mu j} + \sum_{\mu\nu k} \frac{\partial E}{\partial P_{\mu\nu}} \frac{\partial P_{\mu\nu}}{\partial c_{\nu k}} \delta c_{\nu k} -$$

$$- \sum_{jk} \delta \varepsilon_{jk} \left\{ \sum_{\mu\nu} c^*_{\mu j} c_{\nu k} S_{\mu\nu} - \delta_{jk} \right\} -$$

$$- \sum_{\mu\nu jk} \varepsilon_{jk} \left[\delta c^*_{\mu j} \cdot c_{\nu k} \cdot S_{\mu\nu} + c^*_{\mu j} \cdot \delta c_{\nu k} S_{\mu\nu} \right] = 0 \quad (G.9)$$

To study this, we first let $\delta \varepsilon_{jk}$ assume arbitrary values while $\delta c^*_{\mu j} = 0$ and $\delta c_{\nu k} = 0$. Since $\delta \mathcal{G} = 0$ all time, it is necessary that $\{\ \} = 0$. This guarantees that the molecular orbitals ψ_j according to (A.5) are normalized and orthogonal.

Next, we let $\delta c^*_{\mu j}$ take arbitrary values while $\delta c_{\nu k} = 0$ and $\delta \varepsilon_{jk} = 0$. Since $\delta \mathcal{G} = 0$ all time, it is necessary that we have

$$\sum_{\nu} \frac{\partial E}{\partial P_{\mu\nu}} \frac{\partial P_{\mu\nu}}{\partial c^*_{\mu j}} - \sum_{\nu k} \varepsilon_{jk} c_{\nu k} S_{\mu\nu} = 0 \quad (G.10)$$

We will now calculate $\partial P_{\mu\nu} / \partial c^*_{\mu j}$ by taking the derivative of

$$P_{\mu\nu} = \sum_j q_j c^*_{\mu j} c_{\nu j} \quad (G.11)$$

Here, q_j must be $= 1$ from our normalization equation above. This gives

$$\frac{\partial P_{\mu\nu}}{\partial c^*_{\mu j}} = c_{\nu j} \quad (G.12)$$

Further, we put

$\varepsilon_{jk} = 0$ for $j \neq k$

It was shown already by Koopmans [1] that this can be achieved if the molecular orbitals are transformed in a suitable way (see also Pople and Beveridge, [2] p. 39).

We obtain finally

$$\sum_{\nu}\left(\frac{\partial E}{\partial P_{\mu\nu}} - \varepsilon_j S_{\mu\nu}\right)\cdot c_{\nu j} = 0 \tag{9.13}$$

or, if we introduce the Fock matrix element $F_{\mu\nu}$

$$F_{\mu\nu} = \frac{\partial E}{\partial P_{\mu\nu}} \tag{9.14}$$

$$\sum_{\nu}\left(F_{\mu\nu} - \varepsilon_j S_{\mu\nu}\right)\cdot c_{\nu j} = 0 \tag{9.15}$$

which are called Roothaan's equations [3].

Since usually the unit of E is eV and since $P_{\mu\nu}$ is a number, the unit of $F_{\mu\nu}$ is also eV.

Below we write the Roothaan equations explicitely for μ and ν being 1, 2 or 3:

$$\begin{cases}(F_{11}-\varepsilon_j)\cdot c_{1j} + (F_{12}-\varepsilon_j\cdot S_{12})\cdot c_{2j} + (F_{13}-\varepsilon_j\cdot S_{13})\cdot c_{3j} = 0\\ (F_{21}-\varepsilon_j\cdot S_{21})\cdot c_{1j} + (F_{22}-\varepsilon_j)\cdot c_{2j} + (F_{23}-\varepsilon_j\cdot S_{23})\cdot c_{3j} = 0\\ (F_{31}-\varepsilon_j\cdot S_{31})\cdot c_{1j} + (F_{32}-\varepsilon_j\cdot S_{32})\cdot c_{2j} + (F_{33}-\varepsilon_j)\cdot c_{3j} = 0\end{cases} \tag{9.16}$$

In these equations $c_{\nu j}$ are the unknown. To find a non-trivial solution of this homogeneous system of equations, the secular determinant must be zero

$$\begin{vmatrix} F_{11}-\varepsilon_j & F_{12}-\varepsilon_j\cdot S_{12} & F_{13}-\varepsilon_j\cdot S_{13}\\ F_{21}-\varepsilon_j\cdot S_{21} & F_{22}-\varepsilon_j & F_{23}-\varepsilon_j\cdot S_{23}\\ F_{31}-\varepsilon_j\cdot S_{31} & F_{32}-\varepsilon_j\cdot S_{32} & F_{33}-\varepsilon_j \end{vmatrix} = 0 \tag{9.17}$$

This is an equation in ε_j of third order, which gives three solutions for ε_j . If now ε_1 is inserted into Roothaan's equations, c_{11} and c_{21} and c_{31} can be computed, which means that ψ_1 has been determined. The procedure is then repeated for ψ_2 and ψ_3 .

G.3. The Fock matrix elements

We obtain different Fock matrix elements for $\mu_A \mu_A$, $\mu_A \nu_A$ and $\mu_A \nu_B$ when we use eq. (G.14) together with the expression for the orbital exponent

$$\zeta_\mu = \frac{1}{n_\mu}\left[Z_A + \sigma_{\mu\mu} - \sum_\lambda^A N_\lambda \sigma_{\lambda\mu}\right] \tag{G.18}$$

First: $F_{\mu\mu}$ with μ on atom A.

$$F_{\mu_A\mu_A} = -\frac{1}{2}\zeta_\mu^2 + \sum_\nu^A N_\nu \zeta_\nu \frac{1}{n_\nu}\sigma_{\mu\nu} + ML(A) \tag{G.19}$$

Here, $ML(A)$, the Madelung potential, describes the potential in atom A from the charges in atoms B.

$$ML(A) = -\sum_{B \neq A}\left[Z_B V_{AB} - \sum_\nu^B N_\nu \gamma_{AB} + \mu_{t_B} \cdot R_{AB}^{-2}\right] \tag{G.20}$$

Second: $F_{\mu\nu}$ with both μ and ν on atom A.

$$F_{\mu_A\nu_A} = P_{\mu\nu}\cdot\left[\frac{1}{4}(\mu\mu|\mu\mu) + \frac{1}{4}(\nu\nu|\nu\nu) - \frac{1}{2}(\mu\mu|\nu\nu) + \frac{3}{2}(\mu\nu|\mu\nu)\right] + l.d. \tag{G.21}$$

where $l.d.$ means the contribution from the local dipoles, e.g. in the s-x case

$$l.d. = \sum_{B \neq A}\left(Z_B - \sum_\lambda^B N_\lambda\right)\cdot R_{AB}^{-2}\cdot\cos(x, R_{AB})\cdot\int \phi_{2s}^* \, x \, \phi_{2px}\, d\tau \tag{G.22}$$

Third: $F_{\mu\nu}$ with μ on A and ν on B.

$$F_{\mu_A\nu_B} = \frac{1}{2} S_{\mu\nu}\left[F_{\mu\mu} + F_{\nu\nu}\right] - \beta_{\mu\nu} + \\ + P_{\mu\nu}\cdot\left[\frac{1}{4}(\mu\mu|\mu\mu) + \frac{1}{4}(\nu\nu|\nu\nu) - \frac{1}{2}(\mu\mu|\nu\nu)\right] \tag{G.23}$$

Explanation: Since according to eq. (A.10)

$N_\mu = \sum_\nu \frac{1}{2}(P_{\mu\nu}S_{\mu\nu} + P_{\nu\mu}S_{\nu\mu})$ and therefore $\frac{\partial E}{\partial P_{\mu\mu}} = \frac{\partial E}{\partial N_\mu}$, we have

$$\frac{\partial E}{\partial P_{\mu_A\nu_B}} = \frac{\partial E}{\partial N_\mu}\cdot\frac{\partial N_\mu}{\partial P_{\mu\nu}} + \frac{\partial E}{\partial N_\nu}\cdot\frac{\partial N_\nu}{\partial P_{\mu\nu}} = F_{\mu\mu}\cdot\frac{1}{2}S_{\mu\nu} + F_{\nu\nu}\cdot\frac{1}{2}S_{\mu\nu} \tag{G.24}$$

Those parts of E which are not functions of N give a further contribution.

The Fock matrix for ethylene, calculated by use of the HAM/3 approximation, is shown below.

ETHYLENE
FOCK MATRIX

			C1				C2			
			1	2	3	4	5	6	7	8
1		S	-14.002320	-0.000770	0.0	0.0	-9.688454	-10.709740	0.0	0.0
2	C1	X	-0.000770	-8.425504	0.0	0.000033	10.709717	9.082558	0.0	0.0
3		Y	0.0	0.0	-8.231854	0.0	0.0	0.0	-4.742459	0.0
4		Z	0.0	0.000033	0.0	-8.071794	0.0	0.0	0.0	-4.712932
5		S	-9.688454	10.709717	0.0	0.0	-14.002227	0.000769	0.0	0.0
6	C2	X	-10.709740	9.082558	0.0	0.0	0.000769	-8.425502	0.0	0.000035
7		Y	0.0	0.0	-4.742459	0.0	0.0	0.0	-8.231837	-0.000019
8		Z	0.0	0.0	0.0	-4.712932	0.0	0.000035	-0.000019	-8.071746
9		H3	-14.460190	-6.865254	-11.187248	0.0	-1.889218	-3.163151	-1.533270	0.0
10		H4	-14.460182	-6.865247	11.187241	0.0	-1.889218	-3.163148	1.533268	0.0
11		H5	-1.889223	3.163152	-1.533271	0.0	-14.460152	6.865243	-11.187233	0.0
12		H6	-1.889223	3.163153	1.533271	0.0	-14.460159	6.865245	11.187241	0.0

			H3	H4	H5	H6
			9	10	11	12
1		S	-14.460190	-14.460182	-1.889223	-1.889223
2	C1	X	-6.865254	-6.865247	3.163152	3.163153
3		Y	-11.187248	11.187241	-1.533271	1.533271
4		Z	0.0	0.0	0.0	0.0
5		S	-1.889218	-1.889218	-14.460152	-14.460159
6	C2	X	-3.163151	-3.163148	6.865243	6.865245
7		Y	-1.533270	1.533268	-11.187233	11.187241
8		Z	0.0	0.0	0.0	0.0
9		H3	-8.475576	-5.951550	-2.644294	-1.052784
10		H4	-5.951550	-8.475558	-1.052783	-2.644291
11		H5	-2.644294	-1.052783	-8.475569	-5.951556
12		H6	-1.052784	-2.644291	-5.951556	-8.475581

G.4. Solving the Roothaan equations

When the Fock matrix elements $F_{\mu\nu}$ and the overlap integrals $S_{\mu\nu}$ have been calculated, we have found all elements in the secular determinant. If this is put $= 0$, it forms an equation in ε_j of the same order as the number of atomic orbitals in the molecule (in ethylene 12). After solving the equation, we obtain a number (in ethylene 12) of solutions ε_j which are called eigenvalues.

If we insert one of these eigenvalues (ε_j) in Roothaan's equations we can solve for the coefficients $c_{\mu j}$ (called eigenvectors) which define the molecular orbital ψ_j. Since ε_j is an energy, belonging to a molecular orbital, it is often called orbital energy.

We repeat this procedure for all ε_j and instruct the computer to write the eigenvalues at the top of a table with the lowest eigenvalue to the left and then in order of increasing magnitude. We instruct also the computer to write the eigenvectors below their eigenvalue.

The difficulty in this description is that the calculation of the Fock matrix elements requires knowledge of the $c_{\mu j}$ which are the result of the calculation.

We perform therefore a new iteration.

First, we calculate $P_{\mu\nu}$ from $c_{\mu j}$ using

$$P_{\mu\nu} = \sum_j q_i c_{\mu j} c_{\nu j}$$

This means that we must have fixed the values of q_j in advance. In an ordinary molecule with N electrons the $N/2$ lowest energy orbitals are filled with $q_j = 2$ electrons each, but the higher orbitals are empty.

Then, new values of ζ_μ and $P_{\mu\nu}$ are calculated, and we hope that the changes of the eigenvalues and eigenvectors from the new iteration mean an improvement.

The iterations are repeated until no longer any improvement or change are achieved. We have then reached "self-consistency". Normally, in HAM/3 about 10 iterations are sufficient.

According to the description above the calculation is done in two steps: First, ε_j is obtained from the secular equation. Then, $c_{\mu j}$ are obtained from Roothaan's equations. The computer, however, is able to do this in <u>one</u> step.

We will illustrate this by an example, taken from a textbook by Flurry, "Molecular orbital theories" [4].

Let us consider the Roothaan equations below in which all $S_{\mu\nu} = 0$ for $\mu \neq \nu$.

$$\begin{cases} c_{j1}\cdot(\alpha - E_j) + c_{j2}\cdot\beta = 0 \\ c_{j1}\cdot\beta + c_{j2}\cdot(\alpha - E_j) + c_{j3}\cdot\beta + c_{j4}\cdot\beta = 0 \\ c_{j2}\cdot\beta + c_{j3}\cdot(\alpha - E_j) + c_{j4}\cdot\beta = 0 \\ c_{j2}\cdot\beta + c_{j3}\cdot\beta + c_{j4}\cdot(\alpha - E_j) = 0 \end{cases}$$

The determinant is

$$\begin{vmatrix} \alpha - E & \beta & 0 & 0 \\ \beta & \alpha - E & \beta & \beta \\ 0 & \beta & \alpha - E & \beta \\ 0 & \beta & \beta & \alpha - E \end{vmatrix} = 0$$

which can be written (if we divide with β and put $\frac{\alpha - E}{\beta} = X$)

$$\begin{vmatrix} X & 1 & 0 & 0 \\ 1 & X & 1 & 1 \\ 0 & 1 & X & 1 \\ 0 & 1 & 1 & X \end{vmatrix} = 0$$

Let us call this matrix $\mathbb{F}$ and let us guess a matrix $\mathbb{C}$ so that $\mathbb{C}^{+}\mathbb{F}\mathbb{C}$ is diagonal, i.e. $= \mathcal{E}$. We have then solved the problem! $\mathbb{C}$ gives the eigenvectors, i.e. the wavefunctions, and $\mathcal{E}$ gives the eigenvalues, i.e. the orbital energies.

The result is thus:

$$\mathbf{C}^\dagger \begin{vmatrix} X & 1 & 0 & 0 \\ 1 & X & 1 & 1 \\ 0 & 1 & X & 1 \\ 0 & 1 & 1 & X \end{vmatrix} \mathbf{C} =$$

$$= \underbrace{\begin{bmatrix} 0.2819 & 0.6116 & 0.5227 & 0.5227 \\ 0.8152 & 0.2536 & -0.3687 & -0.3682 \\ 0 & 0 & 0.7071 & -0.7071 \\ -0.5059 & 0.7494 & -0.3020 & -0.3020 \end{bmatrix}}_{\mathbb{C}^\dagger} \underbrace{\begin{vmatrix} X & 1 & 0 & 0 \\ 1 & X & 1 & 1 \\ 0 & 1 & X & 1 \\ 0 & 1 & 1 & X \end{vmatrix}}_{\mathbb{F}} \underbrace{\begin{bmatrix} 0.2819 & 0.8152 & 0 & -0.5059 \\ 0.6116 & 0.2536 & 0 & 0.7494 \\ 0.5227 & -0.3682 & 0.7071 & -0.3020 \\ 0.5227 & -0.3682 & -0.7071 & -0.3020 \end{bmatrix}}_{\mathbb{C}}$$

$$= \underbrace{\begin{vmatrix} X+2.1701 & 0 & 0 & 0 \\ 0 & X+0.3111 & 0 & 0 \\ 0 & 0 & X-1.0000 & 0 \\ 0 & 0 & 0 & X-1.4812 \end{vmatrix}}_{\mathcal{E}} \qquad (5.25)$$

If this result is presented in the usual way in quantum chemistry we have

Eigenvalues and eigenvectors

j =	1	2	3	4
ε_j =	-2.1701	-0.3111	+1.0000	+1.4812
	0.2819	0.8152	0	−0.5059
	0.6116	0.2536	0	0.7494
	0.5227	−0.3682	0.7071	−0.3020
	0.5227	−0.3682	−0.7071	−0.3020

How the computer handles the problem to make an intelligent guess of $\mathbb{C}$ is, however, another story.

G.5. Some useful relations for the eigenvalue.

The calculation of the eigenvalues ε_j is illustrated by the numerical example in eq. (G.25). If we study one of the diagonal elements here, we see that

$$\varepsilon_j = \sum_{\mu,\nu} c_{\mu j} F_{\mu\nu} c_{\nu j} \tag{G.26}$$

This expression for the eigenvalue can also be deduced from theory (see p.45 in ref. [2]).

We will now study the derivative $\partial E/\partial q_j$

$$\frac{\partial E}{\partial q_j} = \sum_{\mu\nu} \frac{\partial E}{\partial P_{\mu\nu}} \cdot \frac{\partial P_{\mu\nu}}{\partial q_j} = \sum_{\mu\nu} F_{\mu\nu} \frac{\partial P_{\mu\nu}}{\partial q_j} = \sum_{\mu\nu} F_{\mu\nu} c_{\mu i} c_{\nu j} = \varepsilon_j \tag{G.27}$$

where we have used

$$P_{\mu\nu} = \sum_i q_j c_{\mu j} c_{\nu j} \tag{G.28}$$

This is a simple way to find the eigenvalue ε_j.

It must be stressed that eq.(G.27) has no physical meaning. The charge q_j in orbital ψ_j in a molecule is constant and we are not allowed to vary it during the derivation.

Instead, eq.(G.27) has the following meaning. According to eqs. (F.19–24) E is a mathematical function of q_j and usual mathematical operations are therefore allowed, e.g. derivation.

The notation "differential ionization potential" for $\partial E/\partial q_j$ should be avoided.

In Hartree-Fock theory the calculation of ε_j by use of eq.(G.27) is straight-forward. The result is given in eq.(B.13).

G.6. Comparison with the Hartree-Fock method

We will now compare the Hartree-Fock eigenvalue with the HAM eigenvalue. To be able to compare with the HAM results, which include correlation, we use the expressions, given in Sec.C.7., in which the pair-correlation energies have been added to the Hartree-Fock expressions.

Since the total energies in HAM and Hartree-Fock, with or without correlation, according to Sec.F.3. are identical, we would expect that also the eigenvalues, which are obtained from the total energies, will be equal. This is, however, not the case as demonstrated by Davidson [5] and others [6].

We will therefore in the next section form the HAM total energy from the Hartree-Fock total energy by addition of the idempotency zero eq.(D.12) and then study the eigenvalue obtained from this energy expression.

G.7. The eigenvalue ε_j in Hartree-Fock and HAM

The total energy E^{HAM} in HAM is the sum of the Hartree-Fock total energy E^{HF}, given by eqs.(B.10) + (B.11), and the idempotency zero eq.(D.12)

$$E^{HAM} = E^{HF} + \frac{1}{2}\sum_{\mu\nu\lambda\sigma}\left[P^{\alpha}_{\mu\lambda}P^{\alpha}_{\sigma\nu} + P^{\beta}_{\mu\lambda}P^{\beta}_{\sigma\nu}\right]S_{\nu\mu}S_{\lambda\sigma}(\mu\mu|\mu\mu) - \frac{1}{2}\sum_{\mu\nu}\left[P^{\alpha}_{\mu\nu}+P^{\beta}_{\mu\nu}\right]S_{\nu\mu}(\mu\mu|\mu\mu) \tag{G.29}$$

We intend now to calculate the eigenvalue from the relation $\varepsilon^{\alpha}_{j} = \partial E^{HAM}/\partial q^{\alpha}_{j}$ and find it then necessary first to introduce q^{α}_{j} by use of

$$P^{\alpha}_{\mu\nu} = \sum_{j} q^{\alpha}_{j}\, c^{\alpha *}_{\mu j}\, c^{\alpha}_{\nu j} \tag{G.30}$$

After complicated algebra (see below) we obtain from eq.(G.29)

$$E^{HAM} = E^{HF} + \frac{1}{2}\sum_{\alpha\beta}\sum_{j} q^{\alpha}_{j}\left(q^{\alpha}_{j}-1\right)L_{jj} \tag{G.31}$$

where we have introduced the abbreviation

$$L_{jj} = \frac{1}{2}\sum_{\mu\nu} c_{\mu j}\, c_{\nu j}\, S_{\nu\mu}\cdot\left[(\mu\mu|\mu\mu)+(\nu\nu|\nu\nu)\right] \tag{G.32}$$

We find now the HAM eigenvalue ε^{α}_{j} by taking the derivative of eq.(G.31), using eq.(B.13), which gives

$$\varepsilon^{\alpha}_{j} = \varepsilon^{HF}_{j} + \left(q^{\alpha}_{j} - \frac{1}{2}\right)L_{jj} \tag{G.33}$$

It is obvious that the idempotency zero influences the eigenvalues in HAM so that they differ from the eigenvalues in Hartree-Fock. The difference depends upon the type of orbital.

For an occupied orbital $q_i^{\alpha} = 1$ and we have

$$\varepsilon_i^{\alpha} = \varepsilon_i^{HF} + \frac{1}{2} L_{ii} \tag{G.34}$$

but for an unoccupied $q_a^{\alpha} = 0$ and we get

$$\varepsilon_a^{\alpha} = \varepsilon_a^{HF} - \frac{1}{2} L_{aa} \tag{G.35}$$

For an organic molecule the difference $\frac{1}{2} L_{ii}$ or $\frac{1}{2} L_{aa}$ is of the order of 5 eV.

We remark finally that the eigenvalues ε_i and ε_a have no direct physical meaning, neither in Hartree-Fock nor in HAM, although they are closely connected to ionization energies, excitation energies and electron affinities. Such applications will be discussed separately.

We stated above that the total energies in HAM and Hartree-Fock are identical. This seems to be contradicted by the extra term in eq.(G.31). However, this term is always zero, since in a molecule $q_j^{\alpha} = 1$ or $q_j^{\alpha} = 0$.

Algebra

From (G.29) and (G.30) we obtain

$$E = E^{HF} +$$

$$+ \frac{1}{8} \sum_{\alpha\beta} \sum_{\mu\nu\lambda\sigma} \sum_{jk} q_j^{\alpha} c_{\mu j}^{\alpha *} c_{\lambda j}^{\alpha} S_{\lambda\sigma} q_k^{\alpha} c_{\sigma k}^{\alpha *} c_{\nu k}^{\alpha} S_{\nu\mu} \cdot \tag{G.36}$$

$$\cdot \left[(\mu\mu|\mu\mu) + (\nu\nu|\nu\nu) + (\lambda\lambda|\lambda\lambda) + (\sigma\sigma|\sigma\sigma) \right] -$$

$$- \frac{1}{4} \sum_{\alpha\beta} \sum_{\mu\nu} \sum_{j} q_j^{\alpha} c_{\mu j}^{\alpha *} c_{\nu j}^{\alpha} S_{\nu\mu} \cdot \left[(\mu\mu|\mu\mu) + (\nu\nu|\nu\nu) \right] \tag{G.37}$$

We simplify eq.(G.36) by partitioning it into two terms. We use eq.(A.5) in each of these terms and find then that they are equal.

(S.36) =

$$= \frac{1}{8}\sum_{\alpha\beta}\sum_{jk} q_j^\alpha q_k^\alpha \Bigg\{ \underbrace{\sum_{\lambda\sigma} c_{\lambda j}^\alpha S_{\lambda\sigma} c_{\sigma k}^{\alpha *}}_{\delta_{jk}} \cdot \sum_{\mu\nu} c_{\mu j}^{\alpha *} c_{\nu k}^\alpha S_{\nu\mu} \cdot \left[(\mu\mu|\mu\mu) + (\nu\nu|\nu\nu)\right] +$$

$$+ \underbrace{\sum_{\mu\nu} c_{\mu j}^{\alpha *} S_{\nu\mu} c_{\nu k}^\alpha}_{\delta_{jk}} \cdot \sum_{\lambda\sigma} c_{\lambda j}^\alpha c_{\sigma k}^{\alpha *} S_{\lambda\sigma} \cdot \left[(\lambda\lambda|\lambda\lambda) + (\sigma\sigma|\sigma\sigma)\right] \Bigg\} =$$

$$= \frac{1}{4}\sum_{\alpha\beta}\sum_{\mu\nu}\sum_{j} (q_j^\alpha)^2 c_{\mu j}^{\alpha *} c_{\nu j}^\alpha S_{\nu\mu} \cdot \left[(\mu\mu|\mu\mu) + (\nu\nu|\nu\nu)\right] \qquad (S.38)$$

Since we can take $c_{\mu j}^\alpha = c_{\mu j}'$ we find

$$(S.36) + (S.37) = \frac{1}{2}\sum_{\alpha\beta}\sum_{j} q_j^\alpha (q_j^\alpha - 1) L_{jj} \qquad (S.39)$$

which gives eq. (S. 31).

An approximate value for L_{jj}' can be obtained if all $(\mu\mu|\mu\mu)$ are assumed to be equal $= \gamma_{AA}$. This gives

$$L_{jj} \approx \gamma_{AA} \cdot \sum_{\mu\nu} c_{\mu j}^* c_{\nu j} S_{\nu\mu} = \gamma_{AA} \qquad (S.40)$$

For the atoms in an organic molecule γ_{AA} is about 11 eV (see e.g. [7]) and therefore $\frac{1}{2} L_{jj} \approx 5$ eV.

G.8. Molecules with a small HOMO-LUMO gap

In our study of atoms we pointed out in Sec.E.1. that the HAM method can be used to study only such atoms for which the deduction is valid. This means atoms with a "pure spin-configuration" which can be described by a one-determinant wavefunction, since with a many-determinant wavefunction the idempotency lacks meaning.

In experimental work, however, atomic states are often studied which do not fulfill these requirements (e.g. ^1D). It is then necessary to be able to find the energies of the states when the energies of the configurations have been calculated. Table E.2. gives this connection for atoms.

In molecules similar problems appear. One important problem will be illustrated with reference to ethylene.

In the print-out for ethylene (Sec.A.4.) the highest occupied molecular orbital (HOMO) has the energy -10.538 eV and the lowest unoccupied molecular orbital (LUMO) has the energy -4.269 eV. The HOMO-LUMO gap is thus only 6.27 eV.

In ethylene there are 12 valence electrons. This is sufficient to fill the 6 lowest molecular orbitals with 2 electrons each. There is now a possibility that instead of having 2 electrons in orbital 6 we have them in orbital 7. This means that we have a doubly excited configuration of the ethylene molecule with energy 12.54 eV above the ground configuration.

These two ethylene configurations have the same symmetry and what we observe in the experiments is therefore a mixture, similar to them in Table E.2. To study the mixing we let the wavefunction be

$$\Psi = c_g \cdot \Psi_g + c_e \cdot \Psi_e \tag{G.41}$$

where Ψ_g means the ground configuration with energy E_g , and Ψ_e means the doubly excited configuration with energy E_e .

The Schrödinger equation is

$$\mathcal{H} \Psi = E \Psi \tag{G.42}$$

or

$$\mathcal{H} \left(c_g \Psi_g + c_e \Psi_e \right) = E \left(c_g \Psi_g + c_e \Psi_e \right) \tag{G.43}$$

Multiply this with Ψ_g^* and integrate

$$c_g \left(E_g - E \right) + c_e \cdot H_{ge} = 0 \tag{G.44}$$

and multiply then with Ψ_e^* and integrate

$$c_g \cdot H_{eg} + c_e \cdot (E_e - E) = 0 \qquad (G.45)$$

Here

$$H_{ge} = H_{eg} = \int \Psi_g^* \mathcal{H} \Psi_e \, d\tau = \int \psi_i^*(11)\, \psi_i^*(12) \frac{1}{r_{11,12}} \psi_a(11)\, \psi_a(12)\, d\tau =$$

$$= (ia|ia) = K_{ia} \qquad (G.46)$$

since it can be shown that the only part of $\mathcal{H}$ of interest is $1/r_{11,12}$. HOMO is denoted i and LUMO a and the two electrons are numbered 11 and 12.

To determine c_g and c_e from eqs.(G.51) and (G.52) we must have a secular determinant, which is zero

$$\begin{vmatrix} E_g - E & H_{ge} \\ H_{eg} & E_e - E \end{vmatrix} = 0 \qquad (G.47)$$

With numerical values for ethylene E_g = 0 and E_e = 12.54 and K_{ia} = 1.68 eV (calculated according rules given in Sec.I.1.), we obtain the secular equation

$$\begin{vmatrix} 0 - E & 1.68 \\ 1.68 & 12.54 - E \end{vmatrix} = 0 \qquad (G.48)$$

The solution gives two eigenvalues: E_1 = -0.22 eV and E_2 = 12.76 eV. These values, related to the numerical values for E_g and E_e, correspond to the relations in Table E.2.

From the eigenvalues we can determine the eigenfunctions

$$\Psi_1 = 0.9914 \cdot \Psi_g + 0,1306 \cdot \Psi_e \qquad (G.49)$$

and

$$\Psi_2 = 0.1306 \cdot \Psi_g + 0.9914 \cdot \Psi_e \qquad (G.50)$$

In conclusion we find that the ground state has been displaced downwards 0.22 eV by this configuration interaction (CI).

This phenomenon is important only when the energy gap is small and the exchange integral is large. This means usually molecules with π electrons.

One way to observe this phenomenon is to measure the ionization energy when one electron is ionized from orbital 6 in the ethylene molecule. Before the ionization the energy is depressed by 0.22 eV, but after the ionization there is only one electron in orbital 6 and the phenomenon is no longer possible since it requires a doubly occupied orbital. The calculated IP is therefore influenced by 0.22 eV.

When the ionization takes place from e.g. orbital 5, no such shift takes place.

The double excitation to a valence orbital discussed here is somewhat similar to the double excitation, discussed in Chap.C. to explain the pair-correlation. The double excitation, discussed here, is therefore sometimes denoted as "correlation" although the term "internal correlation" to indicate that a valence orbital is involved should be preferred.

References

1. T. Koopmans, Physica 1, 104 (1933).
2. J.A. Pople and D.L. Beveridge, Approximate Molecular Orbital Theory, McGraw-Hill, New York (1970).
3. C.C.J. Roothaan, Rev.Mod.Phys. 23, 69 (1951).
4. R.L. Flurry, Molecular Orbital Theories of Bonding in Organic Molecules, Dekker, New York (1968).
5. E.R. Davidson, J.Chem.Phys. 57, 1999 (1972).
6. S. Ljunggren, E. Lindholm and L. Åsbrink, to be published.
7. N.C. Baird and M.J.S. Dewar, J.Chem.Phys. 50, 1262 (1969).

H. Ionization and photoelectron spectroscopy.

H.1. Calculation of ionization energy in the HAM model

Ionization means that one electron leaves the molecule. This can happen after supply of energy, usually from photons or electrons.

We assume that the leaving electron comes from the molecular orbital ψ_i^α. The electron charge in this orbital will then change from q_i^α to $q_i^\alpha - 1$. The ionization energy IP_i^α is then the difference of two total energies

$$IP_i^\alpha = E(q_i^\alpha - 1) - E(q_i^\alpha) \qquad (H.1)$$

The total energies are given in different ways in the preceding sections, but to study ($H.1$) we choose eqs. (B.10) + (B.11) + + (G.31) since they are functions of q_j^α. We remember that in eq. (B.11) the pair-correlation energies can be included by letting J mean J' and K mean K'.

The difference of the two total energies gives then a general expression for the ionization energy

$$(H.1) = -H_{ii}^\alpha - \sum_k q_k J_{ik} + \sum_k q_k^\alpha K_{ik}^\alpha + (1 - q_i^\alpha) L_{ii} \qquad (H.2)$$

In this general expression the terms can be rearranged in different ways, giving different formulations of the ionization energy.

The first formulation is obtained if eq. (H.2) is written in the following way

$$(H.1) = -\varepsilon_i^{HF^\alpha} + (1 - q_i^\alpha) L_{ii} \qquad (H.3)$$

Here, ε^{HF^α} is defined by eq. (B.13). If we remove the last term, which depends upon the idempotency relation, we obtain correctly Roothaan's expression for the ionization energy (eq. (65) in ref. [1]).

The second formulation starts from

$$(H.3) = -\varepsilon_i^{HF^\alpha} - \left(q_i^\alpha - \tfrac{1}{2}\right) L_{ii} + \tfrac{1}{2} L_{ii}$$

which gives by use of eq. (G.33)

$$(H.1) = -\varepsilon_i^\alpha + \tfrac{1}{2} L_{ii} \qquad (H.4)$$

The third formulation starts from

$$(H.1) = -H_{ii}^{\alpha} - \sum_{\substack{k \\ \neq i}} q_k J_{ik} - \left(q_i - \tfrac{1}{2}\right) J_{ii} + \sum_{\substack{k \\ \neq i}} q_k^{\alpha} K_{ik} + \left(q_i^{\alpha} - \tfrac{1}{2}\right) K_{ii}^{\alpha} -$$
$$- \left[\left(q_i^{\alpha} - \tfrac{1}{2}\right) - \tfrac{1}{2}\right] L_{ii}$$

where we have $\left(q_i^{\alpha} - \frac{1}{2}\right)$ instead of q_i^{α}. The expression can therefore be interpreted as

$$(H.1) = - {}^{t}\varepsilon_i^{\alpha} \qquad (H.5)$$

where the eigenvalue ${}^{t}\varepsilon$ has been obtained from a study of a molecule in which we have removed one half electron from orbital ψ_i. This will be discussed below.

The eigenvalue, which represents the energy difference (H.1), contains correlation energies to the same extent as the energies in eq. (H.1) contain correlation energies.

We can thus calculate the ionization energy in two ways: To use eq. (H.1) we calculate the total energy of the molecule first and of the ion afterwards. This requires two SCF calculations. To use eq. (H.4) we perform one SCF calculation only, but we have to add the complicated second term in eq. (H.4). Fortunately, the numerical value of L_{ii} is about the same for all orbitals, according to eq. (S.40). This could diminish the numerical work.

It would, however, be of value to have a simpler method for calculation of the IP's.

<u>The transition state method</u>

Instead of studying the molecule, let us study a molecule from which we have removed one half electron from orbital ψ_i^{α}. It is evident that this will influence the SCF calculation and especially change all eigenvalues.

As shown in eq. (H.5) we have

$$IP_i^{\alpha} = - {}^{t}\varepsilon_i^{\alpha} \qquad (H.6)$$

where t means that the transition state has been studied.

<u>How the computer performs the calculation.</u>

The SCF calculation consists of several iterations. The result of such an iteration is a set of ε_j' and $c_{\mu j}'$.

In the new iteration the density matrix is first calculated using

$$P_{\mu\nu} = \sum_j q_j c_{\mu j} c_{\nu j}$$

In molecular calculations the computer takes $q_j = 2$ for all occupied orbitals except for orbital ψ_i whose ionization energy we are calculating. For this orbital $q_i = 3/2$ is used.

The result of this iteration is then that ε_i means $-IP_i$. The other eigenvalues ε_j lack meaning.

The state with $q_i = 3/2$ is usually called "transition state".

Reorganization.

When the electron goes away, the remaining electrons in the molecule will feel a diminishing shielding. Their orbitals change and we have "reorganization".

If we determine IP from eq.(H.4) we study only the molecule, not the ion. The reorganization is thus neglected and the calculated IP is incorrect, often by 5 eV or 10 eV.

It is therefore better to use eq.(H.1) in which we use one SCF calculation for the molecule and one for the ion. The reorganization is thus handled correctly in eq.(H.1).

Also the transition state method treats the reorganization correctly (in the first approximation), since we calculate something which is between molecule and ion. This is a major advantage of eq.(H.5) over eq.(H.4).

The "diffuse" ionization method

It was pointed out above that the "extra term" $\frac{1}{2} L_{ii}$ is approximately equal for all orbitals in a molecule. This can be used in the following way.

Since the transition state method removes the "extra term" for a special orbital i, giving eq.(H.5), it will therefore simultaneously be removed from all other orbitals also. We said above that "the other eigenvalues lack meaning". This statement is not quite true. In fact, they are reasonable approximations to the IP's.

One possibility to reduce the number of SCF calculations is thus to remove one half electron from the highest orbital before the SCF calculation starts and to use the eigenvalues for all orbitals as values for the IP's. This possibility was observed simultaneously in work on the $X\alpha$ method [2] by numerical arguments. Mathematical arguments are discussed in the "Note" below.

Evidently, it does not matter much from which orbital we remove the half electron.

In HAM/3 we have therefore instructed the computer to remove equally much from every occupied orbital. For ethylene with six occupied orbitals this means an occupancy of $2 - \frac{1}{6} \cdot 0.5 = 1.91667$. In this way all orbitals are treated in a similar way. The method is called "diffuse ionization". It introduces an error which for many molecules is only about 0.3 eV or less but sometimes larger.

In the HAM/3 program this is achieved simply by addition of the codeword "PES".

Correlation

It is obvious that a calculation of an ionization energy in a HAM calculation can be successful only if the correlation energies have been taken into account properly.

According to Sec.C.7. the pair-correlation energies $\varepsilon^{corr}_{\mu\nu\lambda\sigma}$ can be assumed to be independent in the first approximation of the occupation numbers. The transformation of eq.(H.1) to give (H.3) and (H.4) and finally (H.5) means therefore that the correlation energy remains unchanged. A condition is, of course, that in all expressions J', K' and ε' have been used instead of J, K and ε.

The use of eq.(H.6) can therefore be expected to give a reasonable value for the change of correlation energy in the ionization process, exactly corresponding to that in eq.(H.1).

It is well known that the correlation energy of the ion is smaller that that of the molecule, since the number of electrons in the ion is smaller (cf. eq.(C.26)). If the correlation is neglected, the calculated IP is therefore incorrect, sometimes by several eV.

Note on the "diffuse" method [3]

Let us remove one half electron from orbital m before the SCF calculation starts. We wish to see how well the eigenvalue $-{}^{tm}\varepsilon_i$ describes the ionization energy IP_i .

We rearrange the general expression eq.(H.2) to obtain a fourth formulation of the ionization energy

$$(H.1) = -H_{ii}^{\alpha} - \sum_{\substack{k \\ \neq m}} q_k J_{ik} - (q_m - \tfrac{1}{2}) J_{im} + \sum_{\substack{k \\ \neq m}} q_k^{\alpha} K_{ik}^{\alpha} + (q_m^{\alpha} - \tfrac{1}{2}) K_{im}^{\alpha} - (q_i^{\alpha} - \tfrac{1}{2}) L_{ii}$$

$$- \tfrac{1}{2} J_{im} + \tfrac{1}{2} K_{im}^{\alpha} + \tfrac{1}{2} L_{ii}$$

or

$$(H.1) = -{}^{tm}\varepsilon_i \quad - \tfrac{1}{2} J_{im} + \tfrac{1}{2} K_{im} + \tfrac{1}{2} L_{ii} \qquad (H.7)$$

If $-\tfrac{1}{2} J_{im} + \tfrac{1}{2} K_{im} + \tfrac{1}{2} L_{ii}$ is small, the eigenvalue $-{}^{tm}\varepsilon_i$ is a good approximation to the ionization energy IP_i. This is usually the case.

If, however, orbital i or orbital m is strongly localized, J_{im} is smaller and a too small ionization energy is obtained. As an example 1-octene ($CH_2=CH-CH_2-CH_2-CH_2-CH_2-CH_2-CH_3$) will be mentioned. The π-orbital in the ethylene group is strongly localized but all other orbitals (σ-orbitals) are delocalized with contributions from all carbon atoms. The eq.(H.6) value for the ionization energy of the π orbital is 10.389 eV and for one of the σ orbitals it is 11.082 eV. The "diffuse" value for the π orbital is much smaller: 8.928 eV, but for the σ orbital it is approximately unchanged: 10.819 eV.

Usually, however, so called "localized" orbitals such as the lone pair in pyridine are sufficiently delocalized to give reasonable IP's by use of a "diffuse" calculation.

It is always possible to check the quality of the "diffuse" IP's by comparison with the eq.(H.6) IP's.

Which value is the "best" ? The eq.(H.6) value or the "diffuse" value ? It depends upon the way in which the parametrization was performed. In HAM/3 "diffuse" ionization was used during the parametrization process. In HAM/3 the "diffuse" value should therefore normally be preferred.

Note on the difference between ionization energies and orbital energies in HAM/3.

Often, two calculations are performed for every molecule.

First, the molecule itself is studied in an SCF calculation. The eigenvalues appear in the print-out. They are denoted as orbital energies.

Second, the diffusely ionized molecule is studied, giving a new set of eigenvalues, denoted as ionization energies.

The energy difference between the two sets of eigenvalues amounts often to 4 or 5 eV. It follows from eq.(H.4) that this energy difference can be approximated as $\frac{1}{2} L_{ii}$.

The same energy difference appears when electron affinities are calculated (see Chap.J.).

H.2. Treatment of ionization energies using the Hartree-Fock method.

In the Hartree-Fock method the ionization energies can be obtained in two ways.

The first way means use of eq.(H.1) which is valid also in the Hartree-Fock method. Its use is called ΔE_{SCF} method. The reorganization is handled correctly in this way, but since no correlation effects are included in Hartree-Fock, the calculated ionization energies are incorrect, sometimes by several eV.

The second way means use of eq.(H.3) after removal of the last term. This means

$$IP_i^{\infty} = - \varepsilon_i^{\infty} \qquad (H.8)$$

Use of eq.(H.8) means that both correlation and reorganization are neglected. Since the two errors have opposite sign, they compensate each other partly [5] . Eq.(H.8) has therefore been used much ("Koopmans' theorem" [6]). The remaining error has been handled by empirical relations such as

$$\begin{aligned} IP_i &= 0.92 \cdot \varepsilon_i \quad \text{eV} \\ &= 0.85 \cdot \varepsilon_i \quad \text{eV} \\ &= 0.770 \cdot \varepsilon_i + 3.17 \quad \text{eV} \\ &= 0.772 \cdot \varepsilon_i + 2.22 \quad \text{eV} \\ &= \varepsilon_i - 4.0 \quad \text{eV} \end{aligned} \qquad (H.9)$$

These crude empirical rules must be considered as unsatisfactory [7].

H.3. Calculation of ionization energies in ab-initio work.

During recent years it has appeared to be possible to go beyond the Hartree-Fock level to calculate ionization energies. This means that it has been possible to include both reorganization and correlation in the calculations.

Two such methods will be mentioned: The perturbation method, called RSPT (Rayleigh-Schrödinger Perturbation Theory) by Chong and coworkers [8] and the Green's function method by Cederbaum and von Niessen.

We will only describe part of the RSPT method.

The total energy of a molecule is given in eq.(C.43) as $E + E_c$, and E_c is given in eq.(C.45) as a sum of pair-correlation energies, which then are given in eq.(C.46). The coefficients C_{jk}^{ab} in eq.(C.46) can be obtained from perturbation theory. This gives to the second order

$$E_{exact} = E_0 + \sum \frac{H_{01} \cdot H_{10}}{E_0 - E_1} \qquad (H.10)$$

where $_1$ means a doubly excited state, or

$$E_{exact} = E_0 + \sum_{jk} \sum_{ab} \frac{[(ja|kb) - (jb|ka)]^2}{\varepsilon_j + \varepsilon_k - \varepsilon_a - \varepsilon_b} \qquad (H.11)$$

Chong and coworkers extended this to the third order and treated afterwards the molecular ion in the same way. The difference of the energies of the molecule and the ion gives then a very good estimate of the experimental IP.

Chong and coworkers have calculated the PES of about 20 molecules with up to three of four atoms. The results are usually in excellent agreement with experiment and with the results from HAM/3 calculations [9] . Due to computer time no larger molecules have been studied.

Cederbaum, von Niessen and coworkers use approximately the same method although Green's functions are used to solve the mathematical problems. They have studied many molecules, also comparatively large (substituted benzenes), and obtain excellent agreement with experiment and with the results from HAM/3 calculations [10 - 18].

von Niessen's work has required very large computer times.

H.4. Experimental methods for study of ionization.

a) The photoelectron spectrometer.

Molecular photoelectron spectroscopy was described 1963 by D. Turner at a conference in Liege [19].

Photons from a helium discharge hit the gas molecules, the molecules are ionized and an electron is ejected. The energy of the electron is measured.

$$h\nu + M \longrightarrow M^+ + e$$

For each orbital in the molecule a peak is obtained, which has vibrational structure (progression of bands) or is broadened.

Turner tried three methods for the determination of the energy of the electrons, namely retarding grids, magnetic deflection and electrostatic deflection. Only the electrostatic deflection is used today.

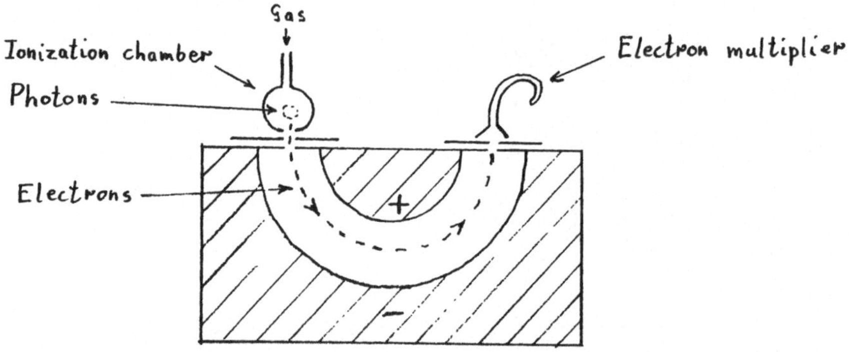

The ionization chamber is a tube, perpendicular to the drawing, and and the photons move in the center of this tube, perpendicular to the drawing.

With helium in the discharge lamp, the resonance line is very strong

58.4 nm 21.21 eV He^* $1s2p \longrightarrow He\ 1s^2$

Other helium lines are fortunately weak

53.7 nm 23.09 eV He^* $1s3p \longrightarrow He\ 1s^2$

52.2 nm 23.74 eV He^* $1s4p \longrightarrow He\ 1s^2$

but it is often necessary to understand that minor features in the spectra may be due to these lines or to impurities in the light source from oxygen or nitrogen.

At higher intensity in the light source, the resonance line of ionized helium can be used

30.4 nm 40.8 eV He^{+*} 2p ⟶ He^{+} 1s

Use of the 58.4 nm line makes possible studies of photoelectron spectra up to about 19 eV, although the relative intensities often are erroneous above 16 eV.

Use of the 30.4 nm line makes possible studies up to about 26 eV where for most molecules the spectra due to the 52.2 nm line begin to predominate. The relative intensity is usually reliable for the whole spectrum, and therefore most spectra shown here have been obtained in this way.

The helium gas in the light source may contaminate the gas in the ionization chamber. Therefore, the 30.4 nm photoelectron spectrum of a gas may exhibit an impurity line from He at 24.55 eV (ionization energy of helium), which is very useful for calibration.

b) ESCA

Some years before Turner, K. Siegbahn and coworkers used X-rays to study photoelectron spectra of atoms and molecules. Their interest to begin with was focussed upon the ionization of 1s electrons, where minor displacements of the ionization energies could be made use of (Electron Spectroscopy for Chemical Analysis). It appeared later that this technique could be used to study also the valence electrons.

The photons Mg $K\alpha$ at 1253.6 eV and Al $K\alpha$ at 1486.6 eV are valuable for studies of the carbon, nitrogen, oxygen and fluorine 2s orbitals with energy 25 ---- 45 eV, which cannot be studied with helium radiation. Unfortunately, only few such studies have been performed.

The experimental equipment will be described in Sec. K.2.

Recently, synchrotron radiation together with a monochromator is used to produce photons at e.g. 50 eV.

c) The (e, 2e) spectroscopy.

In this new method, invented by work in Italy, Australia and Canada, no photons but incident electrons are used. The results are, however, identical with those from photoelectron spectroscopy.

The method is best illustrated with an example:

$$\begin{array}{ccccccccc} e & + & NH_3 & \longrightarrow & NH_3^+ & + & e & + & e \\ 411\ eV & & & & 11\ eV & & 200\ eV & & 200\ eV \end{array}$$

The incident electrons are produced in an ordinary electron gun and are accelerated to 411 eV. The two ejected electrons pass each through an electrostatic analyzer, which transmits electrons of energy 200.0 eV. The deflection angle is 45°. The electrons are measured in coincidence, requiring very long measuring times.

The drawing shows that the apparatus is symmetric with both analyzers in the plane of the drawing. If one of the analyzers is rotated (e.g. ϕ = 10°) around the electron beam, the relative intensities of the spectrum change. The change depends upon the types of the molecular orbitals.

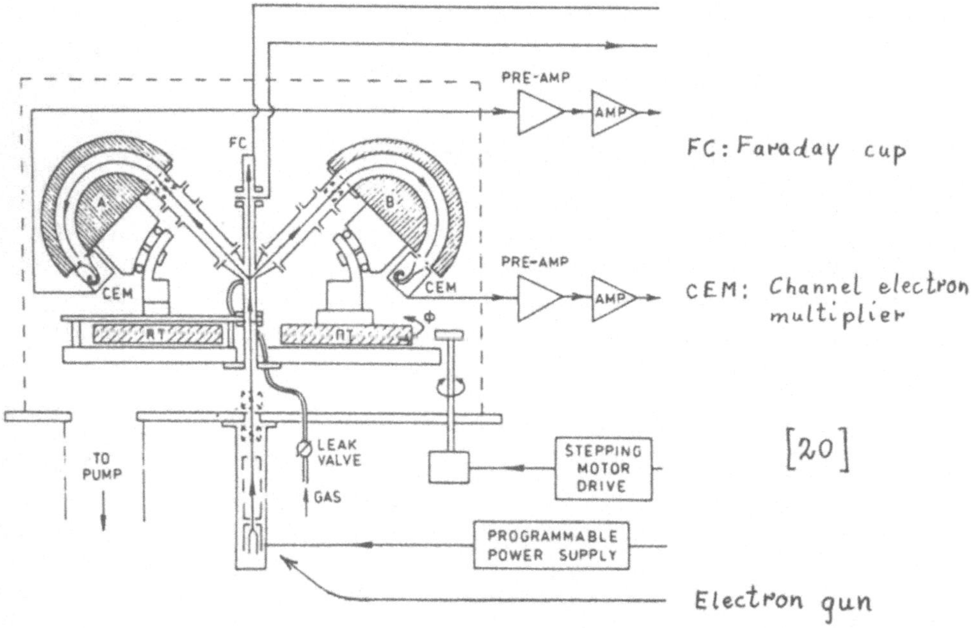

d) Penning ionization electron spectroscopy.

An excited helium atom, He 1s2s ^3S, has a long lifetime. It cannot emit light so that He $1s^2$ ^1S is formed, since both the transition 2s → 1s and the transition triplet → singlet are forbidden. Such metastable atoms with high energy can therefore be used to bombard molecules. The following process takes then place:

$$\underset{19.82\ eV}{\text{He } 1s2s\ ^3S} + M \rightarrow M^+ + \underset{0\ eV}{\text{He } 1s^2} + \underset{\text{kin. energy}}{e}$$

The kinetic energy of the electron depends upon the ionization energy of the orbital in M. Energy analysis gives therefore a curve, similar to the photoelectron spectrum of the molecule [21].

The process takes place when the molecule and the helium atom just touch each other, and when this takes place one of the molecular orbitals, which are localized in the outer part of the molecule, will be ionized. The method can therefore be used to identify π orbitals in a planar molecule, since such orbitals extend a little beyond the main parts of the molecule. Such orbitals get therefore a higher intensity in the spectrum.

To produce the metastable atoms helium gas is bombarded with 60-eV electrons and the ions, which are formed simultaneously, are removed by an electrostatic potential.

e) Experimental difficulties

The main difficulties in studies using electron optics with slow electrons are due to stray fields. Magnetic stray fields are easily compensated using magnetic shielding or compensating magnet coils.

Electrostatic stray fields are due to charging of the surfaces of the electrodes by electrons from the electron beam. Two methods have been used.

The first method is to paint the electrodes with colloidal graphite in alcohol [22]. The surface of the porous graphite takes then the same potential as the metal below, and no surface charges are formed.

The second method is to make the electrodes of molybdenum. Heating to about 200° C in vacuum gives rise to formation of MoS_2, which moves to the surface [23]. The layer of MoS_2 functions then in the same way as a layer of graphite. This method has recently appeared to be very successful (see Sec.I.7.g.).

H.5. Ionization of molecules: some results.

A large number of molecules have now been studied with photo-electron spectroscopy. Earlier only 58.4 nm studies were available. The interpretation of such a study was not easy, since the number of bands (and orbitals) above e.g. 16 eV did not follow from the measurement. Therefore, 30.4 nm studies have been performed for many molecules and compared with theoretical calculations. Only a few molecules will be discussed here.

Ethylene: calculations

It was stated in Sec.H.1. that use of eq.(H.6) and the "diffuse" method give approximately the same result. We illustrate this with a study of orbital 6 in ethylene.

In the first calculation according to eq.(H.6) we have $q_j = 2.000$ for j values 1---5 and $q_6 = 1.500$. This gives ε_6 = -10.912 eV.

	1	2	3	4	5	6	7	8	9	10	11	12
	-24.218	-19.751	-16.255	-14.974	-13.155	-10.912	-4.614	15.441	19.541	25.423	29.892	36.299
q_j →	2.000	2.000	2.000	2.000	2.000	1.500	0.0	0.0	0.0	0.0	0.0	0.0
1 C 1	-0.470	-0.336	0.000	0.071	-0.000	-0.000	-0.000	-0.674	-0.834	0.000	0.006	0.000
2 C 1	0.060	-0.217	-0.000	-0.523	-0.000	0.000	0.000	-0.411	0.387	-0.000	1.001	0.000
3 C 1	0.000	0.000	-0.424	0.000	-0.485	-0.000	-0.000	-0.000	0.000	0.684	0.000	-0.855
4 C 1	-0.000	0.000	0.000	-0.000	0.000	-0.642	0.797	-0.000	0.000	0.000	0.000	0.000
5 C 2	-0.470	0.336	0.000	0.071	0.000	0.000	-0.000	-0.674	0.834	-0.000	-0.006	-0.000
6 C 2	-0.060	-0.217	0.000	0.523	0.000	-0.000	-0.000	0.411	0.387	-0.000	1.001	0.000
7 C 2	0.000	-0.000	-0.424	0.000	0.485	0.000	0.000	-0.000	0.000	0.684	0.000	0.855
8 C 2	0.000	0.000	-0.000	-0.000	0.000	-0.642	-0.797	0.000	0.000	0.000	0.000	-0.000
9 H 3	-0.173	-0.254	-0.245	-0.180	-0.299	-0.000	0.000	0.465	0.286	-0.636	-0.479	0.653
10 H 4	-0.173	-0.254	0.245	-0.180	0.299	0.000	0.000	0.465	0.286	0.636	-0.479	-0.653
11 H 5	-0.173	0.254	-0.245	-0.180	0.299	0.000	0.000	0.465	-0.286	-0.636	0.479	-0.653
12 H 6	-0.173	0.254	0.245	-0.180	-0.299	-0.000	0.000	-0.465	-0.286	0.636	0.479	0.653

In the second calculation we have "diffuse" ionization with 1.91667 electrons in the occupied orbitals. This gives ε_6 = -10.538 eV.

ONE HALF ELECTRON DIFFUSELY REMOVED.

	1	2	3	4	5	6	7	8	9	10	11	12
	-24.292	-19.848	-16.256	-14.883	-13.173	-10.538	-4.269	15.420	19.620	25.496	29.969	36.434
1 C 1	0.466	0.334	-0.000	0.081	0.000	0.000	-0.000	0.676	0.835	0.000	0.003	0.000
2 C 1	-0.055	0.213	0.000	-0.522	0.000	-0.000	-0.000	0.413	-0.381	-0.000	1.004	0.000
3 C 1	-0.000	0.000	-0.420	-0.000	-0.482	0.000	-0.000	-0.000	-0.000	0.687	0.000	-0.857
4 C 1	-0.000	-0.000	-0.000	0.000	-0.000	-0.642	-0.797	0.000	0.000	-0.000	0.000	-0.000
5 C 2	0.466	-0.334	-0.000	0.081	-0.000	0.000	-0.000	0.676	-0.835	-0.000	-0.003	-0.000
6 C 2	0.055	0.213	0.000	0.522	-0.000	0.000	0.000	-0.413	-0.381	-0.000	1.004	0.000
7 C 2	0.000	0.0	-0.420	0.000	0.482	-0.000	-0.000	0.000	-0.000	0.687	0.000	0.857
8 C 2	0.000	0.000	-0.000	0.000	-0.000	-0.642	0.797	-0.000	-0.000	-0.000	0.000	-0.000
9 H 3	0.177	0.257	-0.249	-0.180	-0.301	0.000	-0.000	-0.464	-0.288	-0.635	-0.476	0.652
10 H 4	0.177	0.257	0.249	-0.180	0.301	-0.000	-0.000	-0.464	-0.288	0.635	-0.476	-0.652
11 H 5	0.177	-0.257	-0.249	-0.180	0.301	-0.000	0.000	-0.464	0.288	-0.635	0.476	-0.652
12 H 6	0.177	-0.257	0.249	-0.180	-0.301	0.000	-0.000	-0.464	0.288	0.635	0.476	0.652

When this is repeated for the other orbitals we find:

	1	2	3	4	5	6
eq.(H.6)	24.609	19.905	16.275	15.059	13.093	10.912
"diffuse"	24.292	19.848	16.256	14.883	13.173	10.538
difference	0.317	0.057	0.019	0.176	-0.080	0.374

The result is thus that the "diffuse" IP and the eq.(H.6) IP do not differ much. This result is typical for nearly all molecules studied up till now by HAM/3.

In Sec.M.3. we will finally calculate IP_6 in a third way directly from eq.(H.1) and we will find, as expected, very good agreement with the value 10.912 eV from the eq.(H.6) calculation above.

PES of ethylene

On the next page the ionization of ethylene is shown in three figures.

The photoelectron spectrum obtained using the 40.8 eV line from He II exhibits six bands of approximately equal area. At the top the HAM/3 ionization energies are plotted. We observe that the agreement between theory and experiment is reasonable.

The ESCA study exhibits the same energies but quite different intensities. The molecular orbitals built up from 2s atomic orbitals (orbital 1 and 2) are very strong compared with those from 2p atomic orbitals. The shake-up band at 27.2 eV has a complicated explanation: one electron is ionized from orbital 2 and simultaneously another electron is excited from orbital 6 to orbital 7. The shake-up configuration has intensity = 0 since two-electron transitions are forbidden. However, since it has the same symmetry as orbital 1, interaction will take place and the shake-up configuration will steal intensity from band 1 (see Chap.L.).

The (e,2e) spectrum has bad resolution. It could, however, be shown in this study that the shake-up (satellite) band and band 1 have the same symmetry.

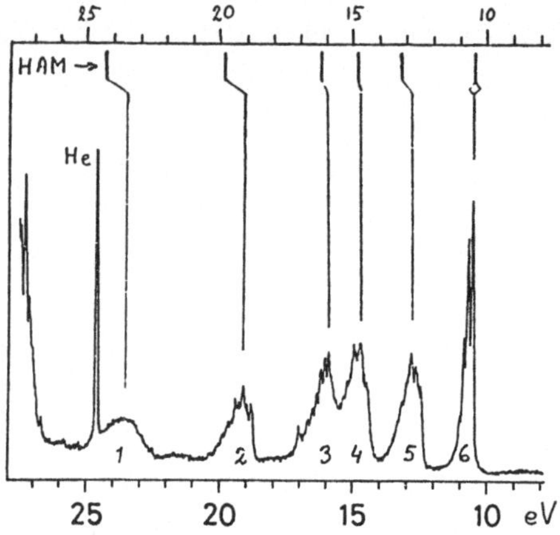

PES
He II
[12]

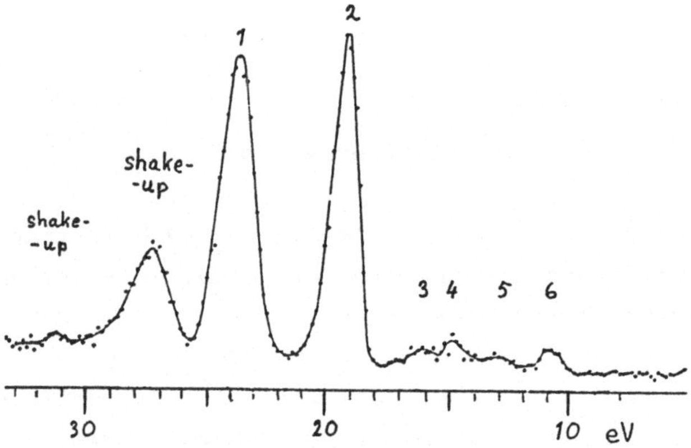

ESCA
Al K
[24]

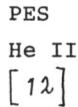

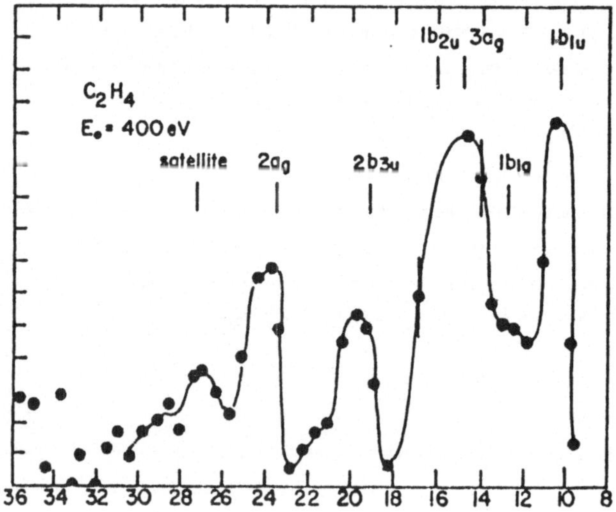

(e,2e)
[25]

PES of butadiene

The geometry, the HAM/3 calculation and the orbitals are shown below. Orbitals 1 --- 11 are occupied and orbitals 12 --- 22 are unoccupied (excited).

ONE HALF ELECTRON DIFFUSELY REMOVED. FILLED ORBITALS GIVE IONIZATION ENERGIES

	1	2	3	4	5	6	7	8	9	10	11
	-25.184	-22.710	-19.400	-18.389	-15.232	-15.176	-13.851	-13.540	-12.581	-11.258	-9.140
1 C 1	-0.288	-0.384	0.281	-0.110	0.019	-0.006	0.090	-0.012	-0.011	0.000	0.000
2 C 1	-0.051	-0.014	-0.154	0.144	-0.042	0.319	0.355	-0.178	0.100	-0.000	0.000
3 C 1	0.003	-0.009	0.011	0.090	-0.351	-0.233	0.023	-0.369	-0.272	-0.000	-0.000
4 C 1	-0.000	-0.000	-0.000	0.000	-0.000	-0.000	-0.000	-0.000	-0.000	0.332	0.544
5 C 2	-0.426	-0.229	-0.244	0.240	0.012	0.140	0.028	0.045	-0.081	-0.000	0.000
6 C 2	-0.003	0.113	-0.153	0.088	0.038	-0.232	-0.394	0.234	-0.139	0.000	-0.000
7 C 2	0.036	-0.061	0.038	0.221	-0.271	-0.087	-0.085	0.131	0.433	-0.000	0.000
8 C 2	0.000	0.000	-0.000	0.000	-0.000	-0.000	-0.000	-0.000	-0.000	0.517	0.383
9 C 3	-0.426	0.229	-0.244	-0.240	0.012	-0.140	0.028	-0.045	-0.081	-0.000	-0.000
10 C 3	0.003	0.113	0.153	0.088	-0.038	-0.232	0.394	0.234	0.139	-0.000	0.000
11 C 3	-0.036	-0.061	-0.038	0.221	0.271	-0.087	0.085	0.131	-0.433	0.000	-0.000
12 C 3	-0.000	-0.000	-0.000	0.000	-0.000	0.000	0.000	-0.000	0.000	0.517	-0.383
13 C 4	-0.288	0.384	0.281	0.110	0.019	0.006	0.090	0.012	-0.011	0.000	-0.000
14 C 4	0.051	-0.014	0.154	0.144	0.042	0.319	-0.355	-0.178	-0.100	-0.000	-0.000
15 C 4	-0.003	-0.009	-0.011	0.090	0.351	-0.233	-0.023	-0.369	0.272	-0.000	0.000
16 C 4	-0.000	-0.000	0.000	-0.000	-0.000	0.000	-0.000	-0.000	0.000	0.332	-0.544
17 H 5	-0.089	-0.175	0.223	-0.066	-0.199	-0.278	-0.087	-0.166	-0.194	-0.000	-0.000
18 H 6	-0.091	-0.163	0.206	-0.185	0.222	0.013	-0.131	0.282	0.167	0.000	0.000
19 H 7	-0.133	0.097	-0.139	-0.296	-0.143	0.059	-0.223	-0.173	0.220	-0.000	0.000
20 H 8	-0.133	-0.097	-0.139	0.296	-0.143	-0.059	-0.223	0.173	0.220	-0.000	-0.000
21 H 9	-0.089	0.175	0.223	0.066	-0.199	0.278	-0.087	0.166	-0.194	0.000	0.000
22 H 10	-0.091	0.163	0.206	0.185	0.222	-0.013	-0.131	-0.282	0.167	-0.000	0.000

	12	13	14	15
	-4.893	-2.005	11.867	15.825
1 C 1	-0.000	0.000	0.080	0.408
2 C 1	0.000	-0.000	-0.099	-0.425
3 C 1	-0.000	0.000	-0.034	0.007
4 C 1	0.643	0.478	0.000	0.000
5 C 2	-0.000	0.000	0.591	0.442
6 C 2	-0.000	0.000	0.392	0.094
7 C 2	-0.000	-0.000	-0.660	0.340
8 C 2	-0.429	-0.691	-0.000	-0.000
9 C 3	0.000	-0.000	-0.591	0.442
10 C 3	0.000	-0.000	0.392	-0.094
11 C 3	0.000	0.000	-0.660	-0.340
12 C 3	-0.429	0.691	0.000	0.000
13 C 4	0.000	-0.000	-0.080	0.408
14 C 4	-0.000	0.000	-0.099	0.425
15 C 4	-0.000	-0.000	-0.034	-0.007
16 C 4	0.643	-0.478	-0.000	0.000
17 H 5	0.000	-0.000	0.092	-0.409
18 H 6	-0.000	-0.000	-0.183	-0.286
19 H 7	-0.000	0.000	0.020	-0.421
20 H 8	0.000	-0.000	-0.020	-0.421
21 H 9	-0.000	0.000	-0.092	-0.409
22 H 10	-0.000	-0.000	0.183	-0.286

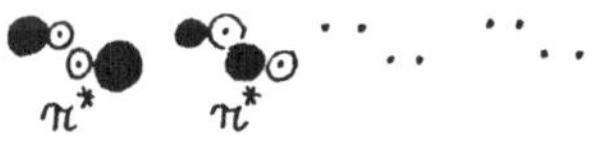

TRANS-BUTADIENE

COORDINATES IN ANGSTROM UNITS

	ATOM	X	Y	Z
1	C	-1.7339	0.6260	0.0
2	C	-0.3970	0.6260	0.0
3	C	0.3970	-0.6260	0.0
4	C	1.7339	-0.6260	0.0
5	H	-2.2710	1.5630	0.0
6	H	-2.2710	-0.3110	0.0
7	H	-0.1400	-1.5630	0.0
8	H	0.1400	1.5630	0.0
9	H	2.2710	-1.5630	0.0
10	H	2.2710	0.3110	0.0

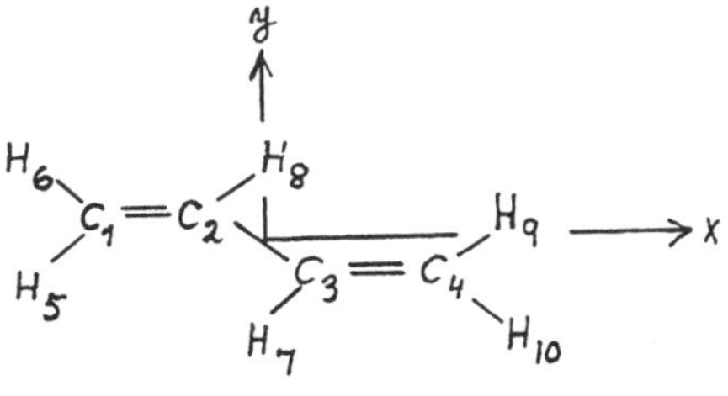

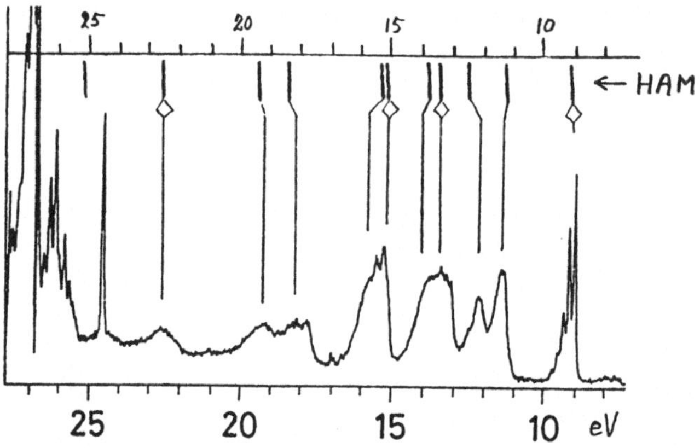

PES

He II

[12]

The agreement between the photoelectron spectrum and the HAM/3 energies is quite reasonable. Also concerning the symmetries of the different orbitals (see the drawings of the orbitals) there is agreement between HAM/3 and more advanced calculations [26].

Using Penning ionization the following result is obtained:

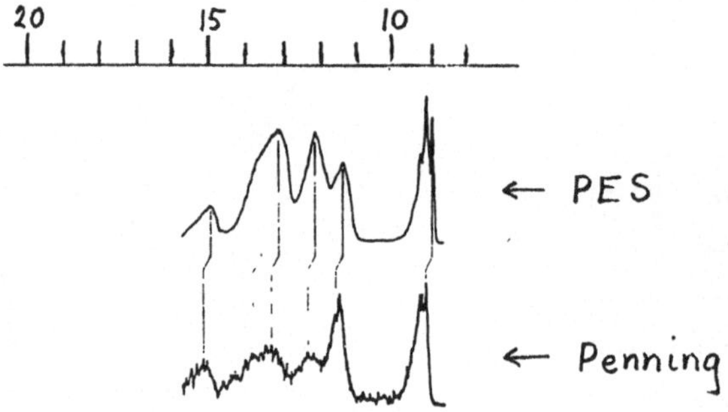

[27]

In Penning ionization spectroscopy the first two bands are strong, since they are π orbitals (No.11 at 9.140 eV and No.10 at 11.258 eV according to the print-out). The third is very weak (No.9 at 12.581 eV), since it is hidden in the inner parts of the molecule.

PES of nitrogen N_2

ONE HALF ELECTRON DIFFUSELY REMOVED. FILLED ORBITALS GIVE IONIZATION ENERGIES

	1	2	3	4	5	6	7	8
	-37.346	-18.398	-16.197	-16.197	-15.422	-7.330	-7.330	26.989
1 N 1	-0.598	0.533	0.0	0.0	0.353	0.0	0.0	0.560
2 N 1	0.0	0.0	-0.658	0.0	0.0	0.0	0.770	0.0
3 N 1	0.0	0.0	0.0	-0.658	0.0	-0.770	0.0	0.0
4 N 1	0.204	0.432	0.0	0.0	0.596	0.0	0.0	-0.785
5 N 2	-0.599	-0.533	0.0	0.0	0.353	0.0	0.0	-0.560
6 N 2	0.0	0.0	-0.658	0.0	0.0	0.0	-0.769	0.0
7 N 2	0.0	0.0	0.0	-0.658	0.0	0.769	0.0	0.0
8 N 2	-0.204	0.432	0.0	0.0	-0.596	0.0	0.0	-0.785
	$2\sigma_g$	$2\sigma_u$	$1\pi_u$		$3\sigma_g$	$1\pi_g$		$3\sigma_u$

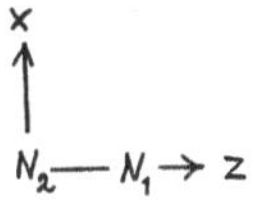

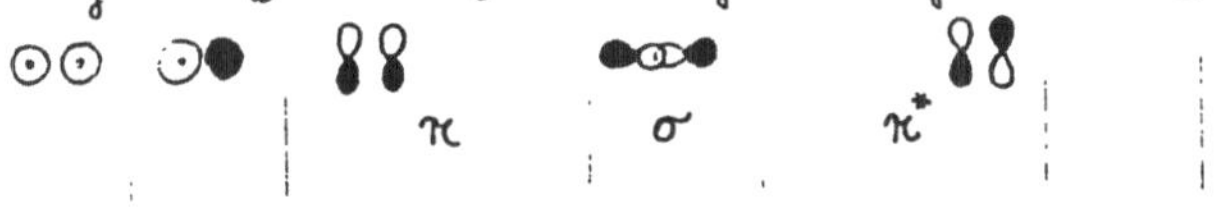

The photoelectron spectra of N_2 are shown below.

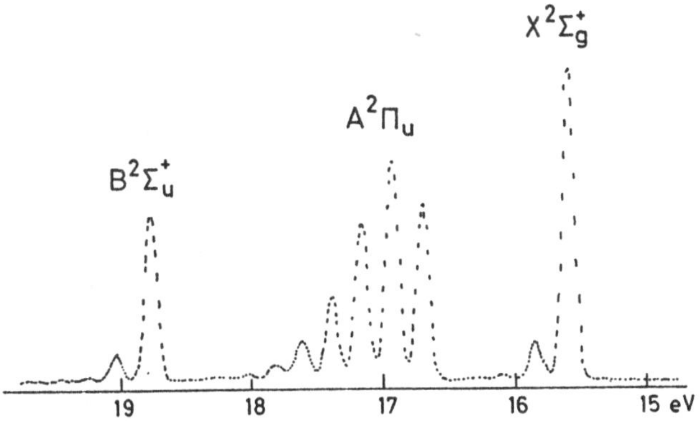

PES

He II

[28]

Fig. 18. Photoelectron spectrum of N_2.

We observe that using He II the X and B bands have about the same intensity, whereas the A band seems to have about the double area. The intensity is thus proportional to the number of electrons ("one band per orbital").

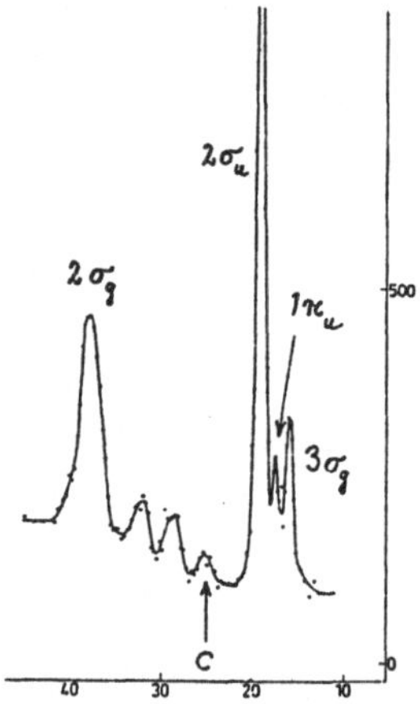

ESCA MgKα [29]

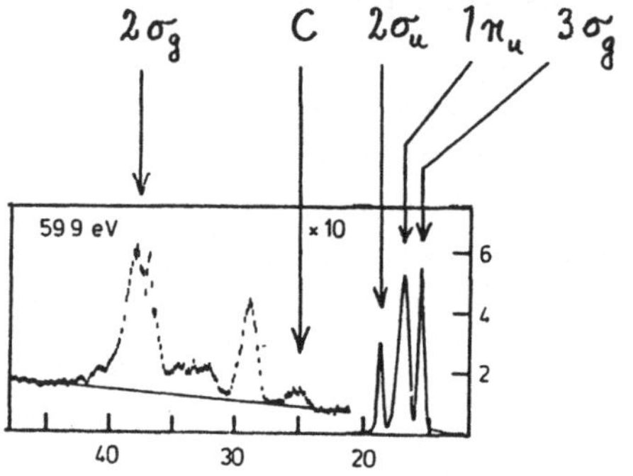

PES
Synchrotron radiation $h\nu$ = 59.9 eV
[30]

We observe that using ESCA the orbitals built up from atomic 2s orbitals, i.e. $2\sigma_g$ and $2\sigma_u$ are very strong. This is true also for the $3\sigma_g$ orbital, which has some 2s character in its lone pairs. On the other hand, the π orbitals have no 2s contribution and therefore a small intensity.

We observe further three shake-up bands at 25 eV, 29 eV and 32 eV. The band at 25 eV is called $C\ ^2\Sigma_u^+$. It is due to ionization of $3\sigma_g$ at 15.7 eV and simultaneous excitation of one electron in $1\pi_u$ to $1\pi_g$. Since its symmetry is the same as for $B\ ^2\Sigma_u^+$ at 18.8 eV, it can steal intensity by configuration interaction.

The shake-up band at 25 eV has been studied by Åsbrink and Fridh [31], using He II.

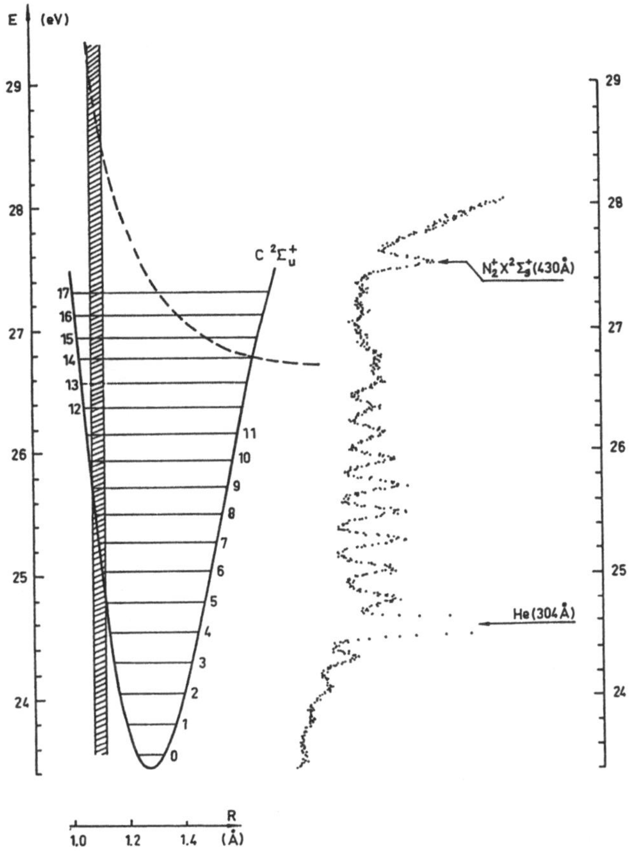

The photoelectron spectrum of N_2 agrees well with the potential energy diagram, constructed by Gilmore [32] from spectroscopic measurements and shown on the next page. (For a slightly improved version of this diagram, see ref. [33].) The Franck-Condon region intersects the $X\ ^2\Sigma_g^+$ and $B\ ^2\Sigma_u^+$ curves near their minima, which gives a single vibrational band in the photoelectron spectrum. The $A\ ^2\Pi_u$ curve is intersected some distance away from the minimum, which gives a progression with five vibrational bands. The $C\ ^2\Sigma_u^+$ curve is finally intersected far away from the minimum, giving many bands in the progression.

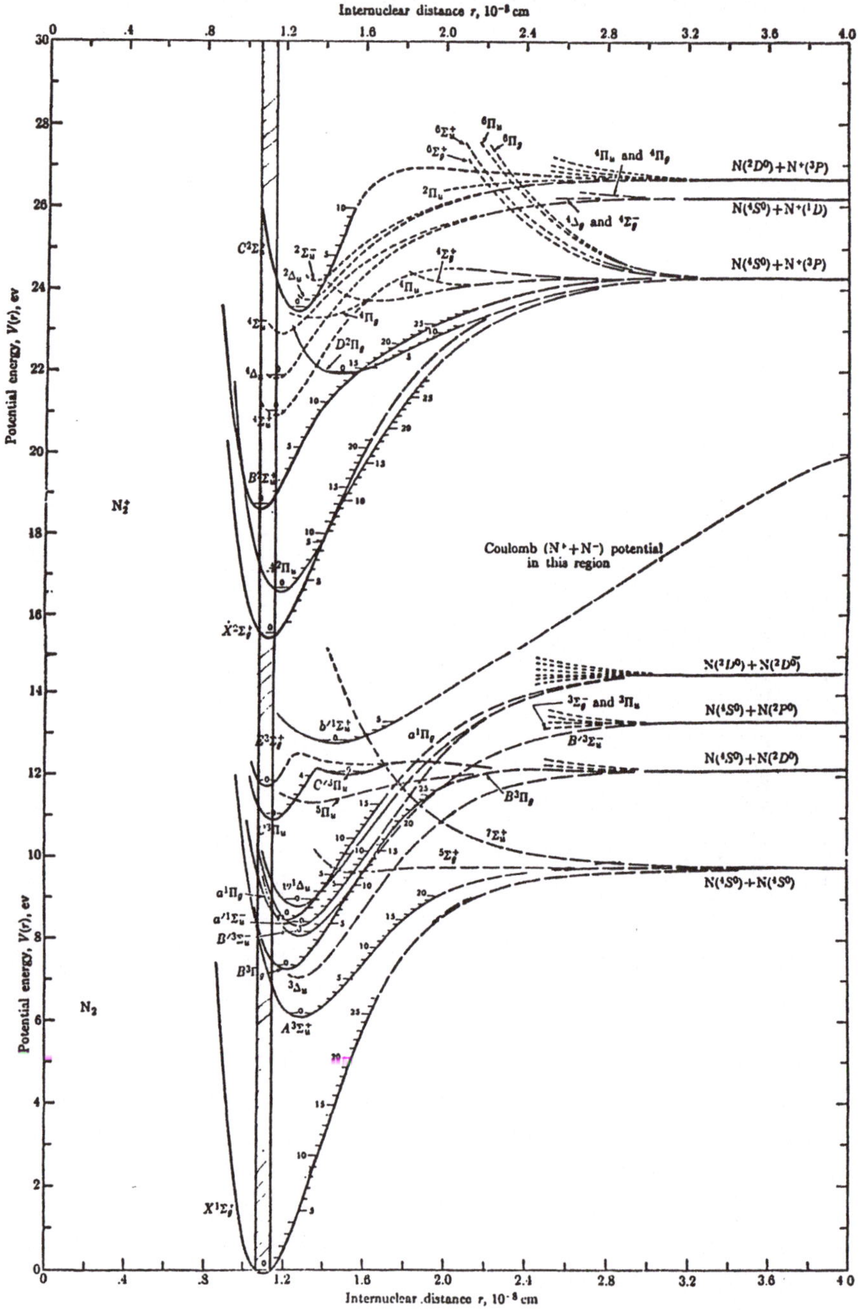

Potential curves for the N_2,   , and N_2^+ molecules based upon spectroscopic and other experimental data.

Gilmore [32]

PES of cyanogen N≡C−C≡N

with internuclear distances C≡N 1.16 Å and C—C 1.37 Å.

ONE HALF ELECTRON DIFFUSELY REMOVED. FILLED ORBITALS GIVE IONIZATION ENERGIES

	3	4	5	6	7	8	9	10	11	12	13	14
	-21.932	-15.069	-15.069	-14.961	-14.527	-13.203	-13.203	-7.037	-7.037	-1.971	-1.971	4.931
1 N 1	0.340	0.0	0.0	0.386	-0.355	0.0	0.0	0.0	0.0	0.0	0.0	-0.209
2 N 1	0.0	0.0	0.382	0.0	0.0	0.0	-0.552	-0.610	0.0	0.0	0.469	0.0
3 N 1	0.0	-0.382	0.0	0.0	0.0	-0.552	0.0	0.0	0.610	0.469	0.0	0.0
4 N 1	0.008	0.0	0.0	0.503	-0.523	0.0	0.0	0.0	0.0	0.0	0.0	0.441
5 C 2	-0.273	0.0	0.0	-0.194	0.059	0.0	0.0	0.0	0.0	0.0	0.0	0.754
6 C 2	0.0	0.0	0.483	0.0	0.0	0.0	-0.379	0.446	0.0	0.0	-0.715	0.0
7 C 2	0.0	-0.483	0.0	0.0	0.0	-0.379	0.0	0.0	-0.446	-0.715	0.0	0.0
8 C 2	0.371	0.0	0.0	-0.108	0.240	0.0	0.0	0.0	0.0	0.0	0.0	-0.131
9 C 3	-0.273	0.0	0.0	0.194	0.059	0.0	0.0	0.0	0.0	0.0	0.0	-0.754
10 C 3	0.0	0.0	0.483	0.0	0.0	0.0	0.379	0.446	0.0	0.0	0.715	0.0
11 C 3	0.0	-0.483	0.0	0.0	0.0	0.379	0.0	0.0	-0.446	0.715	0.0	0.0
12 C 3	-0.371	0.0	0.0	-0.108	-0.239	0.0	0.0	0.0	0.0	0.0	0.0	-0.131
13 N 4	0.340	0.0	0.0	-0.387	-0.354	0.0	0.0	0.0	0.0	0.0	0.0	0.209
14 N 4	0.0	0.0	0.382	0.0	0.0	0.0	0.552	-0.610	0.0	0.0	-0.469	0.0
15 N 4	0.0	-0.382	0.0	0.0	0.0	0.552	0.0	0.0	0.610	-0.469	0.0	0.0
16 N 4	-0.008	0.0	0.0	0.503	0.523	0.0	0.0	0.0	0.0	0.0	0.0	0.441
	$4\sigma_g$	$1\pi_u$		$4\sigma_u$	$5\sigma_g$	$1\pi_g$		$2\pi_u$		$2\pi_g$		
IP_{exp}	22.8	15.6		14.9	14.5	13.4		7.04				

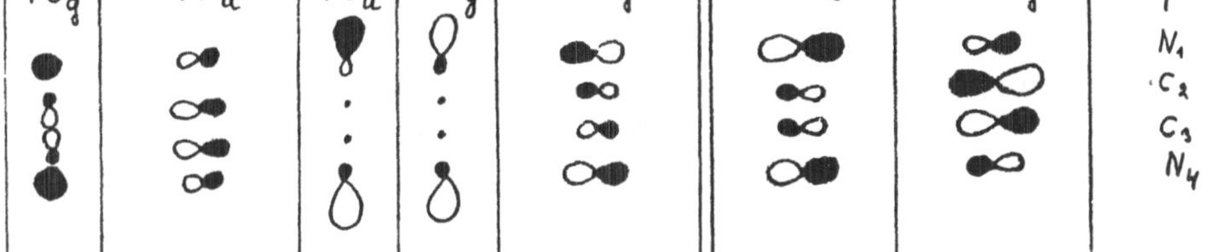

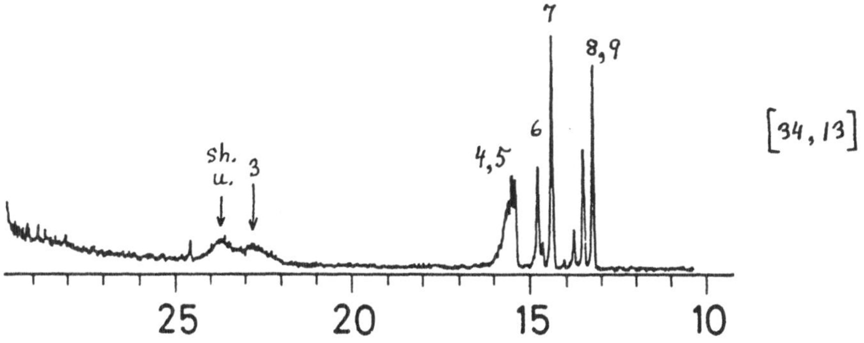

Since the highest experimental IP is 13.4 eV, we have to correct the eigenvalue of orbitals 8 and 9 by −0.2 eV. The "experimental" value for orbitals 10 and 11 is derived from electron affinity studies (see below).

We observe a shake-up band at 23.7 eV which is formed by ionization of one $4\sigma_u$ electron at 14.9 eV together with simultaneous excitation of one $1\pi_g$ electron to $2\pi_u$. The shake up gets intensity by interaction with $4\sigma_g$ at 22.8 eV (see Chap.L.).

PES of acetylene

Internuclear distances are 1.204 Å for $C \equiv C$ and 1.06 Å for $C-H$.

ONE HALF ELECTRON DIFFUSELY REMOVED. FILLED ORBITALS GIVE IONIZATION ENERGIES

	1	2	3	4	5	6	7	8	9	10
	-24.141	-19.733	-17.193	-11.591	-11.591	-3.870	-3.870	15.215	25.598	63.848
1 H 1	0.206	-0.352	-0.293	0.0	0.0	0.0	0.0	0.552	-0.792	-0.725
2 C 2	0.524	-0.336	0.014	0.0	0.0	0.0	0.0	-0.803	0.610	-0.414
3 C 2	0.0	0.0	0.0	0.0	0.627	-0.829	0.0	0.0	0.0	0.0
4 C 2	0.0	0.0	0.0	-0.627	0.0	0.0	-0.829	0.0	0.0	0.0
5 C 2	-0.069	-0.252	-0.486	0.0	0.0	0.0	0.0	0.120	0.553	1.268
6 C 3	0.524	0.336	0.014	0.0	0.0	0.0	0.0	0.803	0.610	0.414
7 C 3	0.0	0.0	0.0	0.0	0.627	0.829	0.0	0.0	0.0	0.0
8 C 3	0.0	0.0	0.0	-0.627	0.0	0.0	0.829	0.0	0.0	0.0
9 C 3	0.069	-0.252	0.486	0.0	0.0	0.0	0.0	0.120	-0.553	1.268
10 H 4	0.206	0.352	-0.293	0.0	0.0	0.0	0.0	-0.552	-0.792	0.725
	$2\sigma_g$	$2\sigma_u$	$3\sigma_g$	$1\pi_u$		$1\pi_g$				
IP_{exp}	23.5	18.7	16.7	11.4		5.2				

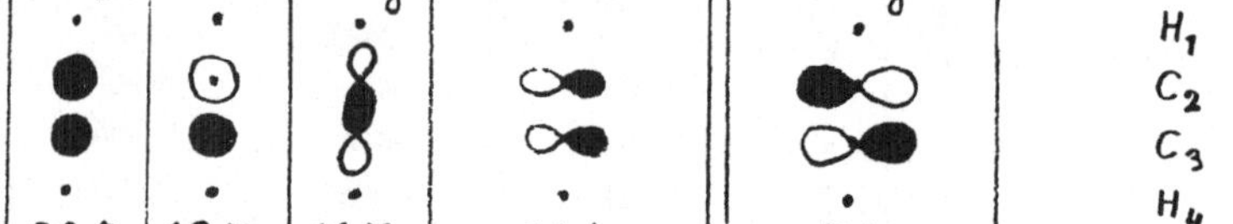

The photoelectron spectrum, observed by Åsbrink [12], is shown below.

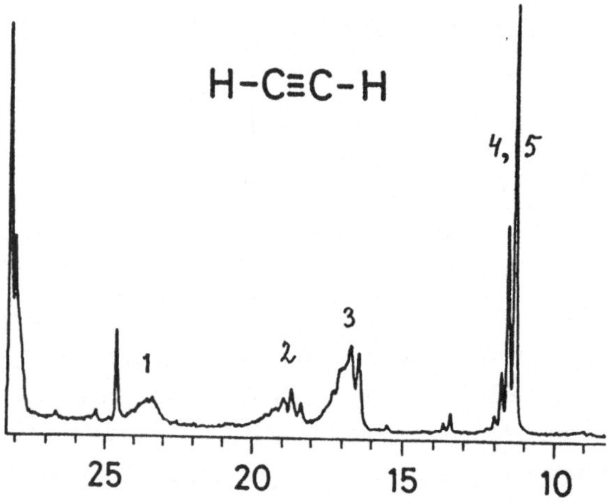

The photoelectron spectrum of acetylene, studied by means of monochromatized X-ray photons with energy 1487 eV [35], is shown below.

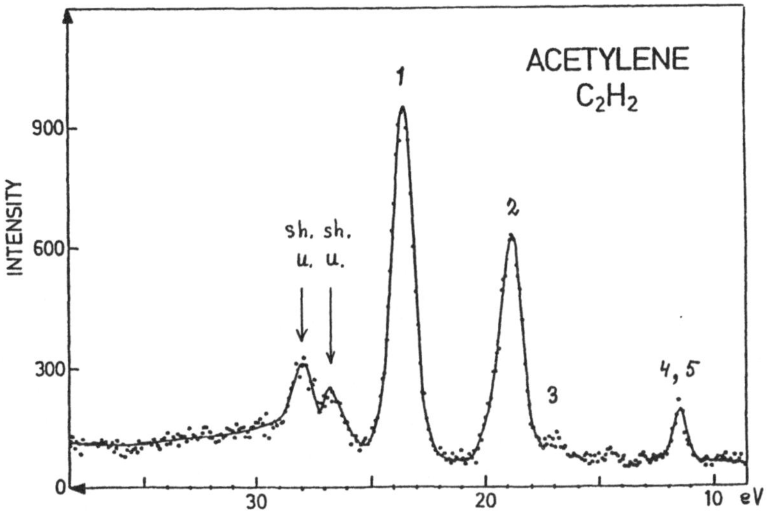

We observe two shake-up bands at 28.0 eV and 26.8 eV (see Sec. L.3. and L.4.). Studies at lower photon energy give similar results [80].

PES of formic acid

FORMIC ACID

COORDINATES IN ANGSTROM UNITS

	ATOM	X	Y	Z
1	C	0.0	0.0	0.0
2	O	1.0111	0.0	0.6970
3	O	-1.2342	0.0	0.4595
4	H	0.0	0.0	-1.0970
5	H	-1.1732	0.0	1.4316

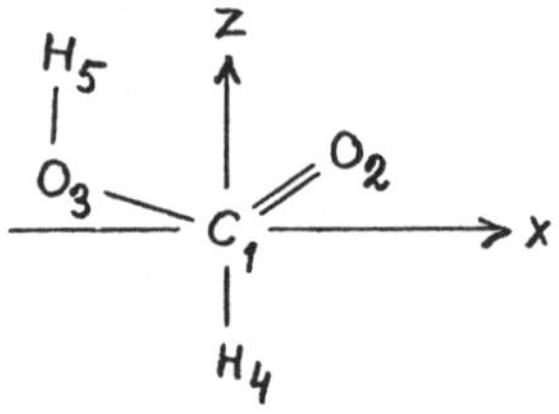

ONE HALF ELECTRON DIFFUSELY REMOVED. FILLED ORBITALS GIVE IONIZATION ENERGIES

			1	2	3	4	5	6	7	8	9	10	11	12
			-35.058	-32.666	-22.030	-17.878	-16.787	-16.194	-14.632	-12.677	-11.105	-5.376	4.465	5.9
1	C	1	0.350	0.205	0.395	0.006	-0.206	0.000	0.010	0.000	-0.071	-0.000	-0.029	-0.7
2	C	1	-0.082	0.349	-0.141	0.211	-0.142	0.000	0.136	0.000	0.039	-0.000	-0.085	0.4
3	C	1	0.000	-0.000	0.000	-0.000	-0.000	-0.514	0.000	-0.244	-0.000	-0.837	-0.000	0.0
4	C	1	0.133	0.113	-0.243	-0.440	0.038	0.000	0.224	0.000	0.102	-0.000	0.044	-0.3
5	O	2	0.326	0.665	-0.153	0.006	0.394	-0.000	-0.216	-0.000	-0.004	0.000	-0.025	0.1
6	O	2	-0.053	-0.064	-0.057	0.181	0.404	-0.000	-0.556	-0.000	0.472	0.000	0.040	-0.3
7	O	2	0.000	0.000	0.000	-0.000	-0.000	-0.353	0.000	-0.743	-0.000	0.581	0.000	-0.0
8	O	2	-0.025	-0.053	-0.101	-0.167	0.306	-0.000	-0.257	-0.000	-0.819	0.000	-0.077	-0.0
9	O	3	0.666	-0.381	-0.101	0.131	0.202	-0.000	0.196	0.000	0.106	0.000	-0.530	0.2
10	O	3	0.060	0.019	0.270	-0.593	-0.244	0.000	-0.447	-0.000	0.190	0.000	-0.084	0.5
11	O	3	-0.000	-0.000	-0.000	-0.000	-0.000	-0.705	-0.000	0.611	0.000	0.373	0.000	-0.0
12	O	3	-0.001	0.002	-0.446	0.103	-0.535	0.000	-0.390	-0.000	-0.071	-0.000	-0.566	-0.2
13	H	4	0.109	0.042	0.273	0.302	-0.101	0.000	-0.184	-0.000	-0.194	0.000	0.013	0.1
14	H	5	0.213	-0.096	-0.297	0.125	-0.196	0.000	-0.149	-0.000	-0.048	-0.000	0.942	0.2

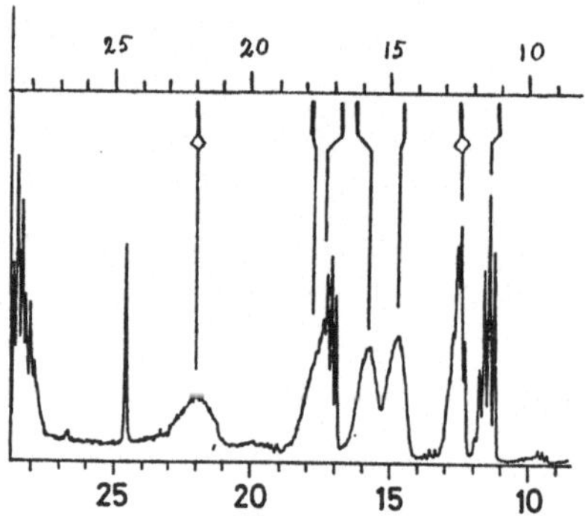

PES

He II

[14]

The carbonyl oxygen atom O_2 has two "lone-pair" (l.p.) orbitals, which do not take part in the chemical bonding. One (orbital 9) is perpendicular to the $C=O$ bond and is of pure 2p type. The other (orbital 7) is in the $C=O$ bond direction and has a strong 2s contribution. Since both orbitals are non-bonding (n), the PES bands are high and narrow.

Also orbital 8 of π type is not much bonding and its PES band is also high and narrow.

PES of ammonia

AMMONIA

COORDINATES IN ANGSTROM UNITS

	ATOM	X	Y	Z
1	H	0.9410	0.0	0.3816
2	H	-0.4706	0.8151	0.3816
3	H	-0.4706	-0.8151	0.3816
4	N	0.0	0.0	0.0

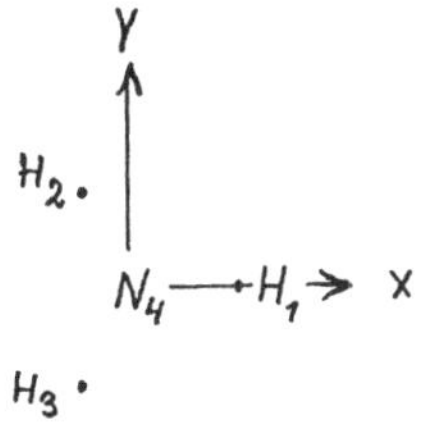

ONE HALF ELECTRON DIFFUSELY REMOVED. FILLED ORBITALS GIVE IONIZATION ENERGIES

		1	2	3	4	5	6	7
		-28.020	-16.526	-16.526	-10.739	5.974	17.914	17.923
1 H	1	0.265	0.453	0.0	-0.084	-0.423	-0.000	-0.959
2 H	2	0.265	-0.226	-0.392	-0.084	-0.423	-0.830	0.480
3 H	3	0.265	-0.226	0.392	-0.084	-0.423	0.830	0.479
4 N	4	0.627	0.0	0.0	0.336	0.812	0.0	0.000
5 N	4	0.0	0.705	0.0	0.0	-0.000	0.000	0.849
6 N	4	0.0	0.0	-0.705	0.0	0.0	0.849	-0.000
7 N	4	0.050	0.0	0.0	-0.919	0.431	0.0	0.000

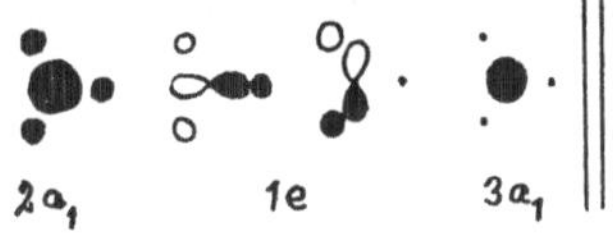

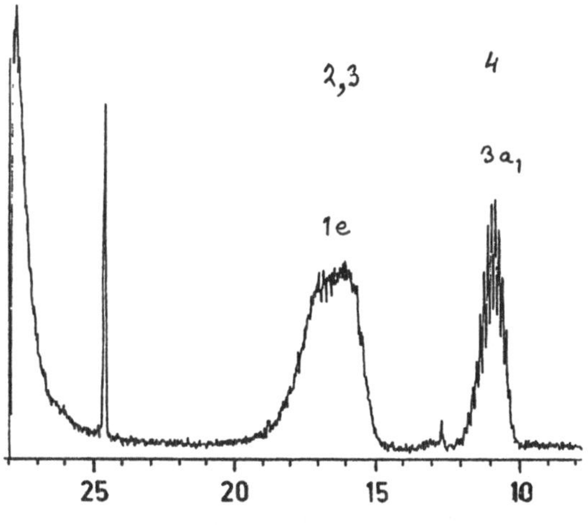

PES He II [18]

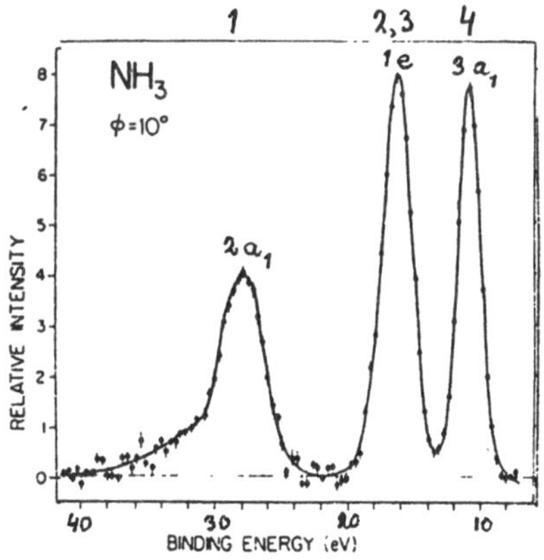

(e,2e) [36]

The (e,2e) study gives the IP also of the 2s type orbital at 28 eV, which is inaccessible using ordinary PES.

PES of benzene

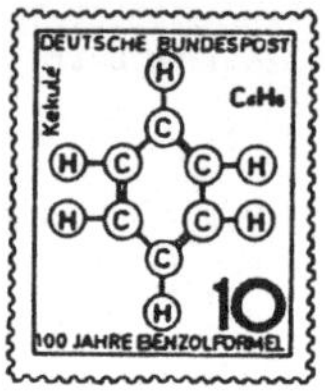

The benzene molecule has symmetry D_{6h} as is evident from the stamp.

If a hydrogen is substituted with e.g. fluorine, the symmetry is C_{2v}.

The figure below shows the photoelectron spectrum [37], the orbitals [38] and their symmetries.

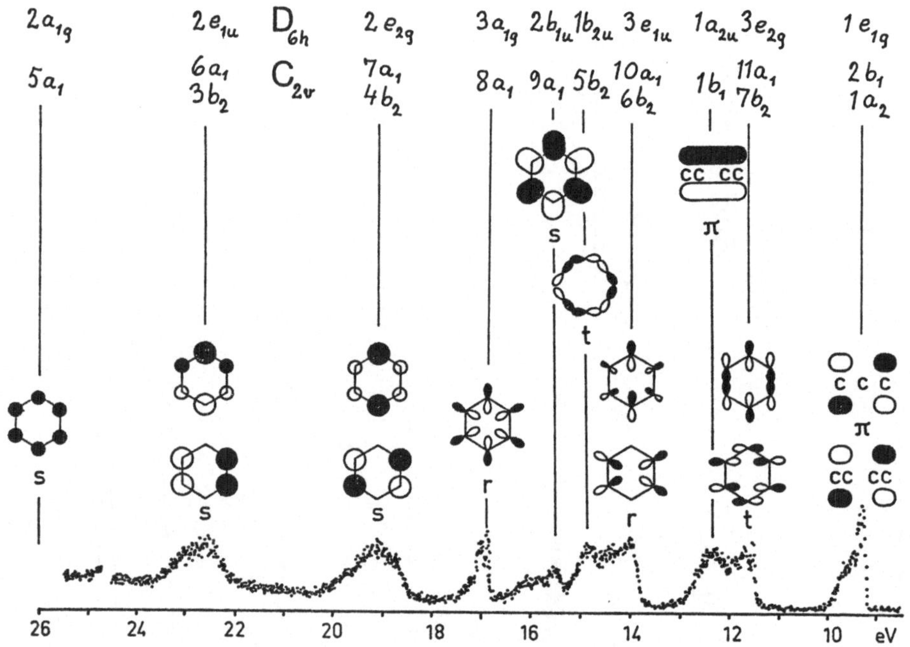

The orbitals have a beautiful symmetry. The orbitals built up from carbon 2s have no node planes ($2a_{1g}$), one ($2e_{1u}$), two ($2e_{2g}$) and three ($2b_{1u}$).. We observe that e denotes a degenerate orbital whereas a and b denote non-degenerate orbitals.

The π orbitals at 9.27 and 12.2 eV are important in chemistry.

The order of the orbitals [39] is now generally accepted. However, the HAM/3 calculation is not quite successful for this molecule. There is an incorrect order between two pairs of orbitals.

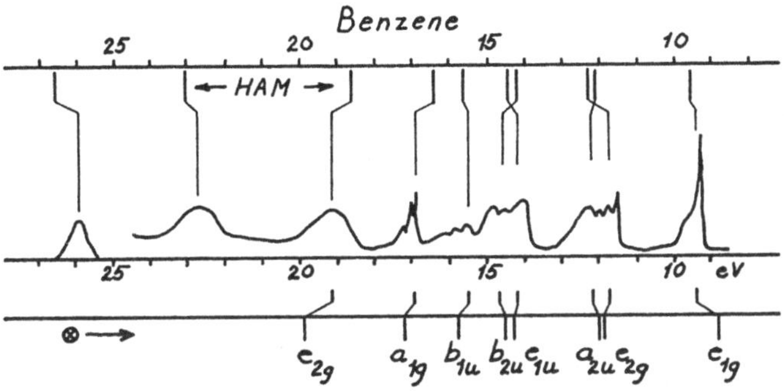

The explanation is simply that when we had performed our last parametrization by use of a steepest descent method, simultaneously studying PES, UV spectra and electron affinities of about 90 molecules, we did not observe this incorrect order. It was observed a year later. In our previous parametrizations we had always had the correct order. The errors depend thus not upon the model, only upon the parameters.

⊗ denotes the results from a Green's function calculation [40].

Using Penning ionization the following result is obtained [21]

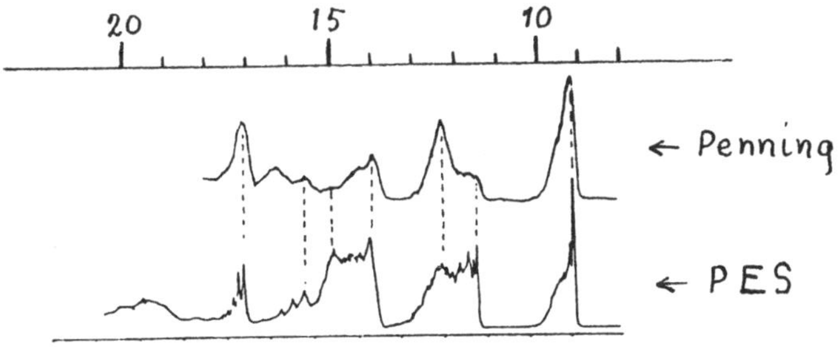

The π bands at 9.3 eV and 12.2 eV are strong but the t-band at 11.8 is weak since it is hidden in the inner parts of the planar molecule (see the orbital picture). The t-band at 14.8 eV is still more hidden and still more weak. The r-band at 17.0 eV is strong since the p-orbitals extend from the molecule.

Using X-ray photons with energy 1487 eV the following photoelectron spectrum of benzene is obtained [41].

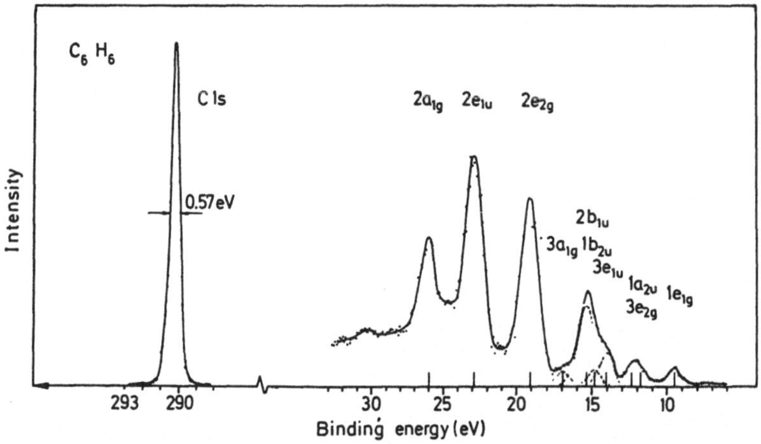

The s-type bands (from orbitals built up mainly from carbon 2s) are much stronger that the other bands (from carbon 2p) as shown already for ethylene.

The explanation is that the size (diameter) of 2s is small compared with that of 2p. Since the wavelength of X-ray photons is also small, such a photon interacts intensely with 2s.

Interpretation of photoelectron spectra

At the beginning of the 1970´s the interpretation of photoelectron spectra was usually unequivocal, but a few molecules were difficult. One of them was benzene, for which it was claimed, that the band at 17.0 eV is of s-type, since it was considered to be natural that all s-type bands are neighbours. The ESCA study showed that this is not the case.

Another problem concerned the π band at 12.2 eV. It was claimed that it has instead the energy 11.7 eV. The Penning study showed that this is not the case.

It is remarkable that so few PES bands have been interpreted by use of experimental results. Besides benzene it has been necessary for only a few molecules. Chemical intuition or studies of similar compounds have sometimes solved the problem. In other cases calculations have been performed by use of the empirically corrected Koopmans´ theorem results (Sec.H.2.) or by true ab-initio calculations (Sec.H.3.) or by SPINDO or HAM. The unsolved problems today concern only a few molecules in which the first orbitals are very close in energy (pyridine, p-benzoquinone, cyclobutane-1,3-dione, quinodimethane, ozone ...).

A third experimental method exploits the symmetries and selection rules for Rydberg transitions. It has been used for a few symmetric molecules (e.g. the 12.2 band in benzene [39]).

A fourth experimental method, which uses the angular dependence of the PES intensitites, will possibly be useful for molecules, adsorbed on surfaces.

High-resolution studies

Only few molecules have been studied by high-resolution photoelectron spectroscopy. Very high resolution was used in Åsbrink's study of H_2 with photons from neon (736 Å; 16.85 eV) [42]

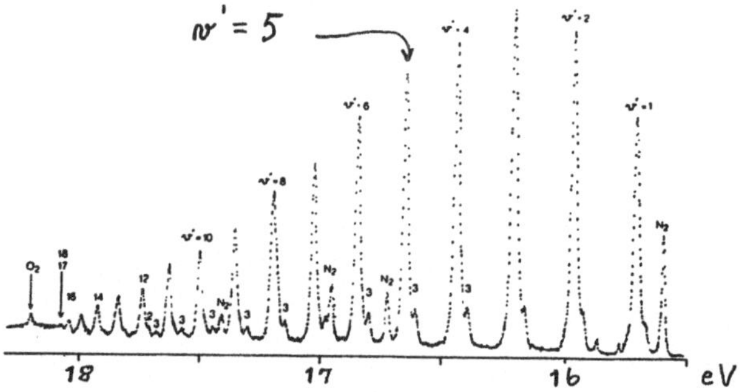

PES of H_2 with He I (584 Å)

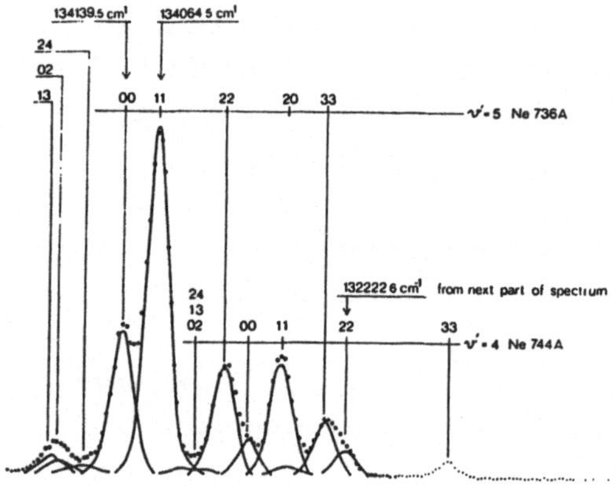

PES of H_2 with Ne (736 Å). The strongest line has in the molecule the vibrational quantum number = 0 and in the ion = 5. The rotational quantum number is = 1 both in the molecule and in the ion.

It is seen that the rotational structure is resolved.

Rotational structure cannot be resolved for any other molecule, but for many small molecules the vibrational structure has been resolved.

For most molecules, however, the PES bands are so broad and diffuse that not even the vibrational structure can be seen. For most organic molecules a photoelectron spectrometer with medium resolving power is therefore sufficient.

H.6. Further studies

The table below includes most studies in which photoelectron spectra have been published together with HAM/3 calculations.

The notations have the following meaning:

acetylene 12,10,43,9 means that PES and HAM/3 calculation are given in refs. [12], [10], [43], [9].

GF12 means that a Green's Function calculation is compared with experiment and HAM/3 calculation in ref. [12].

RSPT9 means the same for a Rayleigh-Schrödinger Perturbation Theory calculation.

UV43 means that the UV spectrum is calculated and discussed in ref. [43].

EA43 means the same for the electron affinities.

Hydrocarbons

CH_4	methane 12,	GF12.			
C_2H_2	acetylene 12,10,43,9,	GF12,	RSPT9,	UV43,	EA43.
C_2H_4	ethylene 12,44,	GF12,		UV45,	EA45.
C_2H_6	ethane 12,	GF12.			
C_3H_4	methyl acetylene 12,9,			UV46.	
	cyclopropene 9.				
	allene 9.				
C_3H_6	cyclopropane 18,	GF18,		UV45,47.	
C_3H_8	propane 12.				
C_4H_2	diacetylene 12,10,	GF12.			
C_4H_6	dimethyl acetylene 12.				
	1,3-butadiene 12,44,	GF12,		UV45,	EA45.
C_4H_8	cyclobutane 48.				
C_4H_{10}	n-butane 12.				
C_5H_6	cyclopentadiene 18,49,	GF18,		UV49,	EA49.
C_6H_2	triacetylene 12	GF12.			
C_6H_6	benzene 12,44,	GF12,		UV50,	EA45.
	dimethyl diacetylene 12.				
	1,5-hexadiyne 18.				
C_6H_8	1,3-cyclohexadiene 51,			UV51,	EA51.
	1,4-cyclohexadiene 51,			UV51,	EA51.
C_7H_8	norbornadiene 18, 51,	GF18,		UV51,	EA51.
C_8H_8	p-quinodimethane 52.				
$C_{10}H_8$	naphthalene 53,			UV53,	EA53.

Compounds with N

N_2	nitrogen 43,9,		RSPT9,	UV43.
NH_3	ammonia 18,	GF18.		
N_2H_2	trans-diazene 54.			
CHN	hydrogen cyanide 43,		RSPT9,	UV43.
CH_2N_2	diazomethane 9.			
	cyanamide 9.			
	diazirine 9.			
CH_3N	methylenimine 54.			
CH_5N	methylamine 18,	GF18.		
C_2N_2	cyanogen 13,34,	GF13,		UV34.
C_2H_3N	acetonitrile 13,9,	GF13,		UV46.
	methylisonitrile 13,9,	GF13,		UV46.
C_2H_5N	ethylenimine 18,	GF18,		UV47.
$C_2H_6N_2$	azomethane 54.			
C_3HN	cyanoacetylene 13,	GF13.		
C_3H_3N	acrylonitrile 18,	GF18.		
$C_3H_3N_3$	1,2,3-triazine 55.			
C_4N_2	dicyano acetylene 13,	GF13.		
$C_4H_4N_2$	trans-1,2-dicyanoethylene 18	GF18.		
	cis-1,2-dicyanoethylene 18,	GF18.		
	1,1-dicyanoethylene 18,	GF18.		
C_4H_3N	methylcyanoacetylene 13,	GF13.		
C_4H_5N	pyrrole 18,49,	GF18,		UV49, EA49.
C_6N_4	tetracyanoethylene 18,	GF18.		
C_6H_5N	pyridine 44,			UV45, EA45.
C_7H_5N	benzonitrile 18.			
C_7H_7N	vinylpyridine 76.			

Compounds with O

H_2O	water 9,		RSPT9.	
CO	carbon monoxide 43,9,		RSPT9,	UV43.
CO_2	carbon dioxide 34,9,			UV34.
CH_2O	formaldehyde 14,	GF14,	RSPT9.	
CH_2O_2	formic acid 14,56,	GF14,		UV57.
CH_4O	methanol 14,	GF14.		
C_2H_2O	ketene 9,		RSPT9.	
$C_2H_2O_2$	glyoxal 14,	GF14.		

C_2H_4O	ethylene oxide 18,	GF18,	UV47.
	acetaldehyde 18,	GF18.	
$C_2H_4O_2$	acetic acid 14.		
C_2H_6O	ethanol 14.		
	dimethyl ether 18,	GF18.	
$C_2H_6O_2$	ethylene glycol 58.		
C_3O_2	carbon suboxide 59,		UV59.
C_3H_2O	propynal 14,15,	GF14,15.	
$C_3H_2O_2$	propiolic acid 14,15,	GF14,15.	
C_3H_4O	propynol 14,15,	GF14,15.	
	acrolein 14,15,60,	GF14,15,	UV60, EA60.
C_3H_6O	acetone 18,	GF18.	
C_4H_4O	furan 18,49,	GF18,	UV49, EA49.
$C_4H_4O_2$	propiolic acid methyl ester 18.		
$C_4H_6O_2$	diacetyl 14.		
$C_4H_8O_2$	1,4-dioxane 18.		
C_5H_8O	cyclopentanone 18.		
$C_6H_4O_2$	p-benzoquinone 61,		UV61, EA61.

Compounds with F

HF	hydrogen fluoride 16,10,	GF16,	RSPT9.
F_2	fluorine 16,10,	GF16,	RSPT9.
CF_2	difluoro methylene 9,		RSPT9.
CF_4	carbon tetrafluoride 16.		
CHF_3	fluoroform 16.		
CH_2F_2	methylene fluoride 16.		
CH_3F	methyl fluoride 16,	GF16.	
C_2F_2	difluoro acetylene 16,10,	GF16.	
C_2F_4	tetrafluoro ethylene 16,11,	GF16.	
C_2HF	fluoro acetylene 16,10,9,	GF16.	
C_2HF_3	trifluoro ethylene 16,11,	GF16.	
$C_2H_2F_2$	1,1-difluoro ethylene 16,11,	GF16.	
	cis-1,2-difluoro ethylene 16,11,	GF16.	
	trans-1,2-difluoro ethylene 16,11,	GF16.	
C_2H_3F	vinyl fluoride 16,11,	GF16.	
C_2H_5F	ethyl fluoride 16.		
C_3HF_3	3,3,3-trifluoro propyne 16.		
C_4F_2	difluoro diacetylene 16,10,	GF16.	

C_4HF	monofluoro diacetylene 10.		
C_5F_4	perfluoro pentadiyne 16.		
C_6F_6	hexafluoro benzene 16.		
$C_6H_xF_y$	fluorinated benzenes, many, 16.		

Compounds with N and O

N_2O	nitrous oxide 34,9,		UV34.
CHNO	hydrogen isocyanate 9.		
CH_3NO	formamide 17,62,	GF17,	UV62, EA62.
	nitrosomethane 63.		
CH_3NO_2	nitromethane 17,	GF17.	
C_2H_5NO	acetamide 17.		
C_3H_3NO	propynoic amide 17.		
C_3H_5NO	acrylamide 17.		
C_3H_7NO	N,N,-dimethyl formamide 18.		
$C_4H_4N_2O$	urea 18,64,		
$C_6H_5NO_2$	nitrobenzene 64,		UV64, EA64.
C_7H_5NO	phenyl isocyanate 50.		

Compounds with N, O and F

N_2F_2	difluoro diazene 54.		
NHF_2	difluoramine 65.		
F_2O	oxygen difluoride 9,		RSPT9.
HOF	hyperfluorous acid 9,		RSPT9.
CNF	cyanogen fluoride 17,9,	GF17.	
C_2NF_3	trifluoro acetonitrile 17,	GF17.	
$C_2N_2F_6$	hexafluoro azomethane 54.		
$C_2HO_2F_3$	trifluoroacetic acid 17.		
$C_2H_2NOF_3$	trifluoro acetamide 17.		
$C_3N_3F_3$	cyanuric fluoride 17,9,	GF17.	
C_3H_2NF	trans-1,2-fluorocyano ethylene 18,		GF18.
C_4NF_3	1-cyano-3,3,3-trifluoro propyne 17.		
$C_4O_3F_2$	difluoro maleic anhydride 17.		
C_5NF_5	pentafluoro pyridine 17.		
$C_6O_2F_4$	tetrafluoro-p-benzoquinone 17,61.		

Large molecules

TCNQ (tetracyanoquinodimethane) 52, UV52, EA52.
uracil 66, UV66.
purine 67.
adenine 67.
carbazole 9.
benzocyclobutadiene 68.
pyridine-N-oxide 69.
azabiphenyls, many different, 70,71,72,81,82.
2-norbornanone 73.
luminol 74.
maleic acid 74.
tri-s-triazine 77.
benzoic acid 78.
quinoid nitronates 79.
cytosine 83.

Radicals

CH_3 75.

References

1. C.C.J. Roothaan, Rev.Mod.Phys. 23, 69 (1951).
2. E.J. Baerends and P. Ros, Chem.Phys. 2, 52 (1973).
3. D.P. Chong, as described in ref. [4].
4. L. Åsbrink, C. Fridh, E. Lindholm, S. de Bruijn and D.P. Chong, Phys.Scripta 22, 475 (1980).
5. W.G. Richards, Int.J.Mass Spectrom.Ion Phys. 2, 419 (1969).
6. T. Koopmans, Physica 1, 104 (1933).
7. R.W. Bigelow, Chem.Phys. 80, 45 (1983).
8. D.P. Chong, F.G. Herring and D. McWilliams, J.Chem.Phys. 61, 78 (1974).
9. D.P. Chong, Theoret.Chim.Acta 51, 55 (1979).
10. G. Bieri, A. Schmelzer, L. Åsbrink and M. Jonsson, Chem.Phys. 49, 213 (1980).
11. G. Bieri, W. von Niessen, L. Åsbrink and A. Svensson, Chem. Phys. 60, 61 (1981).
12. G. Bieri and L. Åsbrink, J.Electron Spectrosc. 20, 149 (1980).
13. L. Åsbrink, W. von Niessen and G. Bieri, J.Electron Spectrosc. 21, 93 (1980).
14. W. von Niessen, G. Bieri and L. Åsbrink, J.Electron Spectrosc. 21, 175 (1980).
15. W. von Niessen, G. Bieri, J. Schirmer and L.S. Cederbaum, Chem.Phys. 65, 157 (1982).
16. G. Bieri, L. Åsbrink and W. von Niessen, J.Electron Spectrosc. 23, 281 (1981).
17. L. Åsbrink, A. Svensson, W. von Niessen and G. Bieri, J.Electron Spectrosc. 24, 293 (1981).
18. G. Bieri, L. Åsbrink and W. von Niessen, J.Electron Spectrosc. 27, 129 (1982).
19. D.W. Turner, C. Baker, A.D. Baker and C.R. Brundle, Molecular Photoelectron Spectroscopy Wiley-Interscience, London (1970).
20. C.E. Brion, S.T. Hood, I.H. Suzuki, E. Weigold and G.R.J. Williams, J.Electron Spectrosc. 21, 71 (1980).
21. K. Ohno, H. Mutoh and Y. Harada, J.Am.Chem.Soc. 105, 4555 (1983).

References (cont.)

22. E. Lindholm, Rev.Sci.Instr. 31, 210 (1960).
23. H.H. Brongersma, private communication.
24. A. Berndtsson, E. Basilier, U. Gelius, J. Hedman, M. Klasson, R. Nilsson, C. Nordling and S. Svensson, Phys.Scripta 12, 235 (1975).
25. M.A. Coplan, A.L. Migdall, J.H. Moore and J.A. Tossell, J.Am.Chem.Soc. 100, 5008 (1978).
26. L.S. Cederbaum, W. Domcke, J. Schirmer, W. von Niessen, G.H.F. Diercksen and W.P. Kraemer, J.Chem.Phys. 69, 1591 (1978).
27. T. Munakata, K. Kuchitsu and Y. Harada, Chem.Phys.Letters 64, 409 (1979).
28. L. Åsbrink, unpublished.
29. K. Siegbahn, C. Nordling, G. Johansson, J. Hedman, P.F. Hedén, K. Hamrin, U. Gelius, T. Bergmark, L.O. Werme, R. Manne and Y. Baer, ESCA Applied to Free Molecules, North-Holland, Amsterdam (1969).
30. S. Krummacher, V. Schmidt and F. Wuilleumier, J.Phys.B: Atom.Molec.Phys. 13, 3993 (1980).
31. L. Åsbrink and C. Fridh, Phys.Scripta 9, 338 (1974).
32. F.R. Gilmore, J.Quant.Spectry.Rad.Transfer 5, 369 (1965).
33. A. Lofthus and P.H. Krupenie, J.Phys.Chem.Ref.Data 6, 113 (1977).
34. C. Fridh, L. Åsbrink and E. Lindholm, Chem.Phys. 27, 169 (1978).
35. J. Müller, R. Arneberg, H. Ågren, R. Manne, P.-Å. Malmquist, S. Svensson and U. Gelius, J.Chem.Phys. 77, 4895 (1982); S. Svensson, P.Å. Malmqvist, M.Y. Adam, P. Lablanque, P. Morin and I. Nenner, Chem.Phys.Letters 111, 574 (1984).
36. S.T. Hood, A. Hamnett and C.E. Brion, Chem.Phys.Letters 39 252 (1976).
37. L. Åsbrink, O. Edqvist, E. Lindholm and L.E. Selin, Chem.Phys.Letters 5, 192 (1970).
38. E. Lindholm, Disc.Faraday Soc. 54, 200 (1973).
39. B.Ö. Jonsson and E. Lindholm, Arkiv Fysik 39, 65 (1969).
40. W. von Niessen, L.S. Cederbaum and W.P. Kraemer, J.Chem.Phys. 65, 1378 (1976).
41. U. Gelius, J.Electron Spectrosc, 5, 985 (1974).
42. L. Åsbrink, Chem.Phys.Letters 7, 549 (1970).

References (cont.)

43. L. Åsbrink, C. Fridh and E. Lindholm, Chem.Phys. 27, 159 (1978).
44. L. Åsbrink, C. Fridh and E. Lindholm, Chem.Phys.Letters 52, 69 (1977).
45. L. Åsbrink, C. Fridh and E. Lindholm, Chem.Phys.Letters 52, 72 (1977).
46. C. Fridh, J.C.S. Faraday II, 74, 2193 (1978).
47. C. Fridh, J.C.S. Faraday II, 75, 993 (1979).
48. L. Åsbrink, C. Fridh, E. Lindholm and G. Ahlgren, Chem.Phys. 33, 195 (1978).
49. L. Åsbrink, C. Fridh and E. Lindholm, J.Electron Spectrosc. 16, 65 (1979).
50. E. Lindholm, J.Mol.Spectrosc. 101, 444 (1983).
51. L. Åsbrink, C. Fridh and E. Lindholm, in preparation.
52. L. Åsbrink, C. Fridh and E. Lindholm, Int.J.Quantum Chem. 13, 331 (1978).
53. L. Åsbrink, C. Fridh and E. Lindholm, Z.Naturforsch. 33a, 172 (1977).
54. D.C. Frost, W.M. Lau, C.A. McDowell and N.P.C. Westwood, J.Mol.Structure (Theochem) 90, 283 (1982).
55. R. Gleiter, J. Spanget-Larsen, R. Bartetzko, H. Neunhoeffer and M. Clausen, Chem.Phys.Letters 97, 94 (1983).
56. F. Carnovale, M.K. Livett and J.B. Peel, J.Chem.Phys. 71, 255 (1979).
57. C. Fridh, J.C.S.Faraday II, 74, 190 (1978).
58. L. Karlsson, L. Åsbrink, C. Fridh, E. Lindholm and A. Svensson, Phys.Scripta 21, 170 (1980).
59. D.P. Chong, Can.J.Chem. 58, 1687 (1980).
60. C. Fridh, L. Åsbrink and E. Lindholm, Phys.Scripta 20, 603 (1979).
61. L. Åsbrink, G. Bieri, C. Fridh, E. Lindholm and D.P. Chong, Chem.Phys. 43, 189 (1979).
62. E. Lindholm, G. Bieri, L. Åsbrink and C. Fridh, Int.J.Quantum Chem. 14, 737 (1978).
63. D.C. Frost, W.M. Lau, C.A. McDowell and N.P.C. Westwood, J.Phys.Chem. 86, 3577 (1982).
64. D.P. Chong, J.Mol.Sci. 2, 55 (1982).
65. D. Colbourne, D.C. Frost, C.A. McDowell and N.P.C. Westwood, Chem.Phys.Letters 72, 247 (1980).

References (cont.)

66. L. Åsbrink, C. Fridh and E. Lindholm, Tetrahedron Letters 52, 4627 (1977).
67. J. Lin, C. Yu, S. Peng, I. Akiyama, K. Li, Li Kao Lee and P.R. LeBreton, J.Am.Chem.Soc. 102, 4627 (1980).
68. T. Koenig, D. Imre and J.A. Hoobler, J.Am.Chem.Soc. 101, 6446 (1979).
69. L. Klasinc, I. Novak, G. Kluge and M. Scholz, Z.Naturforsch. 35a, 640 (1980).
70. V. Barone, C. Cauletti, F. Lelj, M.N. Piancastelli and N. Russo, J.Am.Chem.Soc. 104, 4571 (1982).
71. V. Barone, P.L. Cristinziano, A. Pastore, F. Lelj and N. Russo, Gazz.Chim.Italiana 112, 195 (1982).
72. V. Barone, F. Lelj, C. Cauletti, M.N. Piancastelli and N. Russo, Mol.Phys. 49, 599 (1983).
73. D.C. Frost, N.P.C. Westwood and N.H. Werstiuk, Can.J.Chem. 58, 1659 (1980).
74. S. Ljunggren, G. Merenyi and J. Lind, J.Am.Chem.Soc. 105, 7662 (1983).
75. QCPE Manual: L. Åsbrink, C. Fridh and E. Lindholm, QCPE 12, 393 (1980).
76. V. Barone, N. Bianchi, F. Lelj, G. Abbate and N. Russo, J.Mol.Struct. THEOCHEM 108, 35 (1984).
77. M. Shahbaz, S. Urano, P.R. LeBreton, M.A. Rossman, R.S. Hosmane and N.J. Leonard, J.Am.Chem.Soc. 106, 2805 (1984).
78. Y. Takahata, J.Mol.Struct.THEOCHEM 108, 269 (1984).
79. J. Bergman and P. Sand, J.Chem.Soc.Perkin II, 93 (1985).
80. S. Svensson, P.Å. Malmqvist, M.Y. Adam, P. Lablanquie, P. Morin and I. Nenner, Chem.Phys.Letters 111, 574 (1984).
81. V. Barone, C. Cauletti, L. Commisso, F. Lelj, M.N. Piancastelli and N. Russo, J.Chem.Res. (S), 338 (1984).
82. V. Barone, F. Lelj, L. Commisso, N. Russo, C. Cauletti and M.N. Piancastelli, Chem.Phys.
83. Z.D. Chen, G.X. Xu and K.H. Hsu, Acta Chim.Sinica 41, 791 (1983).

I. Excitation and UV spectroscopy

I.1. Calculation of excitation energy in the HAM model.

Excitation means that one electron moves from an occupied orbital ψ_i to an unoccupied orbital ψ_a. This can happen after supply of energy, usually from electrons or photons.

Configurational energies

We must now observe the spins and get therefore different results when the electron goes from ψ_i^{α} to ψ_a^{α} and when it goes from ψ_i^{α} to ψ_a^{β}. The spin-configurations are different, and the energies of the two spin-configurations are therefore also different.

In the first case after the excitation we have the electron charge $q_i^{\alpha}-1$ in orbital ψ_i^{α} and the electron charge $q_a^{\alpha}+1$ in orbital ψ_a^{α}.

The excitation energy is then the difference of two total energies

$$\text{exc. energy} = E\left(q_i^{\alpha}-1,\ q_a^{\alpha}+1\right) - E\left(q_i^{\alpha},\ q_a^{\alpha}\right) \qquad (I.1)$$

To obtain this energy difference we use as previously (Sec.H.1.) the expression (B.10) + (B.11) + (G.31) for the total energy. This gives the general expression for the excitation energy

$$(I.1) = H_{aa}^{\alpha} + \sum_k q_k J_{ak} - \sum_k q_k^{\alpha} K_{ak}^{\alpha} - H_{ii}^{\alpha} - \sum_k q_k J_{ik} + \sum_k q_k^{\alpha} K_{ik}^{\alpha}$$
$$- J_{ia} + K_{ia}^{\alpha} + \left(1-q_i^{\alpha}\right) L_{ii} + q_a^{\alpha} L_{aa} \qquad (I.2)$$

In this general expression the terms can be rearranged in different ways, giving different formulations for the excitation energy.

The first formulation is obtained if eq.(I.2) is written in the following way

$$(I.1) = \varepsilon_a^{HF^{\alpha}} - \varepsilon_i^{HF^{\alpha}} - J_{ia} + K_{ia}^{\alpha} + \left(1-q_i^{\alpha}\right) L_{ii} + q_a^{\alpha} L_{aa} \qquad (I.3)$$

If we here remove the last two terms, which depend upon the idempotency relation, we obtain correctly Roothaan's expression for the excitation energy (eq.(68) in ref. [1]).

The second formulation starts from

$$(I.3) = \varepsilon_a^{HF\,\alpha} - \varepsilon_i^{HF\,\alpha} - J_{ia} + K_{ia}^{\alpha} + \left(q_a^{\alpha} - \tfrac{1}{2}\right) L_{aa} - \left(q_i^{\alpha} - \tfrac{1}{2}\right) L_{ii} +$$

$$+ \tfrac{1}{2} L_{aa} + \tfrac{1}{2} L_{ii}$$

which gives by use of eq.(S.33)

$$(I.1) = \varepsilon_a^{\alpha} - \varepsilon_i^{\alpha} - J_{ia} + K_{ia}^{\alpha} + \tfrac{1}{2} L_{aa} + \tfrac{1}{2} L_{ii} \qquad (I.4)$$

The third formulation starts from

$$(I.1) = H_{aa}^{\alpha} + \sum_{\substack{k \\ \neq i,a}} q_k J_{ak} + \left(q_i - \tfrac{1}{2}\right) J_{ai} + \left(q_a + \tfrac{1}{2}\right) J_{aa} - \sum_{\substack{k \\ \neq i,a}} q_k^{\alpha} K_{ak}^{\alpha} - \left(q_i^{\alpha} - \tfrac{1}{2}\right) K_{ai}^{\alpha} - \left(q_a^{\alpha} + \tfrac{1}{2}\right) K_{aa}^{\alpha}$$

$$- H_{ii}^{\alpha} - \sum_{\substack{k \\ \neq i,a}} q_k J_{ik} - \left(q_i - \tfrac{1}{2}\right) J_{ii} - \left(q_a + \tfrac{1}{2}\right) J_{ia} + \sum_{\substack{k \\ \neq i,a}} q_k^{\alpha} K_{ik}^{\alpha} + \left(q_i^{\alpha} - \tfrac{1}{2}\right) K_{ii}^{\alpha} + \left(q_a^{\alpha} + \tfrac{1}{2}\right) K_{ia}^{\alpha}$$

$$+ \left[\left(q_a^{\alpha} + \tfrac{1}{2}\right) - \tfrac{1}{2}\right] L_{aa} - \left[\left(q_i^{\alpha} - \tfrac{1}{2}\right) - \tfrac{1}{2}\right] L_{ii}$$

where we have $(q_i^{\alpha} - 1)$ instead of q_i^{α} and $(q_a^{\alpha} + 1)$ instead of q_a^{α}. The long expression can therefore simply be interpreted as

$$(I.1) = {}^{t}\varepsilon_a^{\alpha} - {}^{t}\varepsilon_i^{\alpha} \qquad (I.5)$$

where the eigenvalues ${}^{t}\varepsilon$ have been obtained from a study of a molecule in which we have removed one half electron from orbital ψ_i and added one half electron to orbital ψ_a.

In the <u>second</u> case we have

$$\text{exc. energy} = E\left(q_i^{\alpha} - 1,\ q_a^{\beta} + 1\right) - E\left(q_i^{\alpha},\ q_a^{\beta}\right) \qquad (I.6)$$

This gives a general expression, corresponding to eq.(I.2), in which, however, the eighth term (K_{ia}^{α}) is missing and in which the subscript a is related to β spin. Corresponding to eq.(I.4) we have therefore

$$(I.6) = \varepsilon_a^{\beta} - \varepsilon_i^{\alpha} - J_{ia} + \tfrac{1}{2} L_{aa} + \tfrac{1}{2} L_{ii} \qquad (I.7)$$

In molecular calculations we do not usually distinguish between spins ("restricted" theory). This means that we take

$$\varepsilon_a^{\alpha} = \varepsilon_a^{\beta} = \varepsilon_a \qquad (I.8)$$

which simplifies eq.(I.7).

State energies

We will now study the spins of our two excited configurations and observe that the spinfunction in the first case can be written $\alpha\beta$ and in the second case $\beta\beta$. This means that our excited molecule corresponds the the excited helium atom, which we studied in Sec.E.2., and we find therefore directly from eqs.(I.7) and (E.5) the triplet energy

$$E_T = \varepsilon_a - \varepsilon_i + \frac{1}{2} L_{aa} + \frac{1}{2} L_{ii} - J_{ia} \qquad (I.9)$$

and the singlet energy from eqs.(I.4), (I.7) and (E.6)

$$E_S = \varepsilon_a - \varepsilon_i + \frac{1}{2} L_{aa} + \frac{1}{2} L_{ii} - J_{ia} + 2K_{ia} \qquad (I.10)$$

The transition state method

Let us study a molecule in which we have removed one half electron from orbital ψ_i and added one half electron to orbital ψ_a. Eqs.(I.9) and (I.10) change then into

$$E_T = {}^t\varepsilon_a - {}^t\varepsilon_i - K_{ia} \qquad (I.11)$$

$$E_S = {}^t\varepsilon_a - {}^t\varepsilon_i + K_{ia} \qquad (I.12)$$

as we have shown above.

It is obvious that the transition state method takes care of the reorganization for the same reasons as for ionization (see Sec.H.1.). However, this is less important than for ionization, since the number of electrons in the molecule does not change and the shielding remains constant. This can also be seen from the extra terms in eqs.(I.9) and (I.10)

$$\frac{1}{2} L_{aa} + \frac{1}{2} L_{ii} - J_{ia} + K_{ia} \qquad (I.13)$$

which mean a very small contribution to the excitation energy. It is therefore possible to calculate the excitation energies without use of the transition state. This means

$$E_T \approx \varepsilon_a - \varepsilon_i - K_{ia} \qquad (I.14)$$

$$E_S \approx \varepsilon_a - \varepsilon_i + K_{ia} \qquad (I.15)$$

Also the correlation energy change is taken care of by the transition state method in the same way as for ionization (Sec.H.1.). The correlation energy change is the same either eq.(I.9) or (I.11) are used to calculate the triplet excitation energy, supposed that in all expressions J' , K' and ε' are used instead of J , K and ε . The same can then be shown for the singlet excitation energy.

I.2. A primitive CI method to find the singlet-triplet splitting.

Before the excitation we have two electrons in the occupied orbital ψ_i and during the excitation one of them is excited to ψ_a . We have therefore after the excitation two configurations with the same energy, namely

$$\Psi^1 = \psi_i(1) \cdot \psi_a(2)$$

and

$$\Psi^2 = \psi_i(2) \cdot \psi_a(1)$$

These two configurations mix and the true wavefunction is therefore

$$\Psi = c_1 \cdot \Psi^1 + c_2 \cdot \Psi^2 \qquad (I.16)$$

The Schrödinger equation

$$\mathcal{H}\left(c_1 \Psi^1 + c_2 \Psi^2\right) = E\left(c_1 \Psi^1 + c_2 \Psi^2\right)$$

gives as above (Sec.G.8.)

$$c_1 \left(E^1 - E\right) + c_2 \cdot H_{12} = 0$$

$$c_1 \cdot H_{12} + c_2 \left(E^2 - E\right) = 0$$

with

$$E^1 = E^2 = \varepsilon_a - \varepsilon_i$$

and

$$H_{12} = H_{21} = \int \Psi^{1*} \mathcal{H} \Psi^2 \, d\tau = \int \psi_i^*(1) \psi_a^*(2) \frac{1}{r_{12}} \psi_i(2) \psi_a(1) \, d\tau \tag{I.17}$$

$$= (ia/ia) = K_{ia}$$

The secular equation

$$\begin{vmatrix} \varepsilon_a - \varepsilon_i - E & K_{ia} \\ K_{ia} & \varepsilon_a - \varepsilon_i - E \end{vmatrix} = 0$$

gives two solutions

$$E_S = \varepsilon_a - \varepsilon_i + K_{ia} \tag{I.18}$$

with wavefunction

$$\Psi_S = \frac{1}{\sqrt{2}} \left[\psi_i(1) \psi_a(2) + \psi_i(2) \psi_a(1) \right] \tag{I.19}$$

and

$$E_T = \varepsilon_a - \varepsilon_i - K_{ia} \tag{I.20}$$

with wavefunction

$$\Psi_T = \frac{1}{\sqrt{2}} \left[\psi_i(1) \psi_a(2) - \psi_i(2) \psi_a(1) \right] \tag{I.21}$$

Since the wavefunctions Ψ_S and Ψ_T are symmetric and antisymmetric, respectively, by interchange of the two electrons (1) and (2), they are multiplied with antisymmetric and symmetric spinfunctions so that the product is antisymmetric as required by the Pauli principle. These are the singlet and triplet states.

The exchange integral K_{ia}

The exchange integral K_{ia} is defined as

$$K_{ia} = \int \psi_i^*(1)\, \psi_a(1)\, \frac{1}{r_{12}}\, \psi_a^*(2)\, \psi_i(2)\, d\tau = (ia|ia) \qquad (I.22)$$

To calculate K_{ia} we introduce LCAO

$$\psi_i = \sum_\mu c_{\mu i}\, \phi_\mu$$

and obtain

$$K_{ia} = \sum_{\mu\nu\lambda\sigma} c_{\mu i}\, c_{\nu a}\, c_{\lambda a}\, c_{\sigma i} \int \phi_\mu^*(1)\, \phi_\nu(1)\, \frac{1}{r_{12}}\, \phi_\lambda^*(2)\, \phi_\sigma(2)\, d\tau =$$

$$= \sum_{\mu\nu\lambda\sigma} c_{\mu i}\, c_{\nu a}\, c_{\lambda a}\, c_{\sigma i}\, (\mu\nu|\lambda\sigma) \qquad (I.23)$$

The repulsion integrals $(\mu\nu|\lambda\sigma)$ are usually small except for $\mu=\nu$ and $\lambda=\sigma$. We have therefore four possibilities:

$$(\mu_A\mu_A|\lambda_A\lambda_A) = \gamma_{AA}$$
$$(\mu_A\mu_A|\lambda_B\lambda_B) = \gamma_{AB}$$
$$(\mu_B\mu_B|\lambda_A\lambda_A) = \gamma_{AB}$$
$$(\mu_B\mu_B|\lambda_B\lambda_B) = \gamma_{BB}$$

which gives

$$K_{ia} = \sum_{\mu\lambda} c_{\mu i}\, c_{\mu a}\, c_{\lambda a}\, c_{\lambda i} \cdot \gamma_{AB} \qquad (I.24)$$

For γ_{AB} the Ohno-Klopman formula eq. (F.5) is used.

The choice of γ

Our suggestion to use the Ohno-Klopman formula in the calculation of K_{ia} must be discussed since γ_{AB} is also part of the SCF program (see eq. (F.21)).

It follows from our deductions that both K_{ia} in the calculation of the excitation energies and γ_{AB} in eq. (F.21) originate in the integrals J and K in eq. (B.11). It follows that it is necessary to use the <u>same</u> expression for γ_{AB} in both cases. Since we have presumed that the Ohno-Klopman formula is used in the SCF calculation it must be used also in the calculation of the splitting.

The exchange integral K_{ia} in ethylene

As an example the $\pi\pi^*$ excitation will be studied. To get space in the print-out for the indices a), b), c) and d), we have erased some zeroes.

ONE HALF ELECTRON DIFFUSELY REMOVED. FILLED ORBITALS GIVE IONIZATION ENERGIES

$\mu \downarrow$								i	a					
		$j \rightarrow$	1	2	3	4	5	6	7	8	9	10	11	12
			-24.292	-19.848	-16.256	-14.883	-13.173	-10.538	-4.269	15.420	19.620	25.496	29.969	36.434
1 C		1	0.466	0.334	-0.000	0.081	0.000			0.676	0.835	0.000	0.003	0.000
2 C	A	1	-0.055	0.213	0.000	-0.522	0.000			0.413	-0.381	-0.000	1.004	0.000
3 C		1	-0.000	0.000	-0.420	-0.000	-0.482	a)	b)	-0.000	-0.000	0.687	0.000	-0.857
4 C		1	-0.000	-0.000	-0.000	0.000	-0.000	-0.642	-0.797	0.000	0.000	-0.000	0.000	-0.000
5 C		2	0.466	-0.334	-0.000	0.081	-0.000			0.676	-0.835	-0.000	-0.003	-0.000
6 C	B	2	0.055	0.213	0.000	0.522	-0.000			-0.413	-0.381	-0.000	1.004	0.000
7 C		2	0.000	0.0	-0.420	0.000	0.482	c)	d)	0.000	-0.000	0.687	0.000	0.857
8 C		2	0.000	0.000	-0.000	0.000	-0.000	-0.642	0.797	-0.000	-0.000	-0.000	0.000	-0.000
9 H		3	0.177	0.257	-0.249	-0.180	-0.301	0.000	-0.000	-0.464	-0.288	-0.635	-0.476	0.652
10 H		4	0.177	0.257	0.249	-0.180	0.301	-0.000	-0.000	-0.464	-0.288	0.635	-0.476	-0.652
11 H		5	0.177	-0.257	-0.249	-0.180	0.301	-0.000	0.000	-0.464	0.288	-0.635	0.476	-0.652
12 H		6	0.177	-0.257	0.249	-0.180	-0.301	0.000	-0.000	-0.464	0.288	0.635	0.476	0.652

$$K_{ia} =$$

$$= \underset{a)}{(-0.642)}\underset{b)}{(-0.797)}\underset{a)}{(-0.642)}\underset{b)}{(-0.797)} \cdot \gamma_{AA} + \underset{a)}{(-0.642)}\underset{b)}{(-0.797)}\underset{c)}{(-0.642)} \cdot \underset{d)}{0.797} \cdot \gamma_{AB} +$$

$$+ \underset{c)}{(-0.642)} \cdot \underset{d)}{0.797} \cdot \underset{a)}{(-0.642)}\underset{b)}{(-0.797)} \cdot \gamma_{BA} + \underset{c)}{(-0.642)} \cdot \underset{d)}{0.797} \cdot \underset{c)}{(-0.642)} \cdot \underset{d)}{0.797} \cdot \gamma_{BB} .$$

With $\gamma_{AA} = \gamma_{BB} = 11.1$ eV and $\gamma_{AB} = \gamma_{BA} = 7.7$ eV we obtain $K_{ia} = 1.77$ eV.

Comment: In this calculation the overlaps have been neglected. It is therefore better to normalize the coefficients so that $\sum_{\mu} c_{\mu i}^2 = 1$ (ZDO approximation).

Calculations of matrix elements in other cases [2]

In the example above, $(ia|ia)$, both i and a represent π type orbitals in the z direction. In other cases the same method is used: introduce LCAO, which gives

$$(pq|rs) = \sum c \cdot c \cdot c \cdot c \cdot (\mu\nu|\lambda\sigma) \tag{I.25}$$

As an example we give the expression for the repulsion integral

$$J_{ia} = (ii|aa) = \sum_{\mu\lambda} (c_{\mu i})^2 (c_{\lambda a})^2 (\mu\mu|\lambda\lambda) \tag{I.26}$$

The main difficulty is then to know $(\mu\nu|\lambda\sigma)$ in unusual cases. Convenient formulas have been given by Slater [3] (see [4] p. 81).

$$(ss|ss) = (ss|xx) = F^0 = \gamma_{AA} \tag{I.27}$$

$$(sx|sx) = \frac{1}{3} G^1 \tag{I.28}$$

$$(xy|xy) = \frac{3}{25} F^2 \tag{I.29}$$

$$(xx|xx) = F^0 + \frac{4}{25} F^2 \tag{I.30}$$

$$(xx|yy) = F^0 - \frac{2}{25} F^2 \tag{I.31}$$

Suitable values for the Slater-Condon parameters F^0, F^2 and G^1 and also for γ_{AB} have been given in MINDO/1 [5].

The two-center integrals $(s_A x_A | s_B x_B)$ and $(x_A y_A | x_B y_B)$ are assumed to be zero.

I.3. Calculation of intensities

In the hydrogen atom the transition probability is proportional to the square $|X_{12}|^2$ of the transition moment

$$X_{12} = \int \phi_2^* \, x \, \phi_1 \, d\tau \qquad (I.32)$$

It is easy to show that the transition $2px \rightarrow 1s$ is allowed in a hydrogen atom. Let the X axis be vertical

$$X_{12} = \int \underset{2px}{\overset{\oplus}{\ominus}} \cdot \underset{x}{\frac{+}{-}} \cdot \underset{1s}{\odot} \cdot d\tau = \int \underset{2px \cdot x}{\overset{\oplus}{\oplus}} \cdot \underset{1s}{\odot} \cdot d\tau = \int \overset{\oplus}{\oplus} \cdot d\tau \neq 0$$

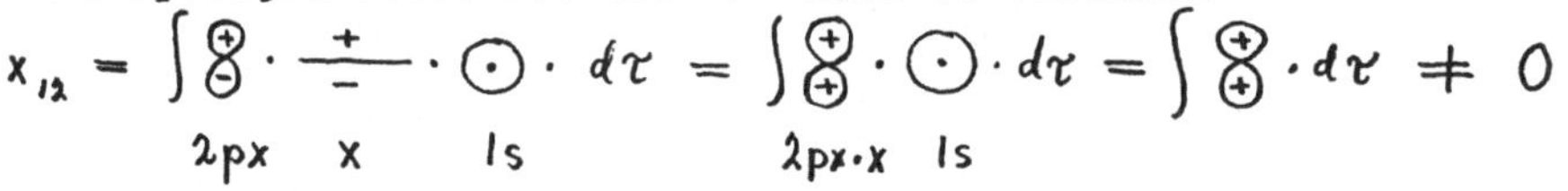

In a molecule we have similarly

$$X_{12} = \int \Psi_2^* \cdot \sum_p x_p \cdot \Psi_1 \cdot d\tau \qquad (I.33)$$

If the transition in the molecule means that one electron goes from ψ_i to ψ_a this is simplified to

$$X_{ia} = \int \psi_i^* \, x \, \psi_a \, d\tau \qquad (I.34)$$

and inserting an LCAO expansion $\psi_i = \sum_\mu c_{\mu i} \phi_\mu$ we get

$$X_{ia} = \sum_\mu \sum_\nu c_{\mu i} c_{\nu a} \cdot \int \phi_\mu \, x \, \phi_\nu \, d\tau \qquad (I.35)$$

For a $\pi\pi^*$ transition in e.g. ethylene we have $\mu = \nu$ and since $\int \phi_\mu^2 \, d\tau = 1$

$$X_{ia} = c_{\mu_A i} c_{\mu_A a} \cdot X_A + c_{\mu_B i} c_{\mu_B a} \cdot X_B \qquad (I.36)$$

With origin in e.g. B, we obtain with the $c_{\mu i}$ values from Sec. I.2.

$$X_{ia} = \overset{a)}{(-0.642)} \cdot \overset{b)}{(-0.797)} \cdot 1.34 = 0.68$$

Since the oscillator strength f_{ia} is calculated using the formula (see Salem, ref. [6], eq. (7-30))

$$f_{ia} = 0.175 \cdot \Delta E \cdot \left[X_{ia}^2 + Y_{ia}^2 + Z_{ia}^2 \right] \qquad (I.37)$$

where ΔE is in eV and X is in Å, we get for the singlet-singlet transition in ethylene

$$f_{ia} = 0.175 \cdot 8.04 \cdot 0.68^2 = 0.65$$

The experimental value is 0.34.

It is a common experience that the calculated transition probability is too large (by a factor 2) using different methods of calculation (see e.g. ref. [6] page 362).

It is easy to show, using qualitative arguments, that the $\pi\pi^*$ transition in ethylene is allowed. In the integral we let the molecular axis be vertical.

$$X_{12} = \int \text{\rotatebox{90}{⬡}} \cdot \frac{+}{-} \cdot \text{⬡} \cdot d\tau = \int \text{⬡} \cdot \text{⬡} \cdot d\tau = \int \text{⬡} \cdot d\tau \neq 0$$

$$\pi^* \quad \cdot \quad x \quad \cdot \quad \pi \qquad\qquad \pi^* \cdot x \qquad \pi$$

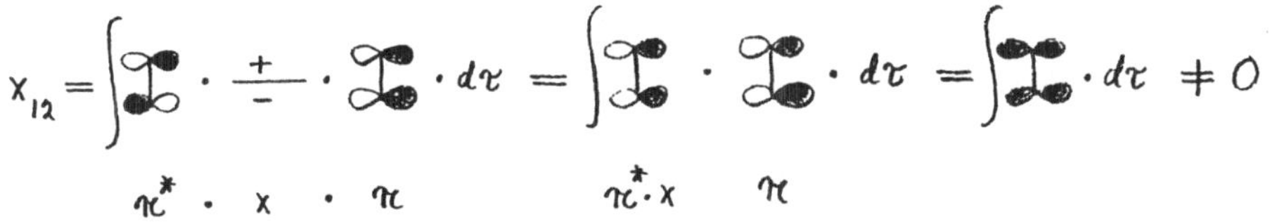

It is clear that our treatment of equ. ($I.35$) is incomplete if $\phi_\mu = 2p_x$ and $\phi_\nu = 2s$. We get therefore two parts in the expression for the transition moment, namely one part corresponding to equ. (I.35) and one part of the $2p_x - 2s$ type. The last part is responsible for the intensity of $n\pi^*$ transitions.

For details we make reference to books by Salem [6] (pp. 347, 358-360) and by McGlynn et al. [7] (pp. 310 and 438, where $Z = n\zeta$ is used).

I.4. Semiempirical methods to calculate excitation

Earlier semiempirical methods to calculate valence excitation have in general been very successful. As examples, the PPP method [8-10] and the CNDO/S method [11] can be mentioned.

Both methods are - like HAM - built up from the general semiempirical energy expression, eq.(C.38), which means that *in principle* the parametrization will result in methods in which all correlation has been taken into account. The PPP method was parametrized for π orbitals only and CNDO/S for planar molecules. The excitation energies were obtained from energy differences (like eq.(I.1) and (I.6)).

When these methods were proposed, the experimental data were rare and uncertain. The parameters are therefore in some studies slightly incorrect. The checks of the methods by comparison of calculated and experimental results were also difficult and had to be presented only in tabular form.

Today, therefore, these methods and their results seem to be partly forgotten, and in many recent papers interpretations can be found which are in conflict with the semiempirical results. We will therefore briefly review some results.

The main possibility today to check the semiempirical results is due to the improved overview of the excitation of molecules from electron-impact energy-loss studies and from synchrotron-radiation absorption results. Today, therefore, the experimental results are presented as spectra, which can be compared with the calculations in a very convincing way.

Pariser and Parr [9] determined their parameters from the spectra of ethylene and benzene. A comparison of the calculated and experimental excitation energies is given in Table I.4.a. (see also Sec.I.8. for the ethylene spectrum and Sec. I.10. for the benzene spectrum).

Table I.4.a. Calculated $\pi\pi^*$ excitation energies compared with experiment (eV).

Ethylene	PPP [9]	Exp [12, 13]
$^1B_{1u}$	7.6	7.6
$^3B_{1u}$	4.5	4.3

Benzene	PPP (1953) [9]	PPP (1956) [14]	Exp (1953) [9]	Exp (1977) [15]
$^1B_{2u}$	4.9 (forbidden)	4.710	4.9	4.9
$^1B_{1u}$	5.3 (forbidden)	5.960	6.0	6.2
$^1E_{1u}$	7.0 (allowed)	6.548	7.0	6.9
$^3B_{2u}$	4.9	4.710	---	5.7
$^3B_{1u}$	4.0	3.590	3.8	3.9
$^3E_{1u}$	4.45	4.149	---	4.8

Using these parameters Pariser and Parr [9] calculated butadiene and obtained very good agreement with recent experimental energies (see also Sec.I.8. for the spectrum).

Table I.4.b.

Butadiene	PPP (1953) [9]	Exp (1953) [9]	Exp (1977) [16, 17]
^1B_u	6.21	6.0	5.9
^1A_g	7.87	---	---
^1A_g	8.51	---	---
^1B_u	9.50	---	9.5
^3B_u	3.92	---	3.3
^3A_g	4.61	---	4.9
^3A_g	8.16	---	---
^3B_u	8.74	---	---

Using the same parameters Pariser [14] calculated the energies and intensities of the transitions in naphthalene and obtained a very good agreement with recent results [18-21]. Koch et al. [19] use synchrotron radiation. Verhaart et al. [20] use a trapped electron method (see Sec.I.7.h.). Their triplet at 2.0 eV is probably uncertain.

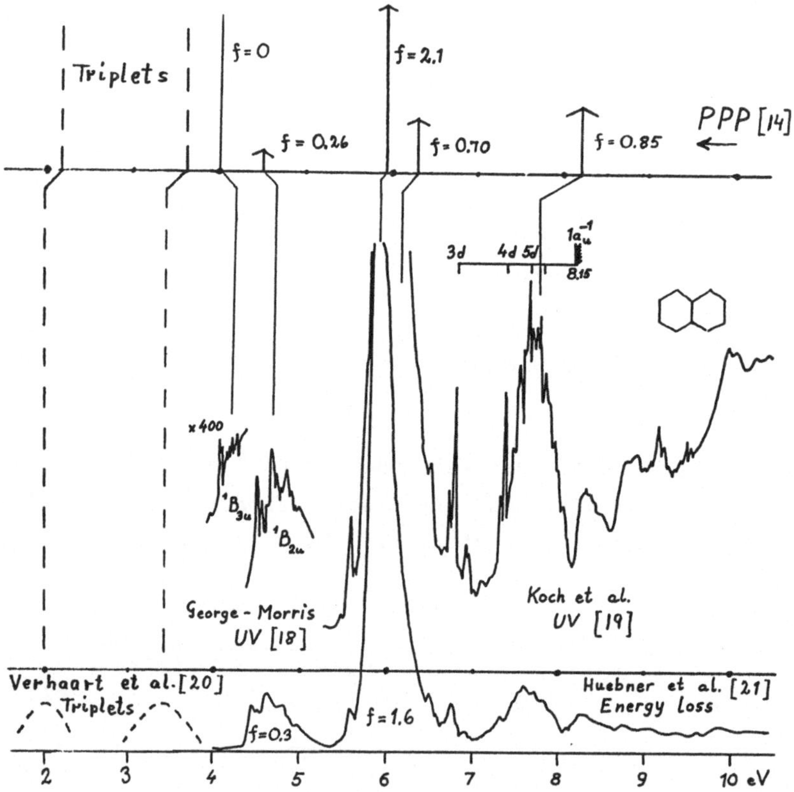

The very good agreement for the excitation of butadiene and naphthalene (also other molecules) shows that PPP calculations are reliable for hydrocarbons and that the strong transitions are due to singlet-singlet $\pi\pi^*$ processes. Rydberg transitions are seen to have a negligible intensity.

These impressions are supported by the fact that if the strong transitions in the UV spectrum are studied in the liquid or solid phase instead, they do not generally change, neither with respect to the energy nor to the intensity [22]. Such experiments prove therefore that the excited orbitals of these transitions have a spatial extension of about the same size as the molecule itself. They must thus be valence orbitals.

For molecules with heteroatoms the parametrisation proved to be more difficult, but many planar six- and five-membered rings have been studied. We show below a few examples of singlet transitions. Only few triplet calculations have been published.

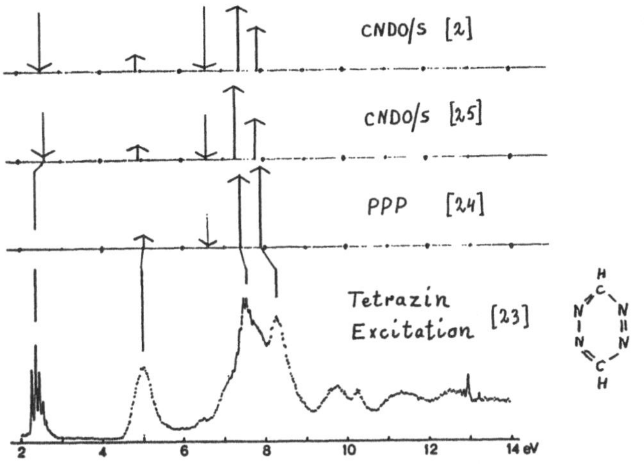

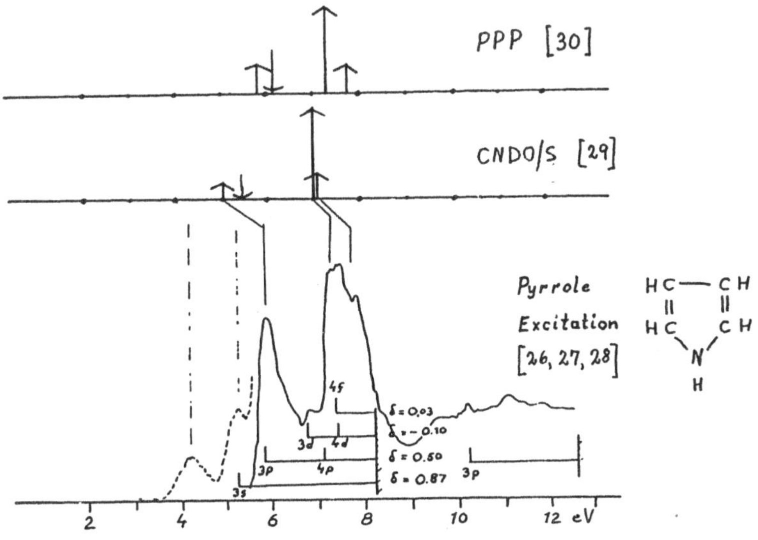

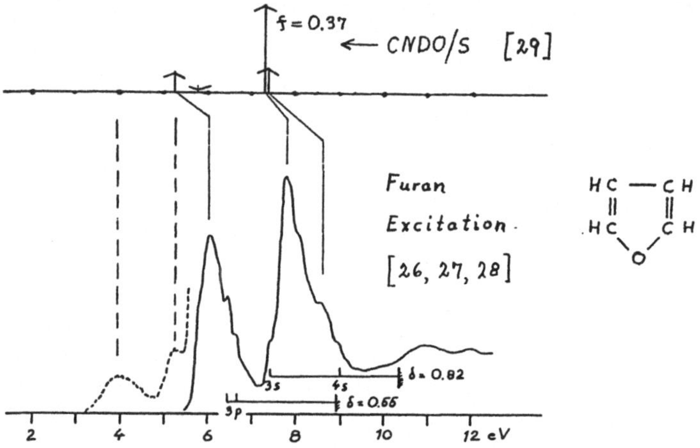

The overall appearance of the excitation is in these and other molecules well described by the semiempirical methods. We observe especially that the Rydberg transitions are very weak.

I.5. Rydberg transitions

Up till now we have considered valence transitions between molecular orbitals built up from carbon 2s and 2p (together with hydrogen 1s) atomic orbitals.

In a Rydberg transition the excited orbital is characterized by the principal quantum number 3 or higher. A Rydberg orbital is therefore much larger than a valence orbital and has a much larger diameter.

Since the electron moves at a large distance from the positively charged molecule ion, the Rydberg orbital of a molecule is similar to that in an atom.

Below we show some Rydberg orbitals in benzene [31] together with recently determined δ-values [68, 69].

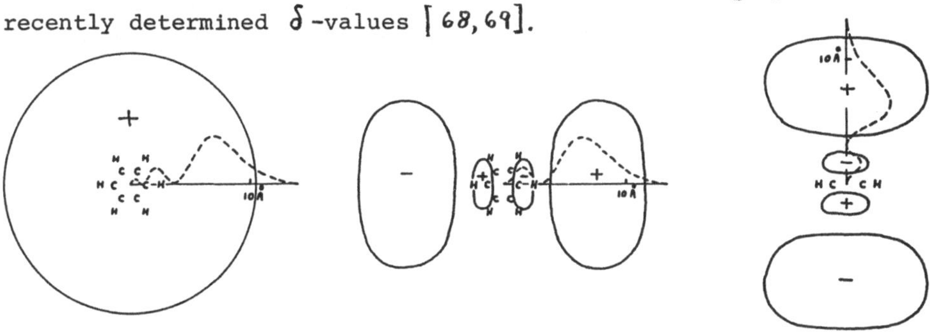

$3sa_{1g}$ $\delta \approx 0.77$ | $3pe_{1u}$ $\delta \approx 0.46$ | $3pa_{2u}$ $\delta \approx 0.16$

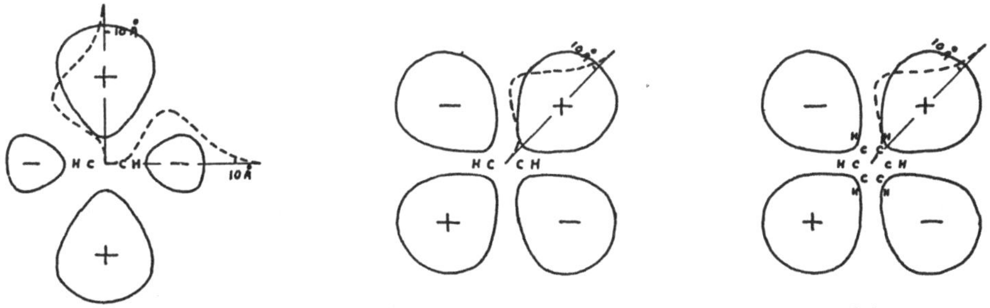

$3da_{1g}$ $\delta \approx 0.175$ | $3de_{1g}$ $\delta \approx -0.08$ | $3de_{2g}$ $\delta \approx 0.24$

The energy of the Rydberg transition is given

$$E = IP - \frac{E_H}{(n-\delta)^2}$$

which is similar to the energy of a hydrogen atom. Here, IP is the ionization energy, E_H = 13.6 eV, n is the principal quantum number and δ the quantum defect.

The vibrational structure of a Rydberg transition is usually completely similar to the vibrational structure of the corresponding band in PES. The explanation is that the Rydberg electron is so remote from the molecule that it does not contribute to the chemical bonding.

It is not easy to find a really instructive example. The best is perhaps carbon suboxide, O=C=C=C=O, for which a beautiful Rydberg series with δ = 1.00 can be seen together with a strong $\pi\pi^*$ singlet transition [32].

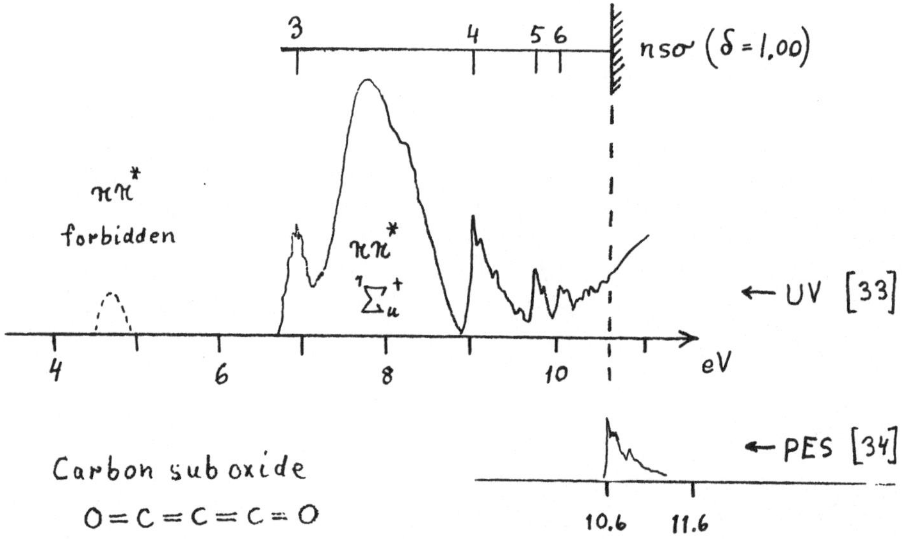

As is often found, the $3s\sigma$ transition has an irregular vibrational structure.

Another example concerns benzene, which was studied in 1972 using continuous light from the synchrotron DESY in Hamburg [35]. It exhibits a large number of very weak Rydberg transitions. The only strong features in this spectrum are the $\pi\pi^*$ transition at 6.9 eV and that at 6.2 eV. This can be seen better in the pyridine spectrum shown in Sec.I.9.

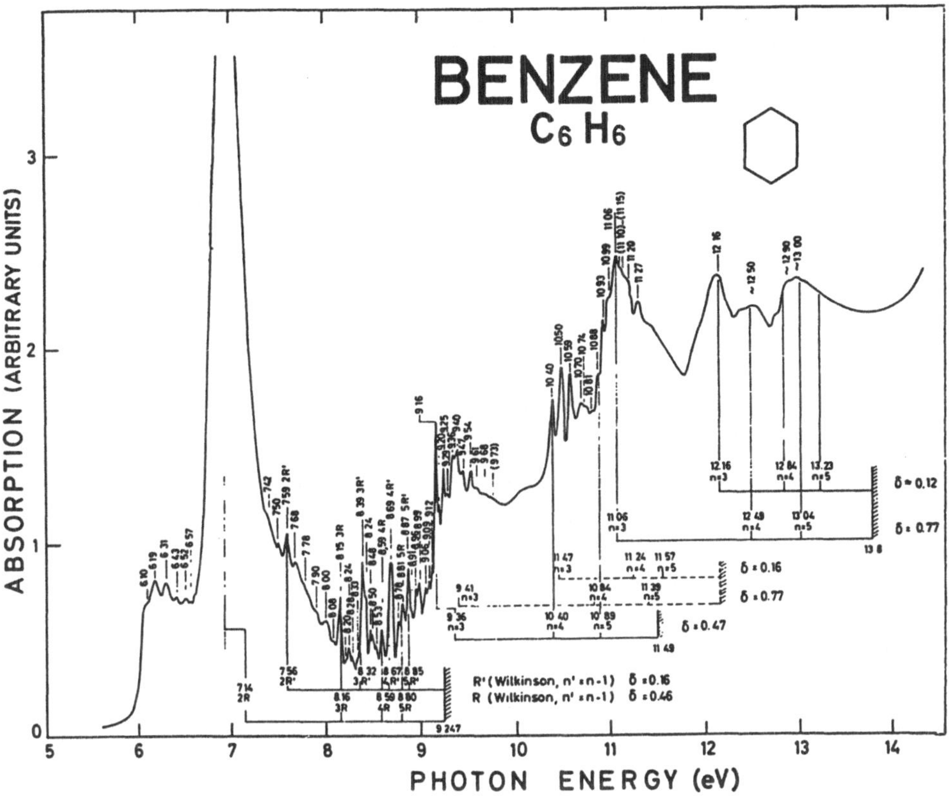

The description, given here, indicates that it is usually very simple to identify the Rydberg transitions in a UV spectrum. The PES, which can be assumed to be well known, gives information concerning IP and vibrational structure, and with a reasonable assumption concerning the quantum defect, one can usually directly pick out a series with a few members. As a contrast, the valence transitions can usually never be identified without access to reliable calculated transition energies and intensities. There are reasons to stress that the calculated intensities are as important as the calculated energies.

A special reason why intensities are important is that in early days the photographic methods made it possible to see the sharp bands in the Rydberg transitions, whereas the broad diffuse valence transitions could not be identified. They were simply denoted as "background". Valence calculations have therefore sometimes been compared with Rydberg experimental results.

I.6. Calculation of excitation energies in ab-initio work.

From the experimentalist's point of view the ab-initio calculations of excitation have not been successful. This concerns the singlet excitations of especially $\pi\pi^*$ type for which the calculated energies are usually much too high.

On the other hand, the calculations of triplet and Rydberg excitations give usually energies in reasonable agreement with experiment.

The methods used in these calculations differ from our use of eqs.(I.11) and (I.12) since the excited singlet and the ground state are treated separately. The intention of the study is then to find the correlated wavefunction $\Psi + \chi$ (see Sec.C.10.) in order to find the correlation energy in the excited singlet. As judged from the calculated energies this is very difficult to achieve.

One may wonder how the correlation energy in the triplet can be calculated so successfully. One explanation might be that in the antisymmetric wavefunction, associated with a triplet (see eq.(I.21)), the two electrons avoid each other because the wavefunction goes to zero when the coordinates of the electrons coincide (Fermi hole). The repulsion between the electrons cannot therefore cause any further avoidance.

But for the singlet no such effect is present and the correlation must be taken care of by χ . It has been proven by mathematics that this is possible.

The molecule, which has been studied mostly, is ethylene.

In ethylene the $\pi\pi^*$ singlet excitation energy is found by experiment to be 7.6 eV. The first calculations gave a much higher energy, but by constructing χ from many Rydberg orbitals the energy has now been depressed to about 8.0 eV. The addition of Rydberg orbitals, which have a large diameter, has, however, caused that the calculated π^* has now a larger diameter than π (by a factor of the order of 1.5 or 2) [36 - 41]. It is possible to improve the energy but then the π^* orbital becomes unphysically large (see Sec.I.4.).

It seems to be generally believed [65] that the ethylene problem has now been solved. However, when the procedure is repeated for a slightly larger molecule, butadiene, the small energy error in ethylene becomes very large.

Buenker, Shih and Peyerimhoff [42] have studied butadiene. To find χ they used a secular equation (see eq.(5.17)) of the size 150 000 x 150 000, which was diagonalized using approximate methods. The triplet energies were in exact agreement with experiment (see Table I.4.b.) and also several Rydberg energies. On the other hand, the singlet ^1B_u energy was calculated as 7.67 eV in pronounced disagreement with the experimental value 5.9 eV. More recent work [66] gives the same results.

The quality of such calculations has therefore been questioned by McDiarmid [43] . Also Buenker and coworkers [65] express pessimism.

Like butadiene, furan has two double bonds (see Sec.I.4. and Sec.I.9.), and similar calculations were therefore performed by Thunemann et al. [44] . Their result is shown in the furan diagram in Sec.I.9. It can be seen that it is in pronounced disagreement with the semiempirical interpretations in Sec.I.4. Unfortunately, the authors present energies for only two of the expected four valence transitions and no calculated intensities.

A similar study of pyrrole [45] gave a very similar result.

A possible explanation of the results for furan and pyrrole is that the errors in the calculations for these molecules are of the same order as the error for butadiene: $\approx$ 1.8 eV.

No other similar calculations of singlet $\pi\pi^*$ energies for molecules of this size seem to have been published.

The difficulties to calculate singlet excitation energies seem to have serious consequences for calculations of shake-up phenomena in PES and for theoretical studies of chemical reactions (see Chap.O.).

The disagreement between the theoretical results for singlet $\pi\pi^*$ excitation and the corresponding experimental results constitutes probably the most important problem today in the front-line research in molecular science.

Comparison with density functional methods

In ab-initio calculations the correlation part of the total energy is, in principle, obtained by use of eq.(C.46):

$$\varepsilon_{jk} = \sum_{ab} \int \Psi \mathcal{H} \Psi_{jk}^{ab} C_{jk}^{ab} \qquad (C.46)$$

Here, a and b are virtual orbitals, and to obtain many terms in eq.(C.46) also Rydberg orbitals have been included. The calculated π^* orbital obtains then a much larger size than the π orbital due to this treatment of the correlation.

The density-functional expression for the corresponding pair-correlation energy is deduced in Sec.C.6. from the exact Gunnarsson-Lundqvist formula for the exchange-correlation energy (eq.(C.24))

$$\varepsilon_{jk} = 2 \cdot \iint \psi_j^*(1)\, \psi_j(1) \cdot \frac{h_c}{r_{12}} \cdot \psi_k^*(2)\, \psi_k(2) \cdot d\tau_1\, d\tau_2 \qquad (C.24)$$

Comparing eqs.(C.24) and (C.46) we observe that in eq.(C.24) the main contribution to the pair-correlation energy comes from those regions in space where simultaneously $\varrho_j \left(= \psi_j^* \psi_j \right)$ and ϱ_k are large, since $h_c \approx 0$ for large r_{12} . The correlation of the electrons is thus largest here, in accordance with our intuition.

The diffuse orbitals, which may enter the calculation of eq.(C.46), have, however, their main density outside the region where the densities ϱ_j and ϱ_k of the valence orbitals are large. It is therefore necessary to have positive and negative contributions in the outer region from these orbitals in order to influence only the inner region, where Rydberg-type orbitals have a small density but where the main correlation takes place. It is obvious that this will require a large number of terms in the CI expansions.

It is possible that this can explain the observation that the number of terms in some CI studies seems to be quite insufficient.

It has recently been estimated that 57 million configurations are quite insufficient to handle a certain two-atom problem. In a recent study of the same problem [67] , the difficulty was circumvented by the introduction of a simple semiempirical correction of PPP origin (eq.(F.5)) into the in other respects "ab-initio" calculation. This solved all problems !

I.7. Experimental methods for study of excitation

a) Absorption of light

Disadvantages: No good light sources in UV with continuous radiation (except the new synchrotron radiation)
Transition probabilities (intensities) difficult to measure
"Forbidden" transitions cannot be studied.

b) Electron impact methods

The modern electron impact methods were initiated by Lassettre and coworkers in 1964.

The principle is simple:

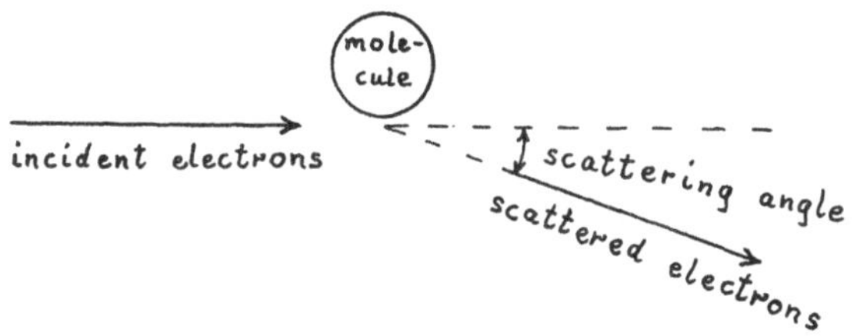

When the electron hits the molecule, the molecule is excited. The energies of the incident and the scattered electrons are measured. Their difference gives directly the excitation energy of the molecule.

Normally (using fast electrons and 0° scattering angle) only optically allowed transitions take place.

Using slow electrons or larger scattering angle both allowed and forbidden transitions occur.

c) Electron impact with high velocity (500 eV) and 0°

Lassettre, Fridh, Geiger, Simpson, Kuyatt

Advantages: Only allowed transitions (as in UV absorption).
Intensity measurements reliable.

Disadvantages: Low resolution

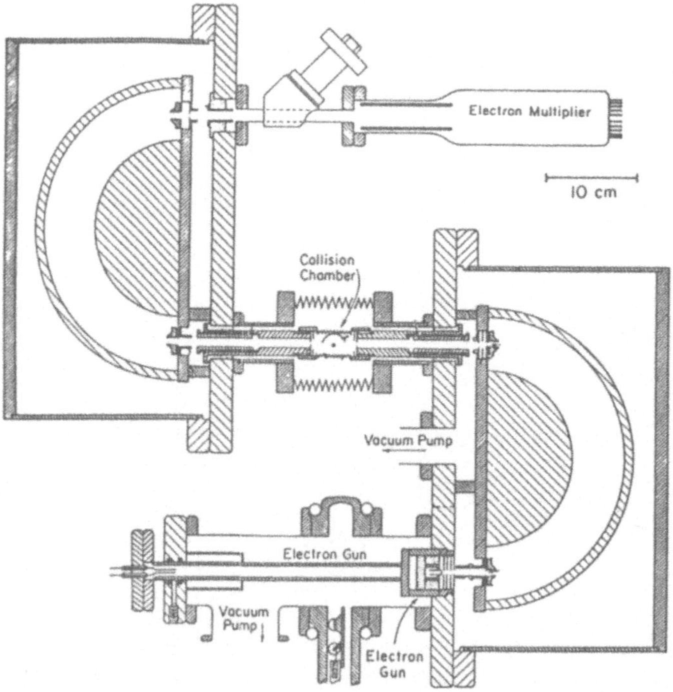

[46]

d) Electron impact with low velocity (10 eV) and 90° or varied angle

Kuppermann, Doering

Advantage: Also forbidden transitions are seen (i.e. singlet ⟶ triplet)

Disadvantage: Weak currents (difficult experiment).

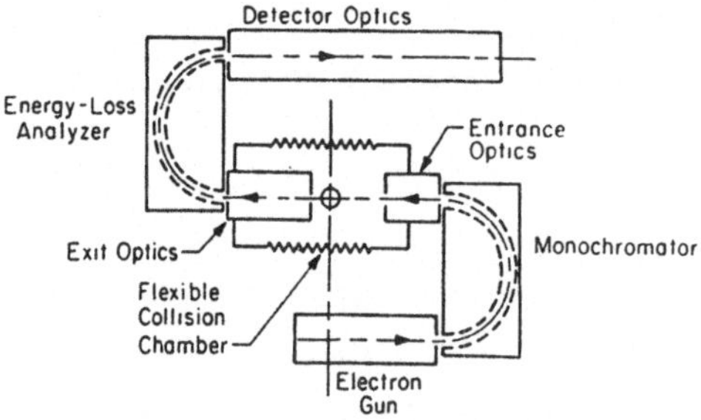

[47]

e) Electron impact with high velocity (2500 eV) and large energy loss

Brion

Advantage: Excitation of 1s electron on one atom to π^* or σ^* or Rydberg orbital in the molecule. Results are often easier to interpret than using c). (See Sec.K.4.)

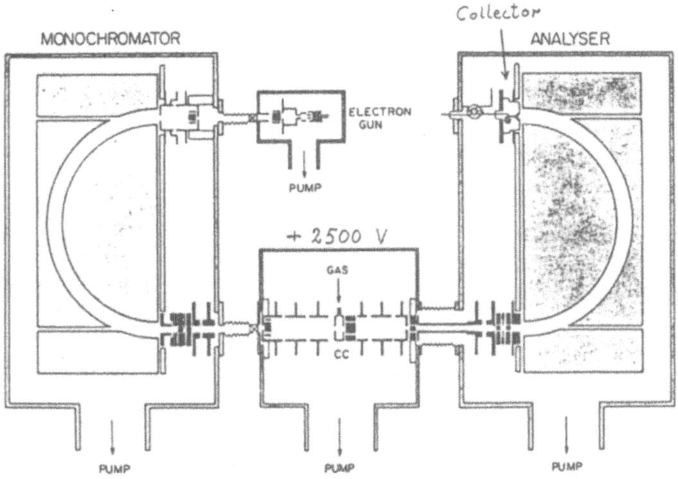

[48]

f) Ion impact with H^+ or He^+ of energy 1000 eV

Moore

Advantage: H^+ gives only singlet → singlet transitions

He^+ gives singlets and triplets

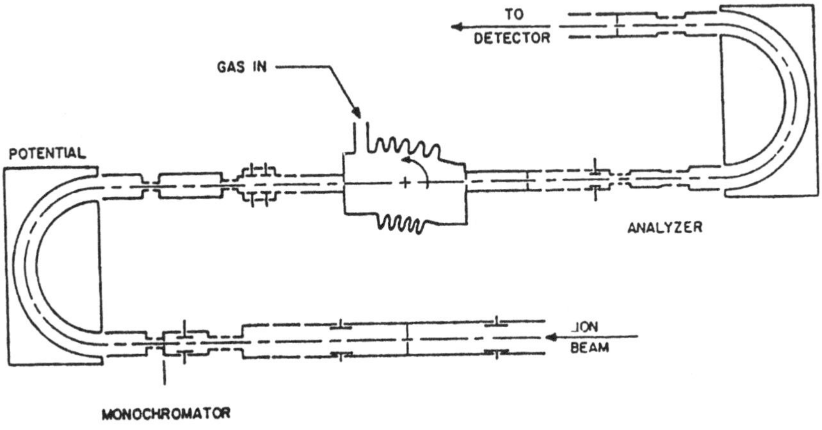

[49]

g) Electron impact with extremely low velocity and 0°

"Trochoidal electron spectrometer"

Allan

Advantages: Extremely slow electrons, still good resolution and good electron intensity.
Both singlets and triplets can be studied

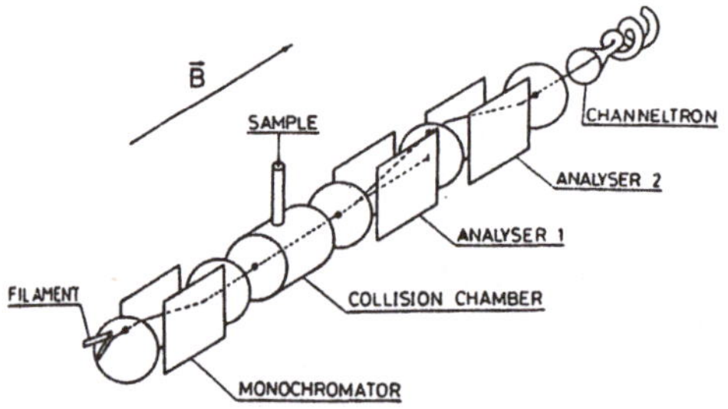

[50]

The trochoidal monochromators were previously used mainly in transmission apparatus to determine electron affinities (see Sec.J.2.)

h) Electron impact with electron velocity = 0 after impact.

"Trapped electron method"

Schulz, Brongersma, van Veen, Hall

Advantage: Often only singlet → triplet transitions

Disadvantage: Low resolution

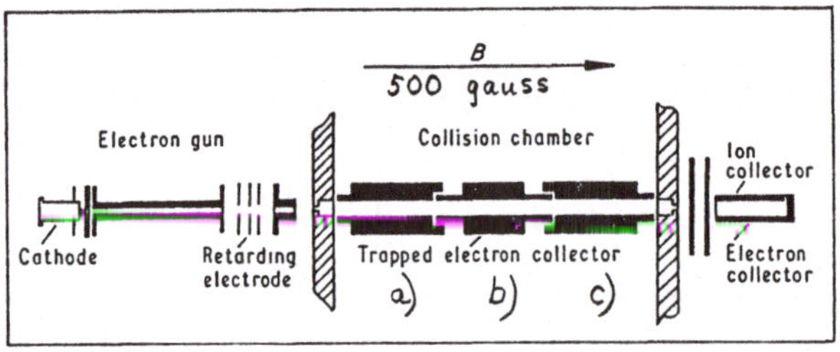

[51]

Most of the electrons have a reasonable velocity (of the order of 10 eV). They follow the magnetic field and leave the collision chamber to the right. But an electron which has given its energy to a molecule and now has the velocity ≈ 0 is repelled by the potential on c) and turns backwards until it is repelled by the potential on a). After some time the electron reaches electrode b) and is measured.

I.8. Excitation of molecules: some results

Excitation of ethylene

It was stated above, that any calculation can be used to find the excitation energy of a molecule. We have therefore calculated ethylene in three ways and find that all ways give about the same value for $\varepsilon_a - \varepsilon_i$.

The first way is the correct one with $q_i = \frac{3}{2}$ and $q_a = \frac{1}{2}$ and gives ${}^t\varepsilon_7 - {}^t\varepsilon_6 = 7.267 - 1.198 = 6.069$ eV. (In this and the following print-outs the orbitals have been normalized according to the ZDO approximation.)

	1	2	3	4	5	6	7	8	9	10	11	12
	-20.996	-16.627	-12.977	-11.556	-10.017	-7.267	-1.198	18.044	22.393	27.403	31.832	38.032
q_j →	2.000	2.000	2.000	2.000	2.000	1.500	0.500	0.0	0.0	0.0	0.0	0.0
1 C 1	0.558	-0.405	0.0	0.100	0.0	0.0	0.0	0.423	0.579	0.0	0.031	0.0
2 C 1	-0.113	-0.311	0.0	-0.631	0.0	0.0	0.0	0.299	-0.248	0.0	0.585	0.0
3 C 1	0.0	0.0	0.552	0.0	0.519	0.0	0.0	0.0	0.0	0.441	0.0	-0.480
4 C 1	0.0	0.0	0.0	0.0	0.0	-0.707	-0.707	0.0	0.0	0.0	0.0	0.0
5 C 2	0.558	0.405	0.0	0.100	0.0	0.0	0.0	0.423	-0.579	0.0	-0.031	0.0
6 C 2	0.113	-0.311	0.0	0.631	0.0	0.0	0.0	-0.299	-0.248	0.0	0.585	0.0
7 C 2	0.0	0.0	0.552	0.0	-0.519	0.0	0.0	0.0	0.0	0.441	0.0	0.480
8 C 2	0.0	0.0	0.0	0.0	0.0	-0.707	0.707	0.0	0.0	0.0	0.0	0.0
9 H 3	0.297	-0.346	0.312	-0.215	0.339	0.0	0.0	-0.341	-0.227	-0.391	-0.281	0.367
10 H 4	0.297	-0.346	-0.312	-0.215	-0.339	0.0	0.0	-0.341	-0.227	0.391	-0.281	-0.367
11 H 5	0.297	0.346	0.312	-0.215	-0.339	0.0	0.0	-0.341	0.227	-0.391	0.281	-0.367
12 H 6	0.297	0.346	-0.312	-0.215	0.339	0.0	0.0	-0.341	0.227	0.391	0.281	0.367

The second way has 2 electrons in all occupied orbitals and gives $\varepsilon_7 - \varepsilon_6 = 7.062 - 0.968 = 6.094$ eV.

	1	2	3	4	5	6	7	8	9	10	11	12
	-20.734	-16.444	-12.858	-11.397	-9.902	-7.062	-0.968	18.268	22.723	27.525	32.079	38.175
1 C 1	0.556	-0.403	0.0	0.103	0.0	0.0	0.0	0.425	0.581	0.0	0.027	0.0
2 C 1	-0.111	-0.311	0.0	-0.631	0.0	0.0	0.0	0.299	-0.243	0.0	0.587	0.0
3 C 1	0.0	0.0	0.551	0.0	0.518	0.0	0.0	0.0	0.0	0.443	0.0	-0.481
4 C 1	0.0	0.0	0.0	0.0	0.0	0.707	0.707	0.0	0.0	0.0	0.0	0.0
5 C 2	0.556	0.403	0.0	0.103	0.0	0.0	0.0	0.425	-0.581	0.0	-0.027	0.0
6 C 2	0.111	-0.311	0.0	0.631	0.0	0.0	0.0	-0.299	-0.243	0.0	0.587	0.0
7 C 2	0.0	0.0	0.551	0.0	-0.518	0.0	0.0	0.0	0.0	0.443	0.0	0.481
8 C 2	0.0	0.0	0.0	0.0	0.0	0.707	-0.707	0.0	0.0	0.0	0.0	0.0
9 H 3	0.299	-0.347	0.313	-0.213	0.340	0.0	0.0	-0.339	-0.228	-0.390	-0.278	0.366
10 H 4	0.299	-0.347	-0.313	-0.213	-0.340	0.0	0.0	-0.339	-0.228	0.390	-0.278	-0.366
11 H 5	0.299	0.347	0.313	-0.213	-0.340	0.0	0.0	-0.339	0.228	-0.390	0.278	-0.366
12 H 6	0.299	0.347	-0.313	-0.213	0.340	0.0	0.0	-0.339	0.228	0.390	0.278	0.366

The third way is diffusely ionized and has 1.91667 electrons in the occupied orbitals and gives $\varepsilon_7 - \varepsilon_6 = 10.538 - 4.269 = 6.269$ eV.

ONE HALF ELECTRON DIFFUSELY REMOVED. FILLED ORBITALS GIVE IONIZATION ENERGIES

	1	2	3	4	5	6	7	8	9	10	11	12
	-24.292	-19.848	-16.256	-14.883	-13.173	-10.538	-4.269	15.420	19.620	25.496	29.969	36.434
1 C 1	0.561	0.407	0.0	0.093	0.0	0.0	0.0	0.421	0.578	0.0	0.030	0.0
2 C 1	-0.118	0.313	0.0	-0.631	0.0	0.0	0.0	0.297	-0.251	0.0	0.582	0.0
3 C 1	0.0	0.0	-0.555	0.0	-0.522	0.0	0.0	0.0	0.0	0.438	0.0	-0.477
4 C 1	0.0	0.0	0.0	0.0	0.0	-0.707	-0.707	0.0	0.0	0.0	0.0	0.0
5 C 2	0.561	-0.407	0.0	0.093	0.0	0.0	0.0	0.421	-0.578	0.0	-0.030	0.0
6 C 2	0.118	0.313	0.0	0.631	0.0	0.0	0.0	-0.297	-0.251	0.0	0.582	0.0
7 C 2	0.0	0.0	-0.555	0.0	0.522	0.0	0.0	0.0	0.0	0.438	0.0	0.477
8 C 2	0.0	0.0	0.0	0.0	0.0	-0.707	0.707	0.0	0.0	0.0	0.0	0.0
9 H 3	0.293	0.344	-0.309	-0.216	-0.337	0.0	0.0	-0.343	-0.227	-0.393	-0.283	0.369
10 H 4	0.293	0.344	0.309	-0.216	0.337	0.0	0.0	-0.343	-0.227	0.393	-0.283	-0.369
11 H 5	0.293	-0.344	-0.309	-0.216	0.337	0.0	0.0	-0.343	0.227	-0.393	0.283	-0.369
12 H 6	0.293	-0.344	0.309	-0.216	-0.337	0.0	0.0	-0.343	0.227	0.393	0.283	0.369

Using the results from the first way, we find

$$E_S = \varepsilon_7 - \varepsilon_6 + K_{67} = -1.198 + 7.267 + 1.77 = 7.84 \text{ eV}$$

$$E_T = \varepsilon_7 - \varepsilon_6 - K_{67} = -1.198 + 7.267 - 1.77 = 4.30 \text{ eV}$$

where E_S and E_T denote the singlet and triplet excitation energies.

The experimental results are shown in the diagram

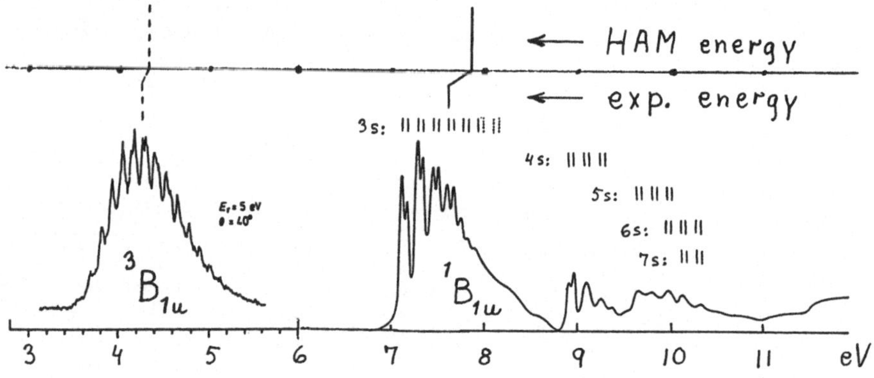

The singlet curve [12] has been obtained using electron impact at 33 000 eV and 0° and the triplet curve [13] at 5 eV and 40°. The C-C stretching vibration (v_2) causes a progression in the triplet curve.

A study with H^+ and He^+ gives similar results [52].

We observe that the HAM/3 calculation agrees well with experiment.

We observe further that also the "third way" above is quite useful, since it introduces an error of only 0.2 eV. It is often practical to obtain the PES and the UV spectrum in <u>one</u> calculation.

Excitation of formic acid

The orbitals in formic acid are shown in Sec. H.3. We expect two strong $\pi\pi^*$ transitions (8 → 10 and 6 → 10) and one weak $n\pi^*$ transition (9 → 10). The other $n\pi^*$ transition (7 → 10) is probably hidden behind stronger transitions.

Only the singlet curve has been observed by Fridh [53] with 500 eV and 0°.

In the diagram below the calculated singlet transitions are marked ↑ or ↓ where the height indicates the intensity. Forbidden transitions are marked ϕ and triplets ¦ .

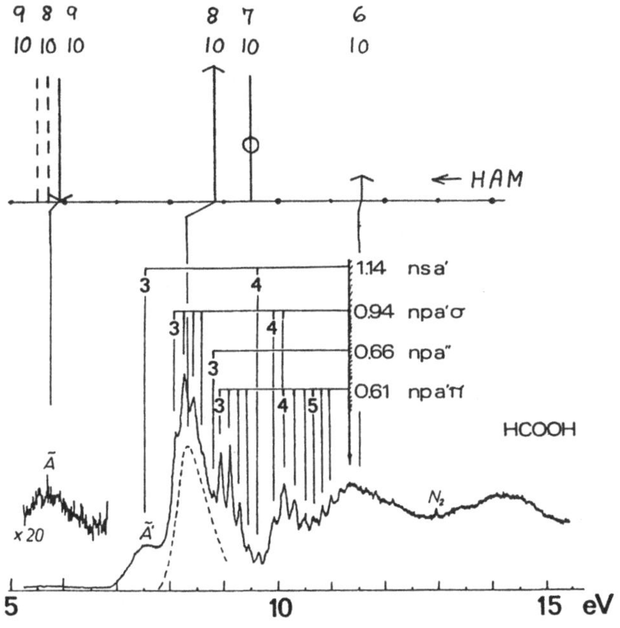

We observe that the $n\pi^*$ transition (9→10) is very weak compared to the $\pi\pi^*$ transition (8 → 10). The calculated f - values are 0.0016 and 0.71, respectively.

Four Rydberg series towards HOMO have been found.

Excitation of butadiene

    The orbitals of butadiene are shown in Sec. H.3. Since there are two π and two π^* orbitals we expect four $\pi\pi^*$ transitions.

    It is easy to see, using the simplified treatment in Sec. I.3. that the transitions $11 \rightarrow 12$ and $10 \rightarrow 13$ are allowed but $11 \rightarrow 13$ and $12 \rightarrow 14$ are forbidden. If we put the molecule in a vertical position we have

$$X_{11 \rightarrow 12} = \int \overset{\circ}{\underset{\bullet}{\circ}} \cdot \frac{+}{-} \cdot \overset{\circ}{\underset{\bullet}{\circ}} \cdot d\tau \; = \int \overset{\circ}{\underset{\circ}{\bullet}} \cdot d\tau \; \neq 0$$

$$\underset{11}{\pi} \;\cdot\; x \;\cdot\; \underset{12}{\pi^*}$$

    The excitation of butadiene is shown below. The singlet curve is obtained by Fridh [16] using 500 eV and $0°$. It is marked ———. The triplet curve is obtained by Kuppermann and coworkers [17] using 20 eV and $85°$. It is marked –·– – – . A study with H^+ and He^+ impact gives similar results [52] .

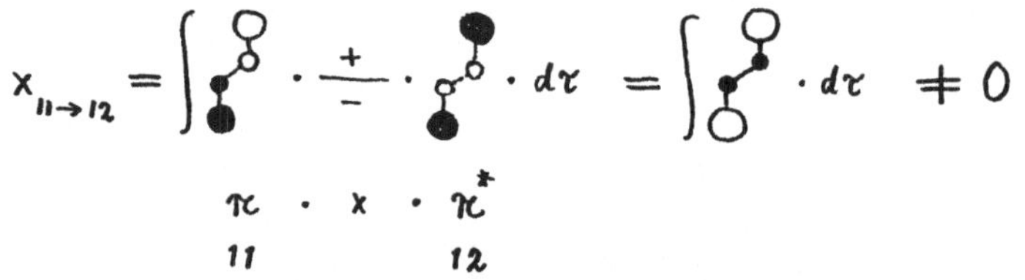

Butadiene excitation

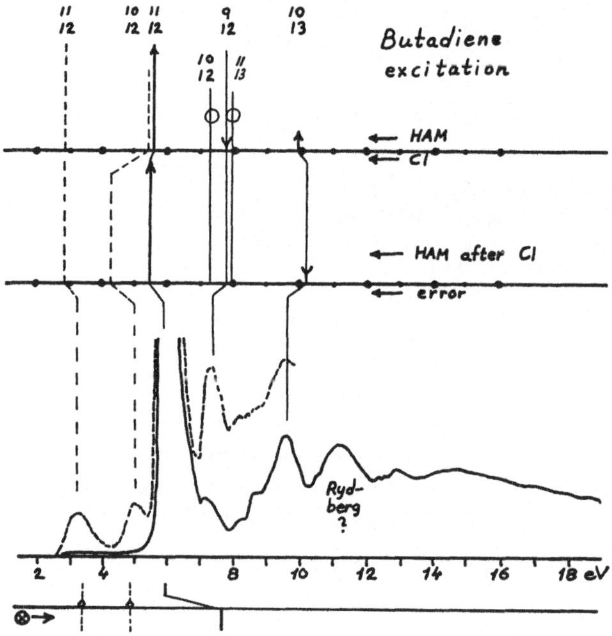

We observe that the very intense band at 5.9 eV is the transition 11 → 12 and that probably the band at 9.5 eV is 10 → 13. Both have symmetry ^1B_u. The two forbidden $\pi\pi^*$ bands have symmetry ^1A_g and both have energies between 7 and 8 eV. It is not probable that anybody has observed the ^1A_g transitions, although they have been objects of several studies. The situation is complicated by three Rydberg transitions in this energy region and also by a $\sigma \rightarrow \pi^*$ transition 9 → 12, which has not been discussed previously. It is very weak with a calculated f-value 0.0009.

⊛ means the result of an ab-initio calculation [42], which has been discussed above (Sec.I.6.).

Below we show the UV spectrum at higher resolution [54].

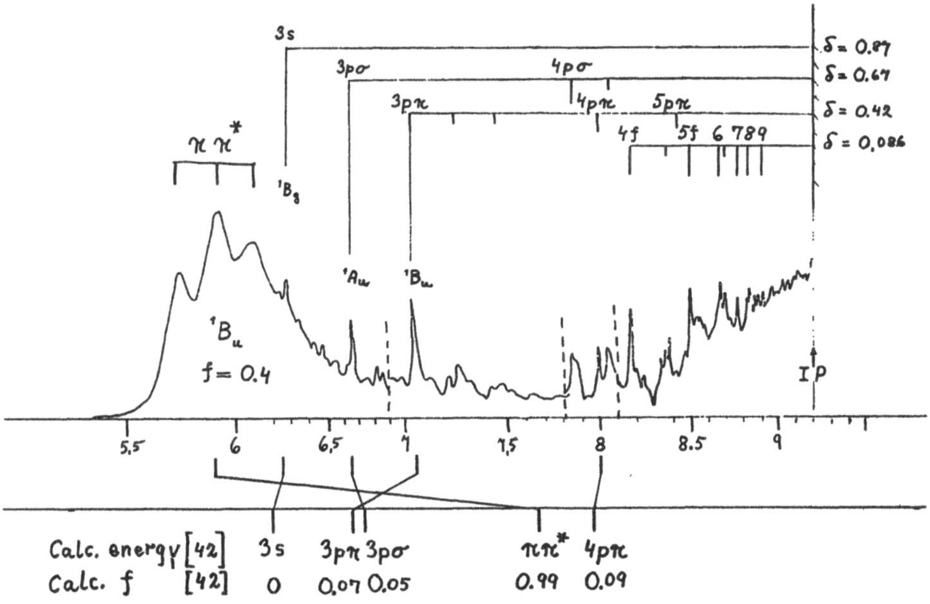

I.9. Degenerate excited configurations will interact: CI

When butadiene has been excited in one of the two allowed processes 11 → 12 or 10 → 13, it exists afterwards in one of the two corresponding excited configurations, which both are ^1B_u. Since they have the same symmetry they interact and with the notations

for transition 11 → 12 or $i \rightarrow a$: Ψ^1 or $\Psi_{i \rightarrow a}$

for transition 10 → 13 or $j \rightarrow b$: Ψ^2 or $\Psi_{j \rightarrow b}$

we find that the true wavefunction is

$$\Psi = c_1 \Psi^1 + c_2 \Psi^2 \qquad (I.38)$$

If we denote the energies of Ψ^1 and Ψ^2 as H_{11} and H_{22} we obtain the secular equation as

$$\begin{vmatrix} H_{11}-E & H_{12} \\ H_{21} & H_{22}-E \end{vmatrix} = 0 \qquad (I.39)$$

It is difficult to deduce expressions for the interaction matrix elements H_{12} when four electrons are involved (see e.g. Pariser and Parr [9], Salem [6] or McGlynn et al. [7]). The result is

for singlets: $H_{12} = \zeta_{iajb} = 2\cdot(ia|jb) - (ij|ab) \qquad (I.40)$

for triplets: $H_{12} = \zeta_{iajb} = \qquad - (ij|ab) \qquad (I.41)$

These repulsion integrals can easily be calculated in the same way as the exchange integral in Sec. I.1.

Solving the secular equation gives always as the result that the higher energy configuration obtains increased energy and the lower energy configuration gets a still smaller energy. The two configurations seem to "repel" each other.

The intensities change also in the CI. It is easy to handle this, starting from

$$x_{12} = \int \Psi_{ground} \cdot x \cdot (c_1 \Psi^1 + c_2 \Psi^2) \cdot d\tau \qquad (I.42)$$

It appears that usually (for positive ζ) the higher energy state gets an increased intensity and the lower energy state gets a decreased intensity. (Butadiene is an exception to this rule.)

The formulas presented here are suitable for calculation by hand. Using a computer it is better to study simultaneously the CI between several excited configurations, e.g. the 20 excited configurations with lowest excitation energies. This is done automatically in the HAM/3 program with the instruction: " CI 20.".

We pointed out above (Sec.I.2.) that the calculation of K_{ia} (the singlet-triplet splitting) must be done in the same way as the calculations in the SCF work, and that the choice of γ_{AB} must be the same in the two cases. This limitation is not valid when the integrals in (I.40) and (I.41) are computed, and other γ_{AB} can therefore be used.

Excitation of furan

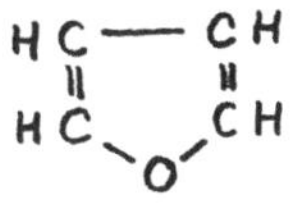

The furan molecule has also two double bonds and gives as butadiene four excitations ($13 \rightarrow 14$, $12 \rightarrow 14$, $13 \rightarrow 15$ and $12 \rightarrow 15$), which all have about the same intensity.

Two of the excitations have $^1A_1$ symmetry. They repel each other, and the high-energy transition takes nearly all intensity. The other excitations are $^1B_2$. Their energy difference is larger, and therefore the influence of the CI is smaller.

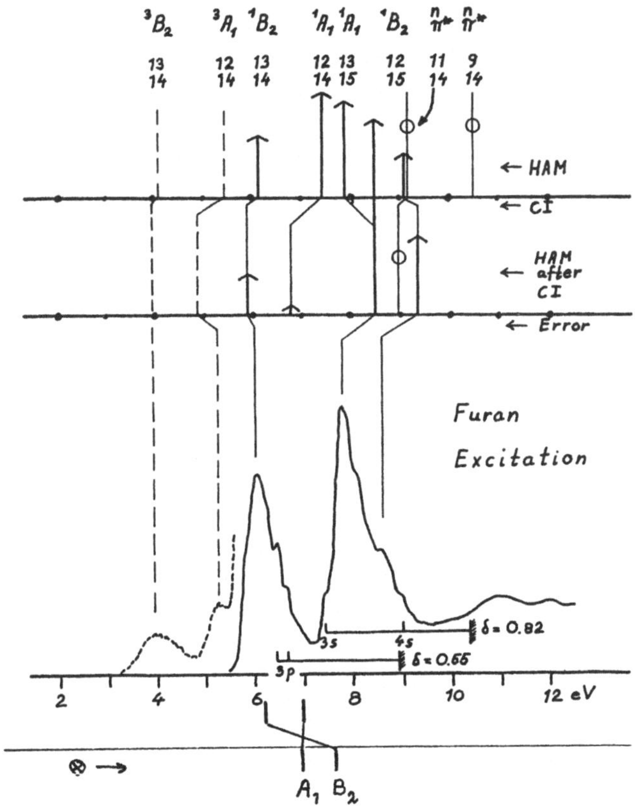

The singlet curve [26] is obtained using 50 eV and 3° and the triplet curve [26] using 50 eV and 50°. For further interpretation see ref. [28].

At the bottom of the diagram the ab-initio results [44] are shown, marked ⊗ , which are discussed in Sec.I.6.

Excitation of pyridine

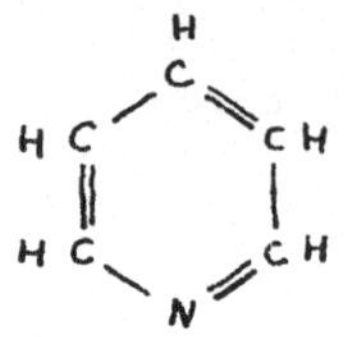

In pyridine there are four transitions of about the same intensity in the energy region 5.5 --- 7.0 eV. The CI moves nearly all intensity to a strong band at 7.3 eV. Two weak bands remain at 4.9 and 6.4 eV, which correspond to the two forbidden bands in benzene (see Table I.4.)

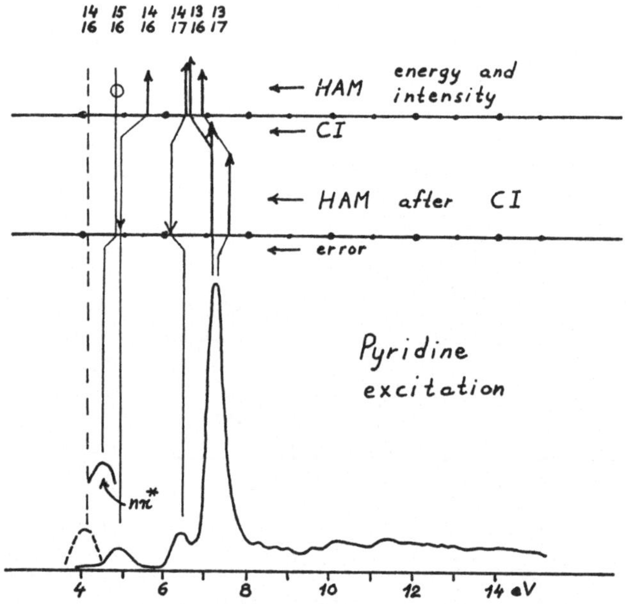

The singlet curve [16] has been obtained by Fridh, using 500 eV and 0° and the triplet curve [55] using 13 eV and 80° or He^{+} impact. The $n\pi^{*}$ transition has been observed using optical spectroscopy.

In the CI calculation the matrix elements have been calculated using the Mataga-Nishimoto γ_{AB}. If instead eq. (F.5) (γ according to Ohno-Klopman) had been used, a large error had been obtained for one of the singlets.

The use of the Mataga-Nishimoto γ for CI calculations is known to be necessary for all aromatic molecules in all semi-empirical methods [56] , and when the HAM/3 program is used to calculate excitations of aromatic molecules, the instruction "CIMAT" has to be used [57] .

I.10. Excitation of linear molecules.

In linear molecules the configuration interaction between excited molecules present special problems.

We take N_2 as an example and use the ZDO print-out. The π orbitals are denoted x and y and the π^* X and Y.

ONE HALF ELECTRON DIFFUSELY REMOVED. FILLED ORBITALS GIVE IONIZATION ENERGIES

	1	2	3	4	5	6	7	8
	-37.346	-18.398	-16.197	-16.197	-15.422	-7.330	-7.330	26.989
1 N 1	-0.641	0.565	0.0	0.0	0.299	0.0	0.0	0.425
2 N 1	0.0	0.0	-0.707	0.0	0.0	0.0	0.707	0.0
3 N 1	0.0	0.0	0.0	-0.707	0.0	-0.707	0.0	0.0
4 N 1	0.299	0.425	0.0	0.0	0.641	0.0	0.0	-0.565
5 N 2	-0.641	-0.565	0.0	0.0	0.299	0.0	0.0	-0.425
6 N 2	0.0	0.0	-0.707	0.0	0.0	0.0	-0.707	0.0
7 N 2	0.0	0.0	0.0	-0.707	0.0	0.707	0.0	0.0
8 N 2	-0.299	0.425	0.0	0.0	-0.641	0.0	0.0	-0.565
	s	s	x	y	σ	Y	X	

As in butadiene and furan we have in N_2 two occupied π orbitals and two unoccupied π^* orbitals and expect therefore four singlet and four triplet transitions. It appears that two of the four are degenerate. The excitation of N_2 can be calculated in two ways.

The simplest way to study the excitation of N_2 or any other linear molecule is to use Recknagel's formulas. They were deduced from the CI equations in 1934 by Recknagel [58] but were forgotten and deduced again in 1955 [59] (for comments see ref. [60]).

$$\begin{aligned} {}^1\Sigma_u^+ &= E_{av} + 4c - a \\ {}^1\Delta_u &= E_{av} + 2b \\ {}^1\Sigma_u^- &= E_{av} + a \\ {}^3\Sigma_u^- &= E_{av} + a \\ {}^3\Delta &= E_{av} \\ {}^3\Sigma_u^+ &= E_{av} - a \end{aligned} \qquad \text{(I.43)}$$

Here, Σ indicates that the angular momentum of the electrons around the molecule axis is = 0 and Δ that it is = $\pm$ 2. The Δ state is thus degenerate and corresponds to two equal eigenvalues. E_{av} is an average excitation energy, approximately $\approx \varepsilon_X - \varepsilon_x$.

The constants in Recknagel's expression are

$$a = 2\cdot(xy|XY) \qquad (I.44)$$

$$b = 2\cdot(xY|xY) \qquad (I.45)$$

$$c = \frac{1}{2}\cdot\left[(xX|xX) + (xX|yY)\right] = \frac{1}{2}\left[K + \overline{K}\right] \qquad (I.46)$$

The first term in c : $(xX|xX)$ is the conventional exchange integral K and the other integrals are of the same type. They can easily be calculated using the rules in Sec.I.2. (for details see ref. [61]) and are obtained in the HAM/3 print-out.

For N_2 the constants are obtained as

$$a = b = 0.71 \text{ (eV)}$$

$$K = 1.61 \qquad \overline{K} = 0.89 \qquad c = 1.25 \text{ (eV)}$$

which gives the energies in Table I.10.

Table I.10. Valence excitation energies of N_2 (eV).

Type	Transition	Energy (calc.)	Method	Exp	HAM CI + corr
$\pi\pi^*$	$^1\Sigma_u^+$	11.90	Recknagel	14.2	11.90+0 = 11.90
	$^1\Delta_u^+$	9.05		10.2	9.05+0 = 9.05
	$^1\Delta_u^-$				9.58−0.54= 9.04
	$^1\Sigma_u^-$	8.33		9.9	8.87−0.54= 8.33
	$^3\Sigma_u^-$	8.33		9.5	8.87−0.54= 8.33
	$^3\Delta_u^-$	7.62		8.6	8.15−0.54= 7.61
	$^3\Delta_u^+$				7.62+0 = 7.62
	$^3\Sigma_u^+$	6.90		7.6	6.90+0 = 6.90
$\sigma\pi^*$	$^1\Pi_g$	8.09+0.57= 8.66	HAM	8.94	
	$^3\Pi_g$	8.09−0.57= 7.52		7.60	
$s\pi^*$	$^1\Pi_u$	11.07+1.13=12.20		12.56	
	$^3\Pi_u$	11.07−1.13= 9.94		10.38	

Comparison with the experimental values in Table I.10. indicates that the calculated values are too small by about 1 eV. The explanation is that our calculated IP for the π orbital, 16.197 eV, is too small by about 0.7 eV as can be seen from the PES in Sec.H.4. If therefore the calculated IP is corrected, the calculated excitation energies will be corrected automatically.

The following molecules have been treated in this way:

ref. [61] : N_2, CO, HCN, HCCH
ref. [62] : CO_2, N_2O, NCCN
ref. [32] : O=C=C=C=O

Proof for Recknagel's formulas.

The configuration interaction (CI) in N_2 consists of four cases, which are all similar. The first case describes the interaction between the singlet $x \rightarrow X$ and the singlet $y \rightarrow Y$.

These two transitions have the same energy which according to eq. (I.15) is

$$E_{xX} = \varepsilon_X - \varepsilon_x + (xX|xX) \tag{I.47}$$

The CI matrix element is according to eq. (I.40)

$$\mathcal{G} = 2 \cdot (xX|yY) - (xy|XY) \tag{I.48}$$

and the energies of the two states is therefore $E \pm \mathcal{G}$. This gives

$$\begin{aligned} {}^1\Sigma^+ &= \varepsilon_X - \varepsilon_x + (xX|xX) + 2(xX|yY) - (xy|XY) = \\ &= \left[\varepsilon_X - \varepsilon_x + \tfrac{1}{2}a - K \right] + 4c - a \end{aligned} \tag{I.49}$$

and

$$\begin{aligned} {}^1\Delta^+ &= \varepsilon_X - \varepsilon_x + (xX|xX) - 2(xX|yY) + (xy|XY) = \\ &= \left[\varepsilon_X - \varepsilon_x + \tfrac{1}{2}a - K \right] + 2(xX|xX) - 2(xX|yY) \end{aligned} \tag{I.50}$$

To simplify the last two terms in $^1\Delta^+$ we study

$$(x+y,\ X-Y \mid x+y,\ X-Y)$$

According to the definition of this integral (see eq.(B.8))
$x+y$ is a π orbital at the angle 45° and $X-Y$ a π^* orbital at -45°. We can therefore rotate all orbitals and obtain

$$4\cdot(xY|xY)$$

since χ is normalized. Direct calculation gives

$$2\,(xX/xX) - 2\,(xX/yY)$$

The last two terms in $^1\Delta^+$ are therefore together $4\,(xY|xY) = 2b$.

The second case describes the interaction between the singlet $x \rightarrow Y$ and the singlet $y \rightarrow X$.

These two transitions have the same energy which approximately equals E_{xX} above. Since eq.(I.13) is slightly different in the first case ($x \rightarrow X$) and the second case ($x \rightarrow Y$), we add the difference of (I.13) in the two cases and find

$$E_{xY} = \varepsilon_Y - \varepsilon_x + (xY|xY) \qquad (I.51)$$

$$-(xx|YY) + (xY|xY) + (xx|XX) - (xX|xX) \qquad (I.52)$$

We simplify first eq.(I.52) by studying

$$(x+y,\ x-y \mid X+Y,\ X-Y)$$

which after rotating and after direct calculation, respectively, gives

$$4\cdot(xy|XY) = 2\,(xx|XX) - 2\,(xx|YY)$$

The result is

$$(I.52) = 2\,(xy|XY) + (xY|xY) - (xX|xX) \qquad (I.53)$$

Since the CI matrix element is

$$\mathcal{G} = 2\,(xY|yX) - (xy|XY)$$

and since $\varepsilon_Y = \varepsilon_X$ and $(xY|xY) = (xY|yX)$ we find

$$^1\Delta_u^- = \varepsilon_X - \varepsilon_x + (xy|XY) - (xX|xX) + 4\,(xY|xY)$$

$$= \left[\, \varepsilon_X - \varepsilon_x + \tfrac{1}{2}\, a - K \,\right] + 2b$$

and

$$^1\Sigma_u^- = \varepsilon_X - \varepsilon_x + (xy|XY) - (xX|xX) + 2\,(xy|XY)$$

$$= \left[\, \varepsilon_X - \varepsilon_x + \tfrac{1}{2}\, a - K \,\right] + a$$

The triplet excitations are handled in the same way.

The Recknagel correction [32]

In the HAM/3 print-out all excitations are handled in the same way. Eq.(I.51) is used and this means that in the study of a linear molecule the small correction eq.(I.52) is neglected.

Since the correction usually amounts to only a few tenths of one eV, it could be omitted, but it is disturbing to have an unsymmetrical solution when the molecule has cylindrical symmetry.

We add therefore eq.(I.53) (the "Recknagel correction") to the energy of the excitation $x \longrightarrow Y$

$$(\text{I}.53) = a + \tfrac{1}{2}\, b - K \qquad (\text{I}.54)$$

The second way to calculate the excitation of N_2 is therefore to use the instruction "CI 10." in the HAM/3 calculation, and afterwards correct all transitions of $x \rightarrow Y$ type by eq.(I.54). The result is shown in the right part of Table I.10.

In his study of the Recknagel correction [32] Chong pointed out that this method can be used for benzene also.

A difficulty is then that the benzene orbitals are in the same plane and therefore not quite equivalent to x, y, X and Y. The method with rotation, used above, is therefore not directly applicable. It is therefore simpler to adjust $\varepsilon_Y - \varepsilon_x$ so that $^1E_{1u}$ and $^3E_{1u}$ become degenerate, separately. This has been done in the figure below.

As pointed out previously, the Mataga γ has to be used for aromatic molecules.

The figure below shows the excitation of benzene, studied in the electron impact energy loss apparatus in Sec.I.7.g [50].

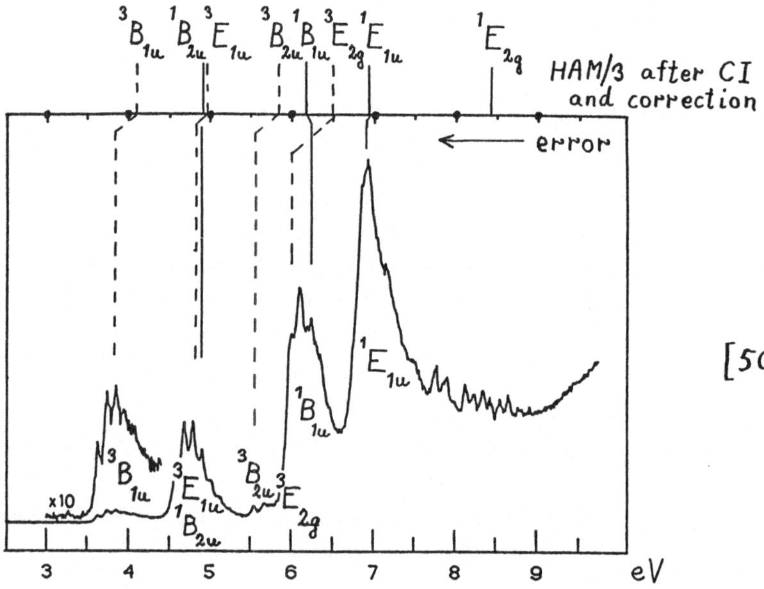

[50]

The electron energy after the collision is only 0.15 eV, and therefore the triplets are strong. We observe that the HAM/3 calculation gives a reasonable result (for a detailed interpretation, see refs.[63, 50]). All transitions are forbidden except that at 6.9 eV (cf. the UV spectrum in Sec.I.5.).

The rule that no correction is necessary for excitations of $x \rightarrow X$ type can be used also for substituted benzenes.

Excitation of nitrogen.

The experimental values in Table I.10. correspond to the energies in the Franck-Condon region in the potential energy diagram in Sec.H.3. They are also shown in the simplified diagram below.

In the left part of the diagram we see how a σ electron ($3\sigma_g$) can be excited to π^* ($1\pi_g$) or to Rydberg ($3p_u$) or to infinity (ionization at 15.58 eV).

In the middle part we see the same for a π electron ($1\pi_u$) and in the right side for an s electron ($2\sigma_u$).

The theoretical energies from Table I.10. are marked $\times$ and the experimental vertical energies with o .

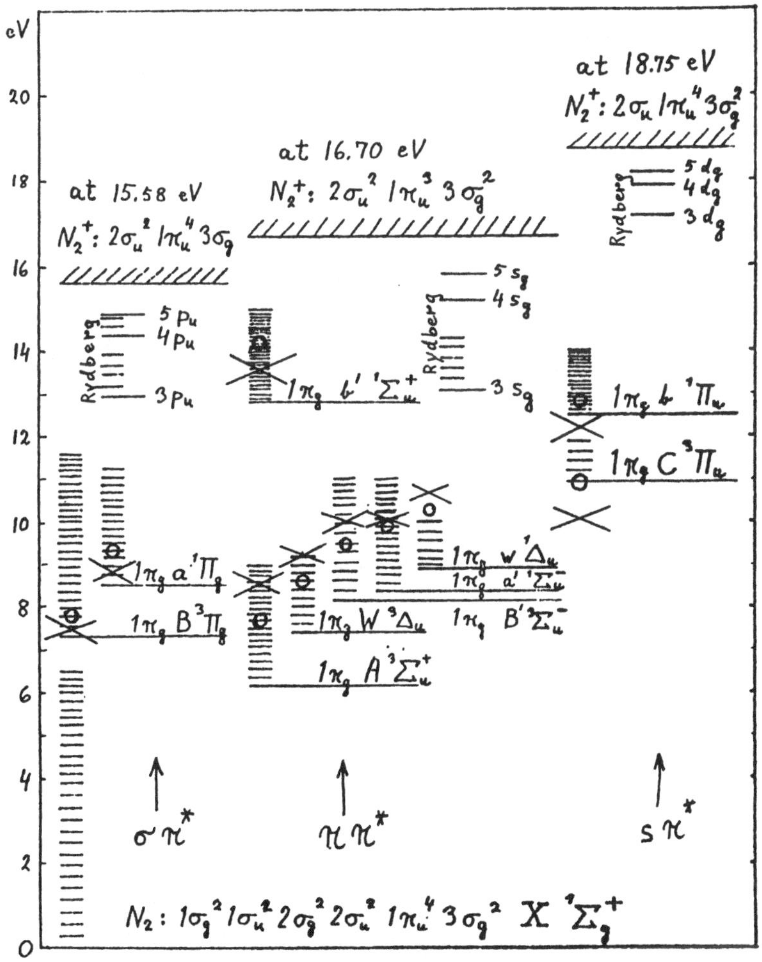

A survey of the spectrum from two electron impact studies of N_2 is given below.

The top picture [46], obtained at 50 eV and 8°, shows mainly singlets. The predominating transition has the energy 12.93 eV. It is due to a $3\sigma_g \rightarrow 3p\pi\ {}^1\Pi_u$ (Rydberg transition) together with $b\ {}^1\Pi_u$. The other transitions are forbidden, but can still be seen at lower intensity.

In the lower picture mainly triplets are shown from an experiment, in which the electrons have energy = 0 after the impact. The apparatus is shown in Sec.I.7.h [51]. The unusual feature around 9.76 eV means probably:

$$e + 9.76\,eV + N_2 \longrightarrow N_2^- \longrightarrow N(^4S) + N(^4S) + e$$

since the dissociation energy of N_2 is 9.76 eV (see the energy diagram in Sec.H.3.).

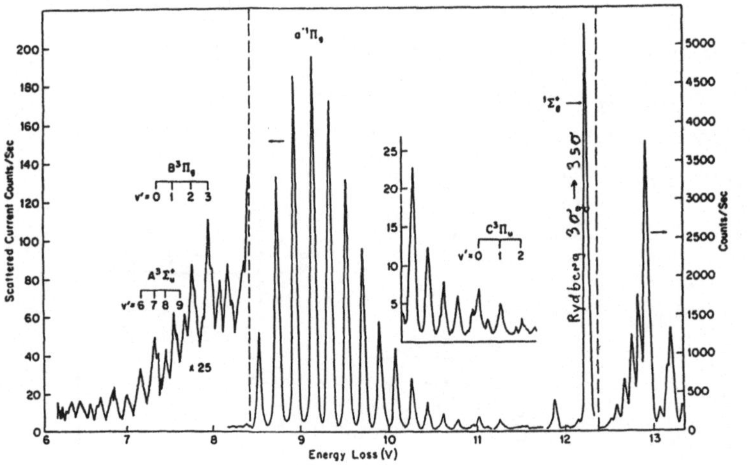

[46]

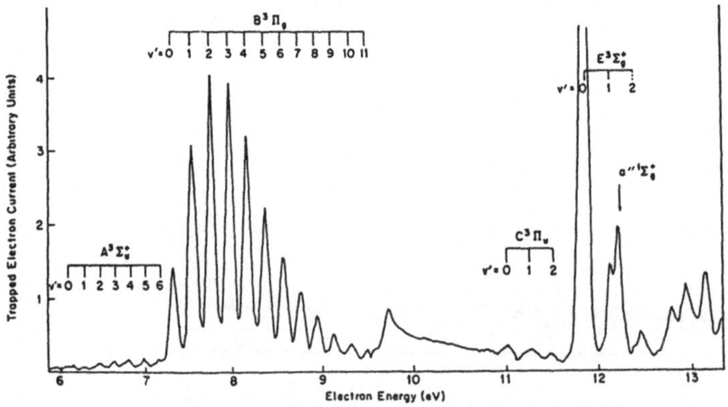

[51]

Excitation of cyanogen N≡C−C≡N

The excitation of cyanogen is of very great interest since the $\pi\pi^* \ {}^1\Sigma_u^+$ band is very intense and not overlapped. It differs in this respect from most other linear molecules except OCCCO (see Sec.I.5.) and is therefore valuable for test of Recknagel´s formulas. It is then important that five of the six transitions are known.

The result from a Recknagel calculation is shown below [62]. The energy of ${}^1\Sigma_u^+$ is given by eq.(I.49) with $\mathcal{E}_x$ corrected by -0.2 eV. The other transitions are obtained in the same way using eqs.(I.43). The agreement is reasonable.

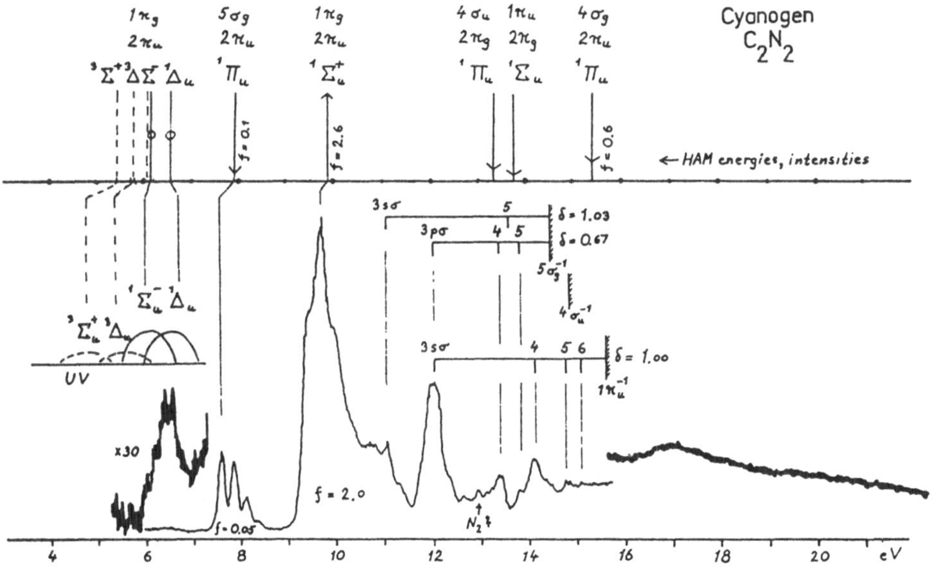

The singlet curve between 6 eV and 22 eV is obtained by Fridh [62] using electron impact spectroscopy. It is marked ———— . The triplet transitions (marked— — — —) between 4 eV and 6 eV and the singlets at 6 eV have been studied spectroscopically [64]. They are marked UV.

References

1. C.C.J. Roothaan, Rev.Mod.Phys. 23, 69 (1951).
2. J. Ridley and M. Zerner, Theoret.Chim.Acta 32, 111 (1973).
3. J.C. Slater, Quantum Theory of Atomic Structure, Vol.1, pp. 339-342, McGraw-Hill, New York (1960).
4. J.A. Pople and D.L. Beveridge, Approximate Molecular Orbital Theory, McGraw-Hill, New York (1979).
5. N.C. Baird and M.J.S.Dewar, J.Chem.Phys. 50, 1262 (1969).
6. L. Salem, The Molecular Orbital Theory of Conjugated Systems, Benjamin, New York (1966).
7. S.P. McGlynn, L.G. Vanquickenborne, M. Kinoshito and D.G. Carroll, Introduction to Applied Quantum Chemistry, Holt, Rinehart and Winston, New York (1972).
8. R. Pariser and R.G. Parr, J.Chen.Phys. 21, 466 (1953).
9. R. Pariser and R.G. Parr, J.Chem.Phys. 21, 767 (1953).
10. J.A. Pople, Trans.Faraday Soc. 49, 1375 (1953).
11. J. Del Bene and H.H. Jaffé, J.Chem.Phys. 48, 1807 (1968).
12. J. Geiger and K. Wittmaack, Z.Naturforsch. 20a, 628 (1965).
13. D.G. Wilden and J. Comer, J.Phys.B: Atom.Molec.Phys. 12, L371 (1979).
14. R. Pariser, J.Chem.Phys. 24, 250 (1956).
15. J.P. Doering, J.Chem.Phys. 67, 4065 (1977).
16. L. Åsbrink, C. Fridh and E. Lindholm, Chem.Phys.Letters 52, 72 (1977).
17. O.A. Mosher, W.M. Flicker and A. Kuppermann, J.Chem.Phys. 59, 6502 (1973).
18. G.A. George and G.C. Morris, J.Mol.Spectr. 26, 67 (1968).
19. E.E. Koch, A. Otto and K. Radler, Chem.Phys.Letters 16, 131 (1972).
20. G.J. Verhaart, P. Brasem and H.H. Brongersma, Chem.Phys.Letters 62, 519 (1979).
21. R.H. Huebner, S.R. Mielczarek and C.E. Kuyatt, Chem.Phys. Letters 16, 464 (1972).
22. M.B. Robin, Higher Excited States of Polyatomic Molecules, Vol.II, Academic Press, New York (1975).
23. C. Fridh, L. Åsbrink, B.Ö. Jonsson and E. Lindholm, Int.J. Mass Spectrom.Ion Phys. 9, 485 (1972).
24. G.W. Pukanic, D.R. Forshey, Br.J.D. Wegener and J.B. Greenshields, Theoret.Chim.Acta 10,240 (1968).

References (cont.)

25. R.L. Ellis, G. Kuehnlenz and H.H. Jaffé, Theoret.Chim.Acta 26, 131 (1972).
26. W.M. Flicker, O.A. Mosher and A. Kuppermann, J.Chem.Phys. 64, 1315 (1976).
27. W.M. Flicker, O.A. Mosher and A. Kuppermann, Chem.Phys.Letters 38, 489 (1976).
28. L. Åsbrink, C. Fridh and E. Lindholm, J.Electron Spectrosc. 16, 65 (1979).
29. J. Del Bene and H.H. Jaffé, J.Chem.Phys. 48, 4050 (1968).
30. I. Fischer-Hjalmars and M. Sundbom, Acta Chem.Scand. 22, 607 (1968).
31. B.Ö. Jonsson and E. Lindholm, Arkiv Fysik 39, 65 (1969).
32. D.P. Chong, Can.J.Chem. 58, 1687 (1980).
33. J.L.Roebber, J.C. Larrabee and R.E. Huffman, J.Chem.Phys. 46, 4594 (1967).
34. J.W. Rabalais, T. Bergmark, L.O. Werme, L. Karlsson and K. Siegbahn, in Electron Spectroscopy, D.A. Shirley (ed.), p.425, North-Holland, Amsterdam (1972).
35. E.E. Koch and A. Otto, Chem.Phys.Letters 12, 476 (1972).
36. L.E. McMurchie and E.R. Davidson, J.Chem.Phys. 66, 2959 (1977).
37. B.R. Brooks and H.F. Schaefer III, J.Chem.Phys. 68, 4839 (1978).
38. R.J. Buenker, S.K. Shih and S.D. Peyerimhoff, Chem.Phys. 36, 97 (1979).
39. J. Rose, T. Shibuya and V. McKoy, J.Chem.Phys. 58, 74 (1973).
40. H. Nakatsuji, J.Chem.Phys. 80, 3703 (1984).
41. K.K. Sunil, K.D. Jordan and R. Shepard, Chem.Phys. 88, 55 (1984).
42. R.J. Buenker, S.K. Shih and S.D. Peyerimhoff, Chem.Phys.Letters 44, 385 (1976).
43. R. McDiarmid, J.Chem.Phys. 79, 9 (1983).
44. K.H. Thunemann, R.J. Buenker and W. Butscher, Chem.Phys. 47, 313 (1980).
45. W. Butscher and K.H. Thunemann, Chem.Phys.Letters 57, 224 (1978).
46. E.N. Lassettre and A. Skerbele, in Methods of Experimental Physics, Vol.3, Molecular Physics (edited by D. Williams), Academic Press, New York (1974), p.868.
47. A. Kuppermann, W.M. Flicker and O.A. Mosher, Chem.Rev. 79, 77 (1979).

References (cont.)

48. S. Daviel, C.E. Brion and A.P. Hitchcock, Rev.Sci.Instrum. 55, 182 (1984).
49. J.H. Moore, J.Chem.Phys. 55, 2760 (1971).
50. M. Allan, Helv.Chim.Acta 65, 2008 (1982).
51. R.I. Hall, J. Mazeau, J. Reinhardt and C. Scherman, J.Phys.B: Atom.Molec.Phys. 3, 991 (1970).
52. J.H. Moore, J.Phys.Chem. 76, 1130 (1972).
53. C. Fridh, J.Chem.Soc., Faraday II, 74, 190 (1978).
54. R. McDiarmid, J.Chem.Phys. 64, 514 (1976).
55. J.P. Doering and J.H. Moore Jr., J.Chem.Phys. 56, 2176 (1972).
56. R.L. Ellis and H.H. Jaffé, in Semiempirical Methods of Electronic Structure Calculation, Part B: Applications, G.A. Segal (ed.), Plenum Press, New York (1977), p.49.
57. E. Lindholm, J.Mol.Spectrosc. 101, 444 (1983).
58. A. Recknagel, Z.Phys. 87, 375 (1934).
59. C.W. Scherr, J.Chem.Phys. 23, 569 (1955).
60. W.C. Ermler and R.S. Mulliken, J.Mol.Spectrosc. 61, 100 (1976).
61. L. Åsbrink, C. Fridh and E. Lindholm, Chem.Phys. 27, 159 (1978).
62. C. Fridh, L. Åsbrink and E. Lindholm, Chem.Phys. 27, 169 (1978).
63. J.P. Doering, J.Chem.Phys. 67, 4065 (1977); 71, 20 (1979).
64. S. Bell, G.J. Cartwright, G.B. Fish, D.O. O'Hare, R.K. Ritchie, A.D. Walsh and P.A. Warsop, J.Mol.Spectrosc. 30, 162 (1969).
65. I.D. Petsalakis, G. Theodorakopoulos, C.A. Nicolaides and R.J. Buenker, J.Chem.Phys. 81, 5952 (1984).
66. V. Bonacic-Koutecky, M. Persico, D. Döhnert and A. Sevin, J.Am.Chem.Soc. 104, 6900 (1982).
67. M.M. Goodgame and W.A. Goddard,III, Phys.Rev.Letters 54, 661 (1985).
68. R.L. Whetten, S.G. Grubb, C.E. Otis, A.C. Albrecht and E.R. Grant, J.Chem.Phys. 82, 1115 (1985).
69. S.G. Grubb, C.E. Otis, R.L. Whetten, E.R. Grant and A.C. Albrecht, J.Chem.Phys. 82, 1135 (1985).

J. Negative ions and electron affinities

J.1. Calculation of electron affinities in the HAM model

Formation of a negative ion means that one electron is added to the molecule. The extra electron enters then an excited, unoccupied orbital ψ_a^α. If energy is released in this process, we say that the electron affinity is positive. Examples are the atoms H (EA = +0.75 eV) and O (EA = +1.46 eV).

The electron affinity E_a^α is then the difference of two total energies

$$EA_a^\alpha = E(q_a^\alpha) - E(q_a^\alpha + 1) \qquad (J.1)$$

In the same way as in Sec.H.1. this gives

$$EA_a^\alpha = -\varepsilon_a^\alpha - \frac{1}{2} L_{aa} \qquad (J.2)$$

or, if we study the transition state

$$EA_a^\alpha = -{}^t\varepsilon_a^\alpha \qquad (J.3)$$

where we have added one half electron to orbital ψ_a^α before starting the SCF calculation.

The electron affinity is thus quite similar to the ionization energy, and the electron affinity can be understood as the ionization energy of the extra electron in ψ_a^α.

It is obvious that reorganization and correlation are taken care of in the same way as in Sec.H.1.

How the computer performs the calculation of EA's.

In the SCF calculation the computer fills all occupied orbitals with two electrons $(q_i = 2)$ and the excited orbital with $\frac{1}{2}$ electron $(q_a = \frac{1}{2})$. The result of the calculation is then that ε_a means $-EA_a$.

It is possible to use the same "diffuse" method for calculation of EA's and for IP's. This means that the half electron is distributed over the occupied orbitals. For ethylene this means an occupancy of $q_i = 2 + \frac{1}{6} \cdot 0.5 = 2.08333$. This somewhat unconventional procedure gives the same accuracy as for IP's (for proof see ref. [1]).

In the HAM/3 program this is achieved simply by addition of the codeword "**EA**" .

Usually, this calculation is performed using the geometry of the molecule. This gives the energy of the vertical transition only.

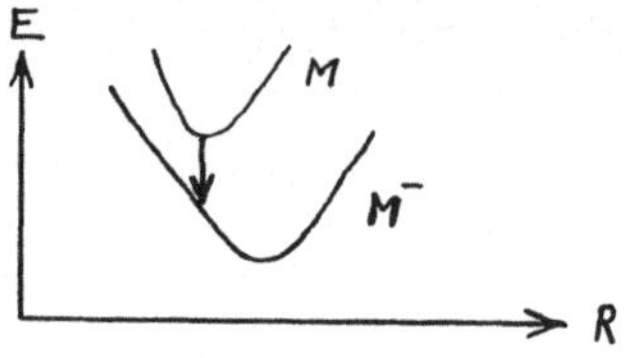

The scientific importance of studies on electron affinities

In the print-out from the computer there are two types of orbitals: occupied and unoccupied. The eigenvalues of the first type correspond to PES and those of the second type to EA. The difference of the eigenvalues corresponds to UV spectra. Only if we have agreement with experiment in all three cases simultaneously can we feel sure that the quantum-chemical calculation describes the molecule properly. Since the first and last cases have been studied during a long time, the recent addition of the EA study is of fundamental scientific importance.

J.2. Experimental methods for determination of EA's

a) Trapped electron methods for negative EA's

Most molecules have negative electron affinities. For e.g. ethylene EA = -1.8 eV. This means that to produce the negative ion it is necessary to supply energy by letting the incident electron have the kinetic energy just 1.8 eV. The negative ion, which is formed, has now excess energy and is unstable. Within typically 10^{-14} secs it ejects the electron.

The kinetic energy of the ejected electron is 1.8 eV if the ethylene molecule is left without vibrations and smaller if the ethylene vibrates. If the vibrational energy is about 1.8 eV, the electron will leave with a very small velocity.

To study the last case the "trapped electron method", described in Sec.I.7.h., may be used [2]. Here, the electrons, which have zero velocity, are measured.

In a study of ethylene [3] the trapped electron current exhibits a maximum at 1.96 eV.

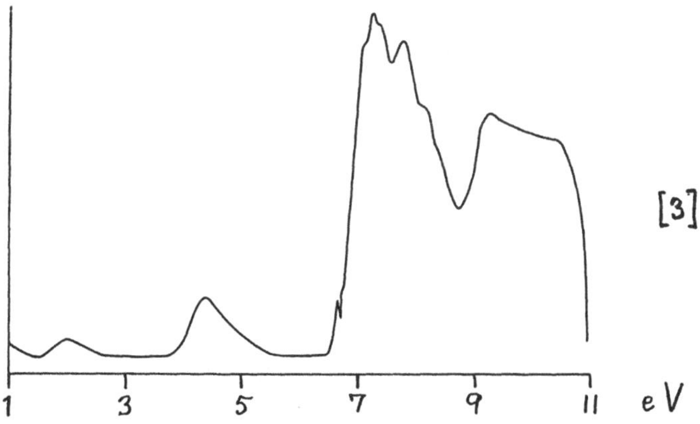

[3]

An unexpected feature is that strong bands indicate the formation of triplet $\pi\pi^*$ at 4.5 eV and singlet $\pi\pi^*$ at 7.7 eV.

b) Transmission methods for negative EA's

In the previous method the electron gives off most of its energy to the molecule.

In the transmission method the electron changes its direction during the interaction with the molecule.

The transmission methods were introduced by Schulz and coworkers in 1972 [4] although Hasted and coworkers had performed similar studies five years earlier.

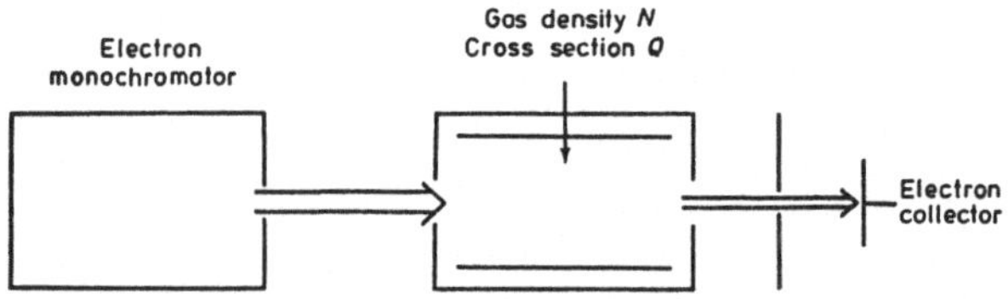

In the transmission apparatus most of the electrons go directly from the electron monochromator to the electron collector. But when the electron energy has been changed to 1.8 eV the electron can be captured by an ethylene molecule in the collision chamber. After about 10^{-14} secs the electron is ejected but probably in such a direction that it cannot reach the collector. The electron current exhibits therefore a minimum at EA.

The trochoidal monochromator has an axial magnetic field B of about 100 gauss and a weak electric field E between two planar electrodes [5] . In this way the electron are deflected upwards in the figure, depending upon their velocity.

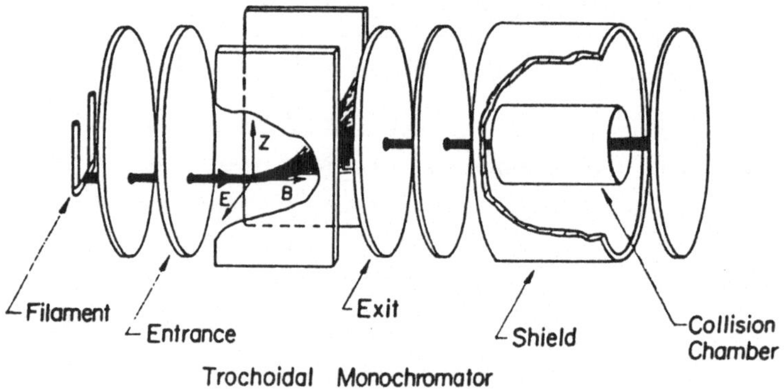

Trochoidal Monochromator

To illustrate this method we study N_2 instead.

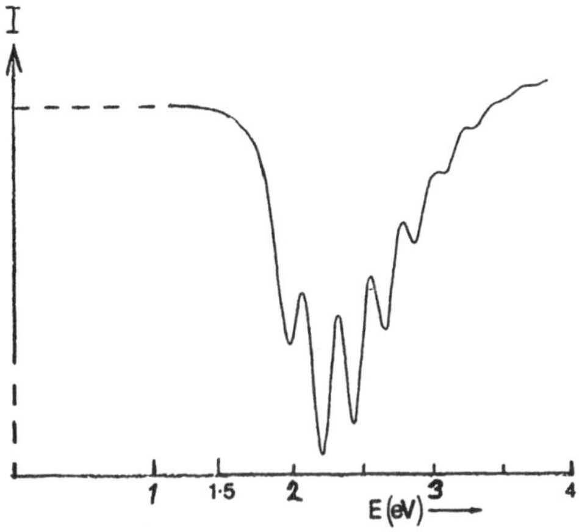

[6]

The curve shows the current to the collector with N_2 in the collision chamber [6]. Usually the derivative is shown instead.

The electron affinity of N_2 is obviously -2.2 eV.

We observe that strong vibrations are introduced when the extra electron enters the antibonding π^* orbital.

For a review of experimental work on organic molecules, see ref. [7].

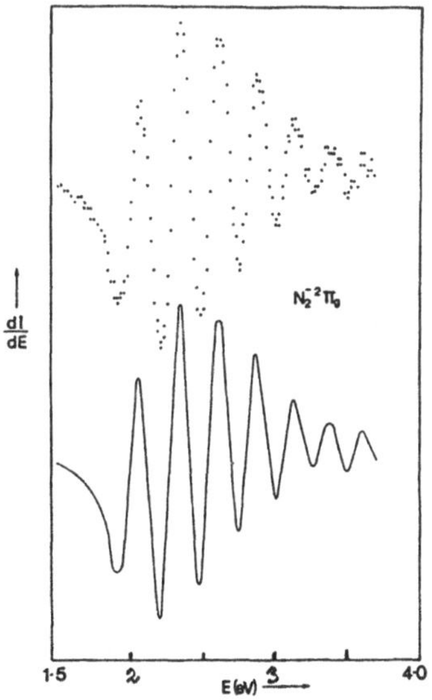

[6]

c) Interpretation of results from transmission methods.

It was previously believed that the bands from the transmission studies give an equally accurate and unequivocal picture of the unoccupied valence orbitals as PES gives a picture of the occupied ones. The "one band per orbital" rule was thus extended to the unoccupied orbitals.

It has recently appeared [8] that the situation is more complicated in larger molecules with several π^* orbitals. Although only few molecules have been studied up till now it seems as if the following tentative rules are valid:

a) When the molecule has several π^* orbitals, shake-up configurations (see Sec.L.5.) interact strongly with the primary attachments. After CI the EA may differ by more than 1 eV from the negative eigenvalue. Since the energy region of interest is only a few eV, this makes the study difficult.

b) Some or all of the higher EA's seem to disappear.

c) Instead, formation of excited triplet or even singlet states seems to replace the higher EA's.

d) In a large molecule there are many triplets. Only those seem to appear which correspond to strong singlet transitions.

As an example the benzene molecule will be discussed.

In benzene, the transmission band at -1.15 eV has a clear vibrational structure. It is attributed to the degenerate π^* orbital e_{2u}.

The band at -4.85 eV is diffuse. It has been attributed to the non-degenerate π^* orbital b_{2g}, but it is demonstrated in Sec.L.8. that it is caused by excitation to $^3E_{1u}$ instead.

This finding is important since the value -4.85 eV was used in the parametrization of HAM/3 (together with EA's from several other molecules). It is also impossible to deduce this value from UV studies.

It is possible that the core-excitation study in Sec.K.4. is a way to determine this value. It is found there that $-\varepsilon_{18}$ = +0.328 eV to be compared with the experimental EA = +1.4 eV. The error is thus 1.1 eV. If we, however, let the primary attachment in orbital 18 interact with shake-up configurations, the calculated EA is displaced by 1.1 eV and the error disappears.

It is unexpected to find that the value, chosen in the HAM/3 parametrization on incorrect grounds, probably is exactly correct.

The CI calculation is completely similar to that in Sec.L.8.

d) Types of negative ions

It is possible to understand these empirical findings from a discussion of the negative ions of a molecule. In the experimental studies four configurations seem to be important. (We will suppose that the molecule has several π and several π^* orbitals.)

1) In the first case the extra electron is in a previously unoccupied orbital, e.g. a π^* orbital. This case, which was discussed in our theoretical treatment in Sec.J.1., is often called "shape resonance". In Sec.L.5. it is called Type B.

2) In the second case the extra electron is in a previously unoccupied valence orbital and simultaneously another electron has been excited from e.g. π to π^*. This case is sometimes called "core excited shape resonance". In Sec.L.5. it is called "shake up" of Type C or Type A.

The first and second configurations interact often and their energies are then displaced.

3) In the third case the extra electron is in a Rydberg orbital, e.g. 3s, and simultaneously another electron has been excited to the same Rydberg orbital. There are thus two electrons in e.g. 2s. Such a negative ion has been observed in N_2 [87]. This "core excited shape resonance" has a rather high energy. We will not discuss it further.

4) The fourth case means a "core excited shape resonance", similar to both case 2) and case 3). The extra electron is in a Rydberg orbital and another electron has been excited from e.g. π to π^*. This possibility does not seem to have been discussed previously in connection with electron affinity spectra.

Since the Rydberg electron around a neutral molecule must be very weakly bound, the energy of this anion is approximately the same as that of the excited state. This explains rule c) above.

When the electron goes away, two products are possible: the excited state or the ground state.

5) Our finding that in e.g. benzene the observed energy corresponds to a triplet, indicates, that the probability for formation in a transmission apparatus can be larger for ions of the fourth type than for ions of the first two types.

e) Photodetachment of small negative ions

Positive electron affinity means that the negative ion is stable. Such negative ions are therefore of importance in nature and in chemistry. Unfortunately, a positive EA is difficult to measure.

The reason is that when the electron is captured by the molecule and the energy is "released", this energy cannot leave the molecule. It stays within the molecule as vibrational energy. It can therefore not be measured.

The usual method to measure positive EA is therefore to produce a beam of negative ions and to ionize them by use of incident photons. The electron affinity is then obtained as the lowest photon energy which produces photoelectrons. This method is complicated and has been used for only few molecules [9].

f) Reduction potentials of large organic molecules

Positive electron affinities of large organic molecules can also be determined in solution by chemical methods. Sjöberg and Eriksen [10, 11] have used pulse radiolysis to study the electron affinity of nitrobenzene $C_6H_5NO_2$ (NB) by comparison with that of duroquinone (DQ) ($C_{10}H_{12}O_2$).

The reaction vessel contains water with a little NB and very little DQ. An electron pulse from an accelerator gives small quantities of NB^- and very small of DQ^-. The amount of DQ^- is too small to be observed. When NB^- hits a DQ molecule the reaction

$$NB^- + DQ \rightleftharpoons NB + DQ^-$$

takes place. The number of DQ^- ions increases therefore rapidly after the pulse and can now be measured. This is done by observing the absorption spectrum of DQ^-.

Since the equilibrium constant of a chemical reaction depends upon the energies, this gives the electron affinity of nitrobenzene, when that of duroquinone is known. A difficulty is that the electron affinities are strongly influenced by the solvent. They are therefore usually described by another name: reduction potentials [12].

Recently, similar reactions have been studied in the gas phase for some medium-size organic molecules [13].

g) Vibrational excitation for observation of temporary negative ions

An important result of the formation of a negative ion with short lifetime is that in the molecule vibrations may be excited. This makes the process important in nature.

Before the process starts, the vibraional quantum number is usually zero. The negative ion is now formed with a certain vibrational quantum number. When then the electron is ejected from the ion, the molecule may be formed in the ground state with a higher vibrational quantum number than zero.

This means that the energies of the incident and the ejected electrons differ by this vibrational energy.To study the resonance we must therefore measure the energies of the two electrons.

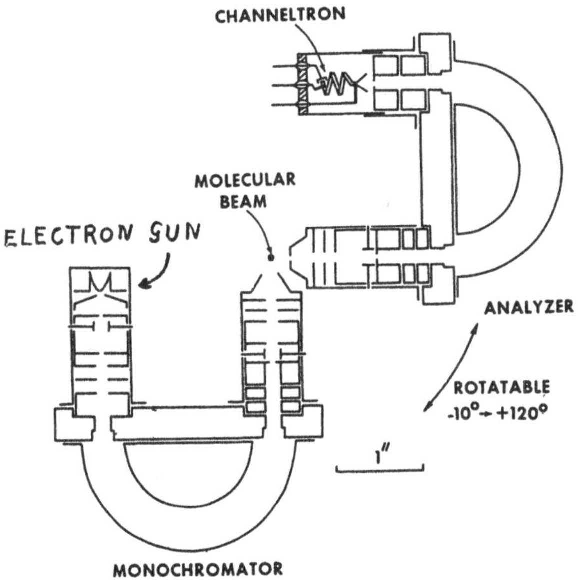

[14]

The importance of this method is that when the resonance process is observed, the energy of the incident electrons gives directly the EA of the molecule at the normal geometry of the molecule. It appears that resonance can be observed also at high energies of the incident electrons.

At low energy the results are the same as with the transmission method.

In the study of CO we expect the same EA for CO as for N_2 (-2.2 eV) since CO and N_2 are isoelectronic. This EA gives a high maximum in the curve below, observed by Tronc and coworkers [15].

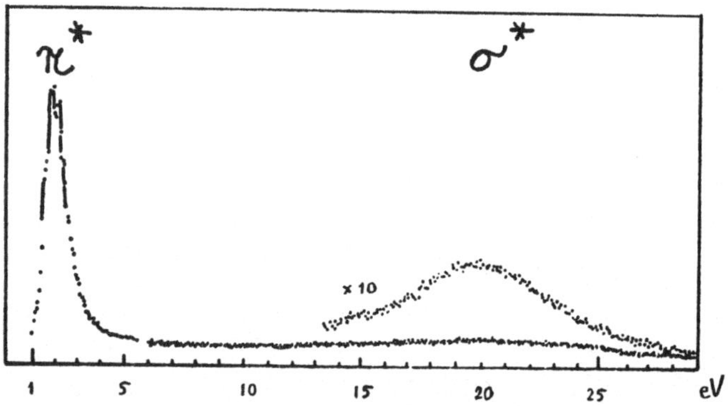

CO

At -20 eV a very broad maximum is due to the σ^* orbital. This interpretation is supported by the study of CO_2 [16] where σ_g^* is seen at -20 + 10 = -10 eV and σ_u^* at -20 - 10 = -30 eV.

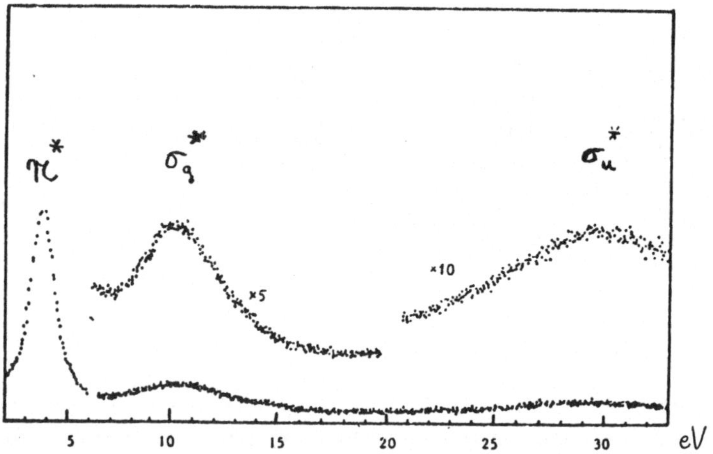

CO_2

The curve for N_2 is nearly identical to that for CO [15].

The results from these studies [15, 16, 54] : -22 eV for N_2, -19.5 eV for CO, -15.5 eV for NO and -9.5 eV for O_2 indicate an electron affinity for C-C beyond that for N_2.

J.3. Electron affinities of molecules: some results

EA of nitrogen

ONE HALF ELECTRON DIFFUSELY ADDED. EMPTY ORBITALS GIVE ELECTRON AFFINITIES

	1	2	3	4	5	6	7	8
	-25.840	-8.021	-5.750	-5.750	-5.049	2.254	2.254	34.782
1 N 1	0.603	0.525	0.0	0.0	0.346	0.0	0.0	0.568
2 N 1	0.0	0.0	0.0	-0.658	0.0	0.0	-0.770	0.0
3 N 1	0.0	0.0	0.658	0.0	0.0	0.770	0.0	0.0
4 N 1	-0.197	0.444	0.0	0.0	0.598	0.0	0.0	-0.778
5 N 2	0.603	-0.525	0.0	0.0	0.345	0.0	0.0	-0.568
6 N 2	0.0	0.0	0.0	-0.658	0.0	0.0	0.770	0.0
7 N 2	0.0	0.0	0.658	0.0	0.0	-0.770	0.0	0.0
8 N 2	0.197	0.444	0.0	0.0	-0.598	0.0	0.0	-0.778
						π^*		σ^*

The print-out shows the result of a calculation in which one half electron has been added to the orbitals 1 --- 5. The eigenvalues ε_6 and ε_7 give then $-EA$, so that EA is calculated as -2.25 eV. This agrees well with the experimental value [6] -2.2 eV.

It is interesting that PES and UV studies of N_2 are in quantitative agreement with the EA study.

The electron affinity for the σ^* orbital 8 has been measured as -22 eV [15]. The calculated value, -34.782 eV, is much higher. The reason for the disagreement is probably the strong local dipoles which have been neglected in HAM/3.

EA of ethylene

ONE HALF ELECTRON DIFFUSELY ADDED. EMPTY ORBITALS GIVE ELECTRON AFFINITIES

	1	2	3	4	5	6	7	8	9	10	11	12
	-17.235	-13.091	-9.519	-7.969	-6.684	-3.653	2.270	21.091	25.780	29.537	34.173	39.901
1 C 1	-0.452	-0.323	0.000	0.101	-0.000	-0.000	0.000	0.683	-0.839	0.000	-0.010	0.000
2 C 1	0.047	-0.204	-0.000	-0.520	0.000	0.000	-0.000	0.417	0.357	-0.000	1.015	0.000
3 C 1	0.000	-0.000	-0.408	0.000	0.468	-0.000	-0.000	0.000	0.000	0.694	0.000	-0.865
4 C 1	0.000	-0.000	-0.000	-0.000	0.000	0.642	0.797	-0.000	-0.000	-0.000	-0.000	-0.000
5 C 2	-0.452	0.323	0.000	0.101	0.000	0.000	0.000	0.683	0.839	-0.000	0.010	-0.000
6 C 2	-0.047	-0.204	0.000	0.520	-0.000	0.000	0.000	-0.417	0.357	-0.000	1.015	0.000
7 C 2	-0.000	0.000	-0.408	-0.000	-0.468	0.000	0.000	0.000	0.000	0.694	0.000	0.865
8 C 2	-0.000	-0.000	0.000	0.000	0.000	0.642	-0.797	0.000	-0.000	0.000	-0.000	0.000
9 H 3	-0.190	-0.268	-0.260	-0.179	0.312	-0.000	-0.000	-0.459	0.294	-0.630	-0.467	0.647
10 H 4	-0.190	-0.268	0.260	-0.179	-0.312	0.000	-0.000	-0.459	0.294	0.630	-0.467	-0.647
11 H 5	-0.190	0.268	-0.260	-0.179	-0.312	0.000	-0.000	-0.459	-0.294	-0.630	0.467	-0.647
12 H 6	-0.190	0.268	0.260	-0.179	0.312	-0.000	-0.000	-0.459	-0.294	0.630	0.467	0.647
							π^*	σ^*_{CH} a_g	σ^*_{CH} b_{1u}	σ^*_{CH} b_{2u}	σ^*_{CC}	σ^*_{CH} b_{3g}

Since ε_7 is +2.27 eV, the electron affinity of ethylene is calculated as -2.27 eV. The experiment [17] gives the vertical EA as about -1.78 eV.

In the figure curve A indicates the derivative of the transmitted current as function of electron impact energy in ethylene. Curve B shows the same after subtraction of a smooth background and magnification.

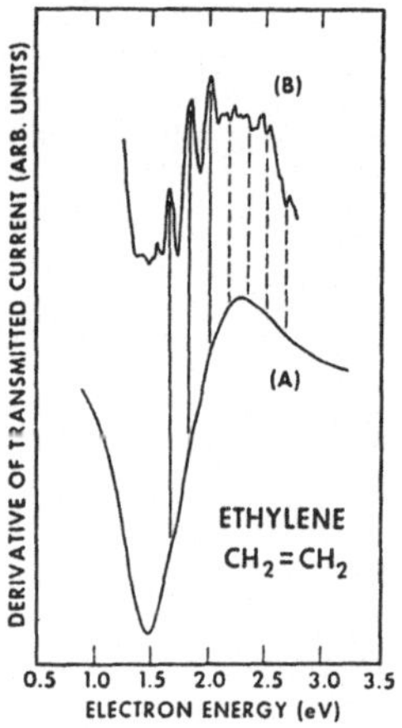

Also here we observe the internal consistency of our calculations with reasonable results simultaneously for IP, UV and EA.

EA of butadiene

In butadiene there are two π^* orbitals and two electron affinities. We reproduce part of the print-out

ONE HALF ELECTRON DIFFUSELY ADDED. EMPTY ORBITALS GIVE ELECTRON AFFINITIES

	4	5	6	7	8	9	10	11	12	13	14	15
67	-12.913	-9.837	-9.717	-8.311	-8.200	-7.101	-5.584	-3.608	0.502	3.408	17.191	20.495
264	-0.097	-0.019	0.008	-0.103	0.020	0.000	0.000	-0.000	-0.000	-0.000	-0.091	-0.420
159	0.144	0.028	0.299	-0.344	0.210	-0.130	0.000	-0.000	-0.000	-0.000	0.094	0.428
011	0.090	0.363	-0.263	-0.048	0.336	0.234	0.000	0.000	0.000	0.000	0.051	-0.014
000	0.000	-0.000	0.000	0.000	0.000	0.000	-0.344	0.552	-0.637	-0.469	-0.000	-0.000
254	0.241	-0.013	0.147	-0.040	-0.026	0.090	0.000	0.000	0.000	-0.000	-0.579	-0.438
142	0.079	-0.024	-0.209	0.378	-0.255	0.180	-0.000	-0.000	-0.000	0.000	-0.394	-0.096
032	0.214	0.236	-0.069	0.115	-0.138	-0.442	-0.000	-0.000	0.000	-0.000	0.666	-0.338
000	-0.000	-0.000	0.000	-0.000	0.000	0.000	-0.510	0.371	0.438	0.697	0.000	0.000
254	-0.241	-0.013	-0.147	-0.040	0.026	0.090	0.000	0.000	0.000	-0.000	0.579	-0.438
142	0.079	0.024	-0.209	-0.378	-0.255	-0.180	-0.000	-0.000	-0.000	0.000	-0.394	0.096
032	0.214	-0.236	-0.069	-0.115	-0.138	0.442	0.000	0.000	-0.000	0.000	0.666	0.338
000	-0.000	-0.000	0.000	0.000	0.000	0.000	-0.510	-0.371	0.438	-0.697	-0.000	-0.000
264	0.097	-0.019	-0.008	-0.103	-0.020	0.000	0.000	-0.000	-0.000	-0.000	0.091	-0.420
159	0.144	-0.028	0.299	0.344	0.210	0.130	0.000	0.000	-0.000	0.000	0.094	-0.428
011	0.090	-0.363	-0.264	0.048	0.336	-0.234	-0.000	-0.000	0.000	-0.000	0.051	0.014
000	-0.000	-0.000	0.000	-0.000	0.000	0.000	-0.344	-0.552	-0.637	0.469	-0.000	-0.000
230	-0.062	0.224	-0.297	0.065	0.144	0.185	-0.000	0.000	0.000	0.000	-0.099	0.414
210	-0.188	-0.238	0.050	0.145	-0.283	-0.144	-0.000	-0.000	-0.000	0.000	0.196	0.284
151	-0.307	0.131	0.040	0.245	0.185	-0.211	-0.000	-0.000	-0.000	0.000	-0.006	0.410
151	0.307	0.131	-0.040	0.245	-0.185	-0.211	-0.000	-0.000	-0.000	0.000	0.006	0.410
230	0.062	0.224	0.298	0.065	-0.144	0.185	0.000	-0.000	0.000	-0.000	0.099	0.414
210	0.188	-0.238	-0.050	0.145	0.283	-0.144	-0.000	-0.000	0.000	-0.000	-0.196	0.284
									π^*	π^*		

Since $\varepsilon_{12} = 0.502$ and $\varepsilon_{13} = 3.409$ the calculated result is $EA_{12} = -0.502$ eV and $EA_{13} = -3.409$ eV.

The electron affinity spectrum [17] shows one EA at -0.6 eV with vibrational structure and one at -2.8 eV. It will be shown in Sec.L.6. that the displacement from 3.409 to 2.8 depends upon interaction with a shake-up configuration. The triplet $\pi\pi^*$ (see Sec.I.8.) has probably too high energy to be important.

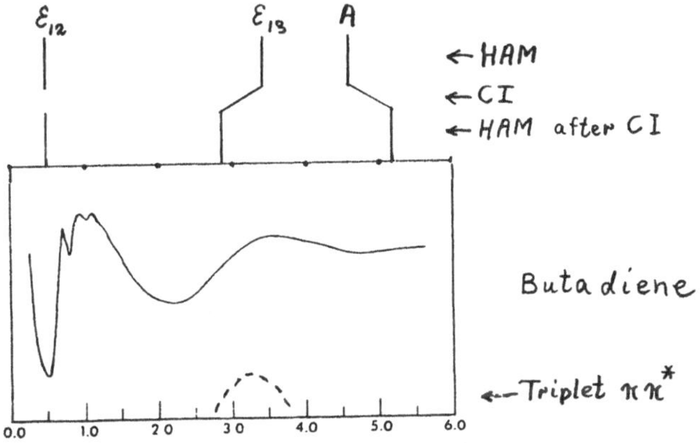

EA of cyanogen

In cyanogen there are two π^* orbitals and two electron affinities. We reproduce the print-out:

ONE HALF ELECTRON DIFFUSELY ADDED. EMPTY ORBITALS GIVE ELECTRON AFFINITIES

	3	4	5	6	7	8	9	10	11	12	13	14
	-14.419	-7.716	-7.716	-7.646	-7.220	-5.905	-5.905	-0.089	-0.089	4.713	4.713	11.844
1 N 1	0.343	0.0	0.0	0.382	-0.349	0.0	0.0	0.0	0.0	0.0	0.0	-0.214
2 N 1	0.0	0.0	-0.378	0.0	0.0	0.551	0.0	0.0	-0.612	0.470	0.0	0.0
3 N 1	0.0	-0.378	0.0	0.0	0.0	0.0	-0.551	-0.612	0.0	0.0	0.470	0.0
4 N 1	0.012	0.0	0.0	0.507	-0.526	0.0	0.0	0.0	0.0	0.0	0.0	0.440
5 C 2	-0.267	0.0	0.0	-0.192	0.057	0.0	0.0	0.0	0.0	0.0	0.0	0.754
6 C 2	0.0	0.0	-0.486	0.0	0.0	0.381	0.0	0.0	0.442	-0.714	0.0	0.0
7 C 2	0.0	-0.486	0.0	0.0	0.0	0.0	-0.381	0.442	0.0	0.0	-0.714	0.0
8 C 2	0.372	0.0	0.0	-0.107	0.239	0.0	0.0	0.0	0.0	0.0	0.0	-0.128
9 C 3	-0.267	0.0	0.0	0.192	0.057	0.0	0.0	0.0	0.0	0.0	0.0	-0.754
10 C 3	0.0	0.0	-0.486	0.0	0.0	-0.381	0.0	0.0	0.442	0.714	0.0	0.0
11 C 3	0.0	-0.486	0.0	0.0	0.0	0.0	0.381	0.442	0.0	0.0	0.714	0.0
12 C 3	-0.372	0.0	0.0	-0.107	-0.239	0.0	0.0	0.0	0.0	0.0	0.0	-0.128
13 N 4	0.343	0.0	0.0	-0.382	-0.349	0.0	0.0	0.0	0.0	0.0	0.0	0.214
14 N 4	0.0	0.0	-0.378	0.0	0.0	-0.551	0.0	0.0	-0.612	-0.470	0.0	0.0
15 N 4	0.0	-0.378	0.0	0.0	0.0	0.0	0.551	-0.612	0.0	0.0	-0.470	0.0
16 N 4	-0.012	0.0	0.0	0.507	0.526	0.0	0.0	0.0	0.0	0.0	0.0	0.440
								π^*		π^*		

The calculated result means thus one EA at + 0.089 eV and another at - 4.713 eV.

The electron affinity spectrum [18] shows two EA's and one

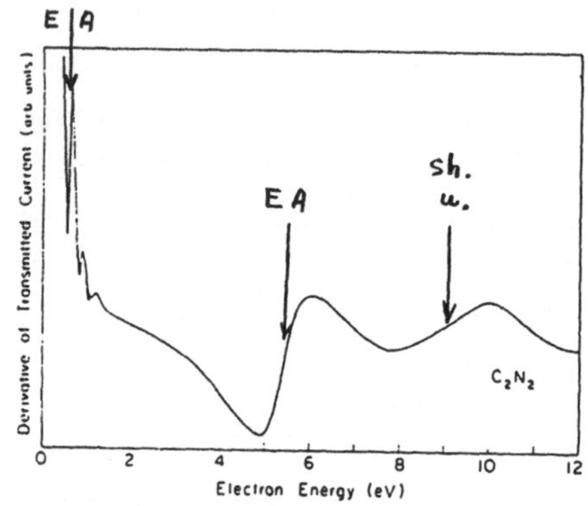

shake up. The curve is rapidly varying at low energies, and it cannot be decided where the first vibrational band is. It is quite possible that the vertical transition is near 0 eV, and we will therefore simply use the HAM/3 value $\varepsilon_{10} = -0.089 \approx -0.10$ eV. This means $EA_{10} = +0.10$ eV. No correction is therefore applied to ε_{10} and ε_{11} also in other types of calculation.

The second EA corresponds roughly to $-\varepsilon_{12}$ and $-\varepsilon_{13}$. This will be discussed in Sec.L.6.

Other molecules

Since there is in every molecule a large number of excited orbitals of different types (π^*, σ^*, Rydberg), every molecule has a large number of negative electron affinities. Few molecules have also one or more positive electron affinities.

Since usually only small electron affinities are of interest, the excited orbitals are usually of π^* type.

The transmission methods are comparatively new, and the number of transmission machines is small. Not many molecules have therefore up till now been measured. Part of them are included in Table J.3., in which the experimental and HAM/3 electron affinities are compared.

Most of the experimental values are from a review article by Jordan and Burrow [7] . No references have been given in these cases.

All calculations have been done using the "diffuse" procedure in which one half electron has been added to the occupied orbitals (see Sec.J.1.). For a small number of molecules CI studies have been performed.

We see from Table J.3. that the agreement between the experimental electron affinity and the HAM/3 result is generally reasonable.

Table J.3. Electron affinities from the transmission method together with negative eigenvalues, obtained from an "EA calculation".

The experimental values are from ref. [7] if no reference is given.

l.m. means "lacks meaning".

The Notes contain more information on special molecules.

Hydrocarbons	Exp	HAM/3	Ref.
ethylene	-1.78	-2.27	
	about -5	l.m. σ^*_{CH} a_g	[19]
	about -7	l.m. σ^*_{CH} b_{1u}	[19]
	about -9	l.m. σ^*_{CH} b_{2u}	[19]
	about-11	l.m. σ^*_{CH} b_{3g}	[19]
	about-30 (extr.)	-34.17 σ^*_{CC}	[20]
propene	-1.99	-2.07	[7,21]
butene-cis	-2.22	-1.97	[7,21]
butene-trans	-2.10	-1.92	[7,21]
butadiene	-0.62	-0.50	[7,22]
	-2.8	-3.40 a)	[7,22]
1,3-cyclohexadiene 	-0.80	-0.36	[7,23]
	-3.43	-3.30	[7,23]
1,4-cyclohexadiene 	-1.75	-1.36 π^*_a	[see below]
	-2.67	-2.26 π^*_b	
norbornadiene 	-1.04	-0.76	
	-2.56	-1.86	
1,3-cyclopentadiene 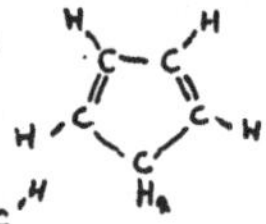	-1.19	-0.95	[24,25]
	-3.3	-3.39	
dimethylfulvene 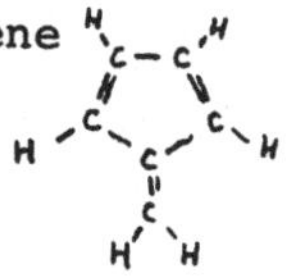	> -0.2	+0.84	[25]
	-3.1	-3.17	
		-3.81	
acetylene	-2.6	-3.29	[7, 26, 27]
	-2.4		[88]
	-6	l.m. σ^*_{CH}	[28]
	about -30 (extr.)	-65.67 σ^*_{CC}	[20]
dimethyl acetylene	-3.43	-3.40	[29]
diacetylene	-1.00	-0.80	[30]
	-5.60	-5.37	

Hydrocarbons (cont.)	Exp	HAM/3	Ref.
allene	-1.9	-2.49	[31]
methane	-7.8	l.m. σ^*_{CH}	[32]
	-5		[33]
	-7.5		[34]
neopentane $C(CH_3)_4$	-6.1	l.m. σ^*_{CH}	[35]
hexatriene-trans		+0.51	[36]
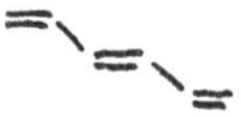	-2.13 b)	-1.82	
	-3.53	-3.69	
hexatriene-cis		+0.46	[36]
	-1.58 b)	-1.80	
	-3.53	-3.77	

Aromatic compounds	Exp	HAM/3	Ref.
benzene	-1.15	{ -1.11, -1.11 } e_{2u}	
	-4.85 c)	-4.86 d)	
toluene	-1.11	{ -1.02, -1.04 }	
	-4.88 e)	-4.77 f)	
phenol	-1.01	-1.00	[7, 37]
	-1.73	-1.38	
	-4.92 e)	-4.96 f)	
aniline	-1.13	-1.17	
	-1.85	-1.51	
	-5.07 e)	-5.07 f)	
fluorobenzene	-0.89	{ -0.73, -1.06 }	
	-4.77 e)	-4.71 f)	
pyridine	-0.62	-0.39	
	-1.20	-1.06	
	-4.58 e)	-4.40 f)	
pyrimidine	>0	-0.03	
	-0.77	-0.63	
	-4.24 e)	-3.69 f)	
pyridazine	>0	-0.24	
	-0.73	-0.55	
	-4.05 e)	-4.20 f)	

Aromatic compounds (cont.)	Exp	HAM/3	Ref.
pyrazine	>0	+0.42	[12]
	-0.87	-0.98	
	-4.10 e)	-3.83 f)	
s-triazine	>0	{ +0.01 0.0	
	-4.0 e)	-2.86 f)	
styrene (planar)	-0.25	+0.08	[7,38]
	-1.05	-0.80	
	-2.48	-2.23	
	-4.67 e)	-5.01 f)	
benzonitrile	>0	+0.22	
	-0.54	-0.49	
	-2.49	-1.70	
	-3.20	-2.55	
	-4.9 e)	-4.98 f)	
naphthalene	-0.19	+0.36 g)	[7, 38-42]
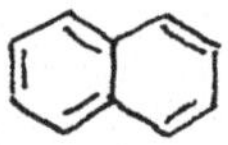	-0.90	-0.49 g)	
	-1.67	-1.68 g)	
	-3.37 h)	-2.94 g)	
	-4.72 h)	-5.60 g)	
	-5.71 h)		
biphenyl (45°)	-0.3	+0.42 i)	[43]
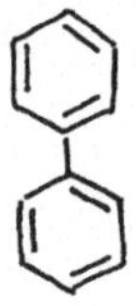	-0.97	-0.32 i)	
	-1.25	-0.44 i)	
	-1.87	-1.28 i)	
	-4.46 e)	-3.76 f)	
	-5.6 e)	-5.06 f)	
fluorene	-0.2	+0.33 i)	[43]
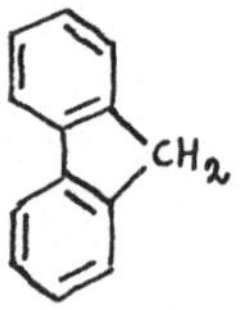	-0.71	-0.48 i)	
	-1.45	-0.58	
	-2.12	-1.37	
	-3.94 e)	-3.84 f)	
	-4.8 e)	-5.10 f)	
	-5.7 e)		

Oxo-compounds	Exp	HAM/3	Ref.
CO	-1.8	-2.31	[15]
	-19.5	-36.83 σ^*	[15]
CO_2	-3.8	-3.00	[44]
	-10.8	-17.30 σ^* $5\sigma_g$	[16]
	-30	-47.58 σ^* $4\sigma_u$	[16]
formaldehyde H_2CO	-0.86	-1.54	
acetone $(CH_3)_2CO$	-1.51	-1.13	
acrolein $CH_2=CH-CHO$	>0	+0.40	
	-2.47	-2.96	
acetaldehyde CH_3CHO	-1.19	-1.20	
p-benzoquinone 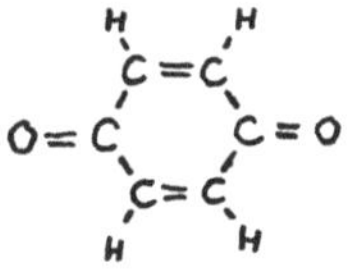	+1.86	+2.71 j)	[45-47, 13]
	-0.72	-0.24 j)	
	-1.43	-1.14 j)	
	-2.15 k)	-4.05	
	-4.2 k)		
maleic anhydride 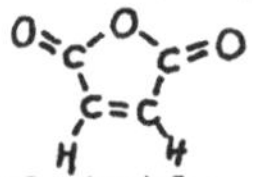	+1.42	+1.98	[48, 13]
	-2.14	-1.60	
	-2.93	-2.44	
maleimide 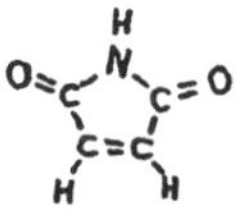	?	+1.66	[48]
	-2.66	-1.93	
	-2.66	-2.70	
4-cyclopentene-3,5-dione 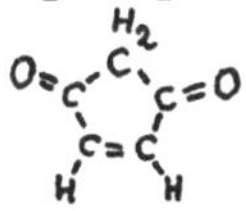	?	+1.86	[48]
	-0.80	-1.10	
	-2.91	-2.64	
Fluoro-compounds			
ethylene	-1.78	-2.27	
fluoroethylene	-1.91	-2.27	[49,50]
difluoroethylene-1,1	-2.39	-2.27	[49,50]
tetrafluoroethylene	-3.00	-1.75 l)	[49,50]
benzene	-1.15	-1.11	
	-4.85 c)	-4.86 d)	
fluorobenzene	-0.91	-0.73	[51, 52]
	-1.40	-1.06	
	-4.66 e)	-4.71 f)	
difluorobenzene-1,4	-0.62	-0.39	[51, 52]
	-1.41	-0.79	
	-4.51 e)	-4.36	

Fluoro-compounds (cont.)	Exp	HAM/3	Ref.
trifluorobenzene-1,3,5	-0.77	-0.39	[51,52]
	-4.48 e)	-4.14	
hexafluorobenzene	-0.42	+0.57 l)	[51,52]
	-4.50 e)	-3.18	
Other compounds			
N_2	-2.2	-2.25	[6,53]
	-22	-34.78 σ^*	[15]
N_2O	-2.2	-2.19	[54]
	-8.4	-14.91 σ^* 8σ	
	?	-39.78 σ^* 9σ	
acetonitrile CH_3CN	-2.84	-3.10	
acrylonitrile $CH_2=CH-CN$	-0.21	-0.27	
	-2.74	-2.27	
		-4.10	
pyrrole 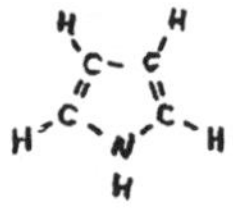	-2.38	-2.14	[24]
	-3.44	-3.59	
furan 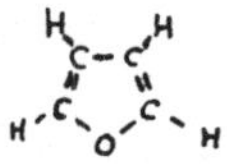	-1.76	-1.72	[24]
	-3.14	-3.38	
nitromethane CH_3NO_2	+0.49	-0.21	[55,13]
nitrobenzene	+0.57 m)	+1.17 m)	
	-0.55	+0.11 m)	[42]
	-1.36	-0.89 m)	
	-2.9 m)	-4.12	
	-3.79 m)		
	-4.69 m)		
imidogen NH	+0.38	+0.49	[56]
formamide $HCONH_2$	-2.0	-2.50	[57]
tetracyanoethylene	+2.9	+2.89	[58]
tetracyano-quinodimethane 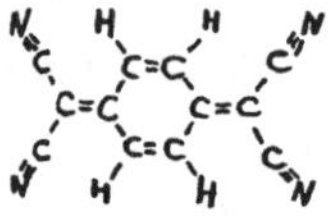	+2.8	+3.52	[59]
		+0.69	
		+0.46	
HCN	-2.26	-3.12	[60]
cyanogen C_2N_2	-0.58	+0.09 o)	[18]
	-5.37	-4.73 o)	
C_2	+3.54 σ	+3.41 σ	[83]

Table J.3. Notes

a) butadiene: CI changes -3.40 to -2.84 eV (see Sec.L.6.).
b) hexatriene: this difference cannot be explained.
c) benzene: -4.85 probably caused by $^3E_{1u}$ (see Sec.L.8.).
d) benzene: CI changes -4.86 to -3.76 eV (see Sec.L.8.).
e) possibly due to triplet, but triplet energies not well known.
f) CI gives displacement towards higher value (cf. benzene).
g) naphthalene: see Sec.L.7. and L.8.
h) naphthalene: -3.37 possibly due to triplet and -4.72 and -5.71 possibly due to singlets.
i) biphenyl and fluorene: PES and UV indicate corrections -0.4 eV to the HAM values, which improves the agreement.
j) p-benzoquinone: PES and UV indicate corrections about -1.0, -0.6 and -1.2 eV, giving HAM values about +1.71, -0.84 and -2.34 eV. CI changes then -2.34 to -1.08 eV, which possibly explains the experimental value -1.43 eV. --- As a further test the UV absorption of the anion can be studied (cf. Sec.L.7.). This shows that a more precise HAM value is (+1.85 ± 0.15) eV.
k) p-benzoquinone: -2.15 is possibly due to excitation to triplets (three triplets have this energy) and -4.2 to a singlet excitation.
l) hexafluorobenzene: PES and UV indicate a large correction, changing +0.57 to -0.82 eV. The parameters for F-F interaction are obviously unsuccessful. --- The same explanation is probably valid also for tetrafluoroethylene although its UV spectrum is not well understood.
m) nitrobenzene: The HAM/3 values can be corrected from PES [80] and UV [81] . The corrections about -0.6, -0.65 and -0.45 give corrected HAM values about +0.57 eV, -0.54 eV and -1.34 eV. The last two values agree well with the experimental ones and the first value, +0.57 eV, is given in the "Exp" column. There are probably no experimental values to compare with.

These results are valid for isolated nitrobenzene molecules in the gas phase.

In the liquid phase the EA +0.57 has to be replaced by a larger value, about + 1.00 eV. This value gives then good agreement if it is used to calculate the UV spectrum of nitrobenzene in methanol or water [81] or the UV spectrum of the anion of nitrobenzene in the solvent methyltetrahydrofuran [82].

The reason for the changed EA is probably the large dipole moment of the negative ion in its ground state [81].

It is not clear why the use of ion-molecule reactions have given the high value EA = +0.97 eV [13] in agreement with our liquid phase value.

n) nitrobenzene: possibly due to triplets.

o) cyanogen: see Sec.L.6.

J.4. The relation between the PES, UV and EA results

It was pointed out in Sec.J.1. that since ε_i is obtained from PES and ε_a from EA and $\varepsilon_a - \varepsilon_i$ from UV, it is desirable that the calculations give agreement in all three cases. If there is disagreement between experiment and theory in one case and agreement in the other two, then we must conclude that the theoretical model is unable to describe nature. On the other hand, if there is disagreement in two cases by the same amount and in the same direction and agreement in the third, the model is probably acceptable and only the numerical values of some parameters require improvement.

In HAM/3 studies the agreement has appeared to be reasonable in all three cases for a number of molecules. Examples: ethylene, butadiene, N_2 (see above), cyclopentadiene, furan, pyrrole [24], naphthalene [39], p-benzoquinone [45], formamide [57], tetracyanoquinodimethane [59], 1,3-cyclohexadiene, 1,4-cyclohexadiene, norbornadiene, benzene, pyridine, pyrimidine, pyridazine, pyrazine, s-triazine, nitromethane [61], acetonitrile [61,62].

Some further examples are given in the Notes.

As far as we know, the simultaneous study of all three cases has up till now been performed only with HAM/3 and, for a few molecules, also with the PPP method [63,64]. Recently it has appeared that also the $X\alpha$ method is successful and a few molecules have been studied [88].

As an example of such a study, however performed without access to calculations, McDiarmid and Doering's study of 1,4-cyclohexadiene [65] can be mentioned. They discuss the UV spectrum, which is reproduced below together with the HAM/3 excitation energies.

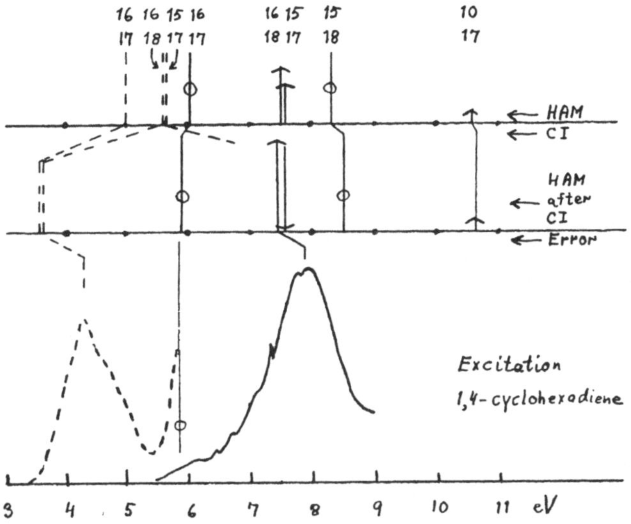

Both curves are obtained by use of electron impact energy loss, the singlet curve [65] using 100 eV and 2° and the triplet curve [66] using 50 eV and 80°.

In the HAM/3 calculation we have first studied the PES and EA and corrected the eigenvalues a little. These eigenvalues have then been used in the UV calculation. We see that the agreement is reasonable with one strong allowed and two forbidden transitions.

It has, however, been claimed by Jordan et al.[67,7] (see also ref. [43]), that the order of the π^* orbitals should be reversed. The consequence would then be that the two forbidden transitions at 5.8 eV and 8.5 eV would be allowed instead. We would thus expect two strong transitions at these energies instead, which would disagree completely with experiment. The order of the π^* orbitals in ref. [65] and HAM/3 and earlier work [68] is therefore certainly correct.

J.5. Other calculations of electron affinities

In the same way as eq.(H.8) was used in Hartree-Fock calculations of ionization energies we can obtain electron affinities according to Koopmans' theorem using

$$E_a^{\alpha} = - \varepsilon_a^{\alpha} \qquad (J.4)$$

Use of eq. (J.4) means in the same way that both correlation and reorganization are neglected. In calculations of electron affinities, unfortunately, both errors have the same sign and do not compensate each other.

For planar π-systems it has, however, been possible to find a purely empirical formula to correct both errors simultaneously (Younkin, Smith and Compton [69])

$$EA_a \approx -0.95\ \varepsilon_a - 1.9 \quad eV \qquad (J.5)$$

This formula gives very reasonable results for such molecules in the same way as Koopmans' theorem can be corrected for ionization energies (see Sec.H.2.). The success of this method depends probably upon the fact that the orbital energies are obtained from a PPP-type SCF method.

When ab-initio SCF calculations are used, the difficulties, which are caused by the fact that correlation error and reorganization error add, become more obvious. In two review articles on molecular anions it is pointed out by Radom [70] and Simons [71] that the correlation energy change approximately equals the electron affinity itself. Therefore, in these papers no results for any molecules in Table J.3. are given.

One can try to introduce corrections, but they become very large and probably uncertain. In a study of norbornadiene [72] the STO-3G orbital energies were corrected

$$EA_a \approx - {}^{STO}\varepsilon_a \quad + \quad \text{about } 7\ eV$$

The correction term, which has to compensate for both correlation and reorganization and perhaps also for other errors, is thus very large. It is in this study assumed to be constant.

In a later study [18] of cyanogen, butadiene and butadiyne the theoretical splitting between two EA's is much larger than the experimental. This proves that the correction term is not a constant.

A further illustration to the difficulties is presented in a recent study [31] of two small molecules: propyne with experimental EA = -2.8 eV and allene with EA = -1.9 eV. The difference of the experimental EA's is thus -0.9 eV, but the corresponding difference from an ab-initio calculation was +3.05 eV. A larger basis set gave +0.15 eV. HAM/3 gives -0.7 eV.

It is, on the other hand, clear that methods such as RSPT and Green's functions, which have been very successful for calculating IP's, must be useful also for EA's due to the similarity of the two phenomena [73]. Unfortunately, it appears to be necessary to use a still larger basis set to obtain reliable EA's than are needed for calculating IP's [73]. The consequence has been that calculations for only a few small molecules have been made.

Of all molecules in Table J.3. it is probable that only two have been studied in ab-initio work: CO_2 for which EA was calculated as -5.14 eV [74] and C_2 for which good EA's were obtained [73, 83].

$X\alpha$ methods have been more successful and several molecules have been studied [88, 89].

The Green's function method has, on the other hand, been very successful for calculation of the vibrational structure in the electron affinity spectrum [75].

A less difficult problem concerns the electron affinity of radicals, which have not been included in Table J.3. Experimental work has been performed by Brauman [76] using cyclotron resonance methods for many organic radicals (which generally have positive electron affinities) and theoretical studies have been made for these closed shell anions.

J.6. Unoccupied σ^* orbitals

When HAM/3 was parametrized in 1976, nothing was known about the energies of unoccupied σ^* orbitals. The parameters, which determine the corresponding eigenvalues, have therefore been determined from the energies of occupied orbitals. From for instance PES of ethylene this can be done, since orbital 2 in ethylene is 2s-2s antibonding and $2p\sigma$-$2p\sigma$ antibonding. We expect therefore that the calculated eigenvalues of unoccupied σ^* orbitals in general are reasonably useful.

An exception concerns bonds with hydrogen, especially C-H bonds, since there are no contributions from σ^*_{CH} in any occupied orbitals in any molecules. The parametrization in HAM/3 of parameters determining σ^*_{CH} had therefore to be performed without use of any experimental results, and it gives therefore eigenvalues for σ^*_{CH} which lack meaning. They are usually much too high. and are denoted l.m. in Table J.3.

It seems probable today that the EA of the σ^*_{CH} orbital is of the order of -6 or -8 eV. This conclusion has been drawn from studies of several molecules (C_2H_4, CH_4, C_2H_2, diacetylene and p-benzoquinone) [30] . The apparatus in Sec.J.2.g. was used in the first studies.

Ethylene has been studied by Walker, Stamatovic and Wong [19] with the following result.

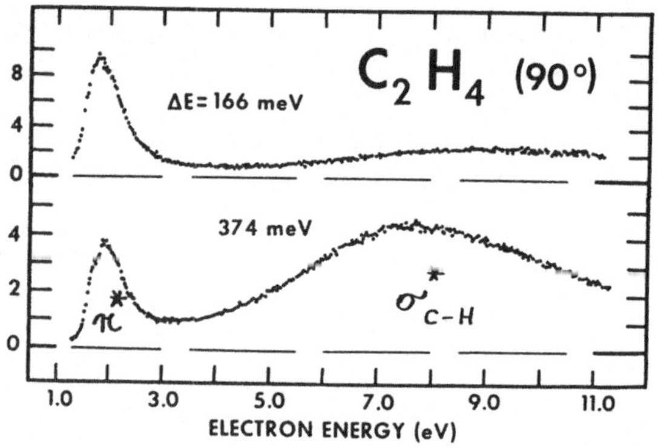

The change of vibrational quantum number, mentioned in Sec.J.2.d., refers in the π^* band mainly to the C-C vibration (vibrational energy 0.166 eV) but in the broad band mainly to the C-H vibration (0.374 eV). This indicates that the broad band is formed when the extra electron enters one of the antibonding C-H orbitals a_g, b_{1u}, b_{2u} or b_{3g}. The antibonding C-C orbital b_{1u} has according to Tronc so high energy, that it does not mix with the hydrogens. The HAM/3 orbital 11 is thus incorrect and should have much smaller hydrogen contributions. Inspection of the broad band indicates therefore that the four σ^*_{CH} EA's could be in order: -5 eV, -7 eV, -9 eV and -11 eV, as suggested in Table J.3.

This interpretation agrees with the results from an $X\alpha$ calculation [77] which shows broad bands at about these energies. Nothing is known, however, about the breadths of the individual bands, and if they are sufficiently narrow, it might be possible to resolve them by improved experimental methods. It will be shown below (Sec.K.4.), that core-excitation is probably such a method.

In most molecules the study of σ^*_{CH} energies is complicated by π^* orbitals and heteroatoms. The ideal molecule for such studies seems therefore neopentane, $C(CH_3)_4$, to be, in which there are eleven σ^*_{CH} orbitals (two t_2, one t_1 and one e).

A transmission study [35] has indicated EA = -6.1 eV.

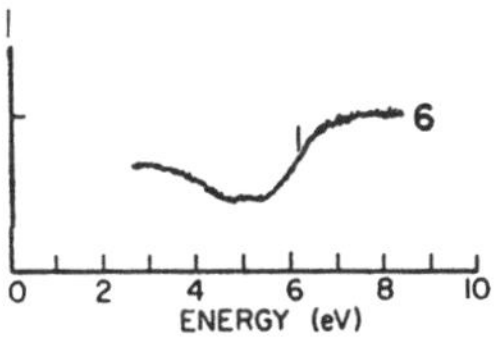

It is possible that this can explain the main part of the UV spectrum of neopentane [78] which has been studied using synchrotron radiation.

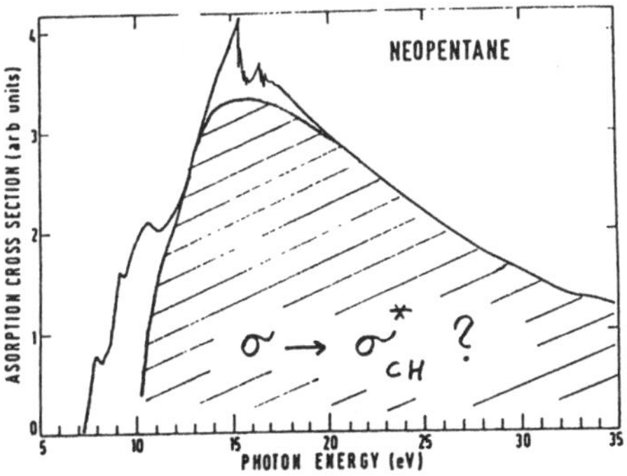

In the spectrum the unscratched parts are known to be of Rydberg type and therefore the scratched area must correspond to valence transitions.

The valence excitation energies can be estimated from a HAM/3 calculation with one half electron diffusely added. The excitation energies are then $(6.1 - \varepsilon_i)$ eV (we neglect K_{ia}). This gives excitation energies about 12.5 eV, 14.2 eV, 14.9 eV, 16.1 eV, 19.0 eV, 22.7 eV and 26.3 eV in reasonable agreement with the UV spectrum.

Since many small hydrocarbons have similar UV spectra [79], it is probable that the explanation is the same for them.

These excitations can also be seen in ESCA spectra [84]. When ethylene is bombarded with X-rays (1254 eV), the C 1s line is obtained an 290.7 eV. At 8.4 eV higher energy (299.1 eV) a weak band is seen, in which the 1s ionization and a $^1(\pi\pi^*)$ excitation occur simultaneously [85,86]. Around 18 eV (309 eV) there is a broad band, rather similar to the neopentane excitation above, and therefore probably due to simultaneous ionization of 1s and excitation of $\sigma \rightarrow \sigma^*_{CH}$. In propane there is no band at 8.4 eV, only a broad band around 18 eV.

References

1. L. Åsbrink, C. Fridh, E. Lindholm, S. de Bruijn and D.P. Chong, Phys.Scripta 22, 475 (1980).
2. E.H. van Veen and F.L. Plantenga, Chem.Phys.Letters 38, 493 (1976).
3. E.H. van Veen, Chem.Phys.Letters 41, 540 (1976).
4. L. Sanche and G.J. Schulz, Phys.Rev. A5, 1672 (1972).
5. A. Stamatovic and G.J. Schulz, Rev.Sci.Instr. 39, 1752 (1968) and 41, 423 (1969).
6. D. Mathur and J.B. Hasted, Chem.Phys. 16, 347 (1976).
7. K.D. Jordan and P.D. Burrow, Acc.Chem.Res. 11, 341 (1978).
8. to be published.
9. J.A.R. Samson, Phys.Reports 28, 303 (1976).
10. L. Sjöberg and T.E. Eriksen, Radiochem.Radioanal.Letters 35, 275 (1978).
11. L. Sjöberg and T.E. Eriksen, J.C.S.Faraday I, 76, 1402 (1980).
12. I. Nenner and G.J. Schulz, J.Chem.Phys. 62, 1747 (1975).
13. G. Caldwell and P. Kebarle, J.Chem.Phys. 80, 577 (1984).
14. M.J.W. Boness and G.J. Schulz, Phys.Rev. A9, 1969 (1974).
15. M. Tronc, R. Azria and Y. LeCoat, J.Phys.B: Atom.Molec. Phys. 13, 2327 (1980).
16. M. Tronc, R. Azria and R. Paineau, J.Physique Lettres 40, L-323 (1979).
17. P.D. Burrow and K.D. Jordan, Chem.Phys.Letters 36, 594 (1975).
18. L. Ng, V. Balaji and K.D. Jordan, Chem.Phys.Letters 101, 171 (1983).
19. I.C. Walker, A. Stamatovic and S.F. Wong, J.Chem.Phys. 69, 5532 (1978).
20. M. Tronc, private communication.
21. K.D. Jordan and P.D. Burrow, J.Am.Chem.Soc. 102, 6882 (1980).
22. S.W. Staley, J.C. Giordan and J.H. Moore, J.Am.Chem.Soc. 103, 3638 (1981).
23. J.C. Giordan, M.R. McMillan, J.H. Moore and S.W. Stuart, J.Am.Chem.Soc. 102, 4870 (1980).
24. L. Åsbrink, C. Fridh and E. Lindholm, J.Electron Spectrosc. 16, 65 (1979).
25. S.W. Staley M.D. Bjorke, J.C, Giordan, M.R. McMillan and J.H. Moore, J.Am.Chem.Soc. 103, 7057 (1981).
26. Unpublished results by L. Åsbrink in a transmission machine.
27. K.H. Kochem, W. Sohn, K. Jung, H. Erhardt and E.S. Chang, J.Phys.B: Atom.Molec.Phys. 18, 1253 (1985).

References (cont.)

28. L. Andric and R.I. Hall, Contributed Papers, Symposium on the Physics of Ionized Gases, Dubrovnik 1982.
29. L. Ng, K.D. Jordan, A. Krebs and W. Rüger. J.Am.Chem.Soc. 104, 7414 (1982).
30. M. Allan, Chem.Phys. 86, 303 (1984).
31. B. Ciommer, K.M. Nguyen, H. Schwarz, G. Frenking, G. Kwiatkowski and E. Illenberger, Chem.Phys.Letters 104, 216 (1984).
32. D. Mathur, J.Phys.B: Atom.Molec.Phys. 13, 4703 (1980).
33. K. Rohr, J.Phys.B: Atom.Molec.Phys. 13, 4897 (1980).
34. H. Tanaka, M. Kubo, N. Onodera and A. Suzuki, J.Phys.B: Atom.Molec.Phys. 16, 2861 (1983).
35. J.C. Giordan and J.H. Moore, J.Am.Chem.Soc. 105, 6541 (1983).
36. P.D. Burrow and K.D. Jordan, J.Am.Chem.Soc. 104, 5247 (1982).
37. A. Modelli, D. Jones, F.P. Colonna and G. Distefano, Chem.Phys. 77, 153 (1983).
38. K.D. Jordan and P.D. Burrow, Chem.Phys. 45, 171 (1980).
39. L. Åsbrink, C. Fridh and E. Lindholm, Z. Naturforsch. 33a, 172 (1977).
40. P.D. Burrow, private information.
41. M. Allan, private information.
42. A. Modelli, private information.
43. A. Modelli, G. Distefano and D. Jones, Chem.Phys. 82, 489 (1983).
44. I. Cadez, F. Grestau, M. Tronc and R.I. Hall, J.Phys.B: Atom.Molec.Phys. 10, 3821 (1977).
45. L. Åsbrink, G. Bieri, C. Fridh, E. Lindholm and D.P. Chong, Chem.Phys. 43, 189 (1979).
46. M. Allan, Chem.Phys.81, 235 (1983).
47. M. Allan, Chem.Phys. 84, 311 (1984).
48. A. Modelli, G. Distefano and D. Jones, Chem.Phys. 73, 395 (1982).
49. N.S.Chiu, P.D. Burrow and K.D. Jordan, Chem.Phys.Letters 68, 121 (1979).
50. M.N. Paddon-Row, N.G. Rondan, K.N. Houk and K.D. Jordan, J.Am.Chem.Soc. 104, 1143 (1982).

References (cont.)

51. J.R. Frazier, L.G. Christophorou, J.G. Carter and H.C. Schweinler, J.Am.Chem.Soc. 69, 3807 (1978).

52. K.D. Jordan and P.D. Burrow, J.Chem.Phys. 71, 5384 (1979).

53. K. Rohr, J.Phys.B: Atom.Molec.Phys. 10, 2215 (1977).

54. M. Tronc, L. Malegat and R. Azria, to be published.

55. R.N. Compton, P.W. Reinhardt and C.D. Cooper, J.Chem.Phys. 68, 4360 (1979).

56. R.J. Celotta, R.A. Bennett and J.L. Hall, J.Chem.Phys. 60, 1740 (1974).

57. E. Lindholm, G. Bieri, L. Åsbrink and C. Fridh, Int.J. Quantum Chem. 14, 737 (1978).

58. E.C.M. Chen and W.E. Wentworth, J.Chem.Phys. 63, 3183 (1975).

59. L. Åsbrink, C. Fridh and E. Lindholm, Int.J.Quantum Chem. 13, 331 (1978).

60. P.D. Burrow, cited in ref. [18].

61. Unpublished work by E. Lindholm and L. Åsbrink.

62. C. Fridh, J.Chem.Soc.Faraday II, 74, 2193 (1978).

63. R. Pariser and R.G. Parr, J.Chem.Phys. 21, 767 (1953).

64. N.S. Hush and J.A. Pople, Trans.Faraday Soc. 51, 600 (1955).

65. R. McDiarmid and J.P. Doering, J.Chem.Phys. 75, 2687 (1981).

66. A. Kuppermann, W.M. Flicker and O.A. Mosher, Chem.Rev. 79, 77 (1979).

67. K.D. Jordan, J.A. Michejda and P.D. Burrow, Chem.Phys.Letters 42, 227 (1976).

68. R. Hoffmann, E. Heilbronner and R. Gleiter, J.Am.Chem.Soc. 92, 706 (1970).

69. J.M. Younkin, L.J. Smith and R.N. Compton, Theoret.Chim.Acta 41, 157 (1976).

70. L. Radom, in Applications of Electronic Structure Theory, H.F. Schaefer III (ed.), Plenum Press, New York (1977), p.333.

71. J. Simons, in Theoretical Chemistry, Advances and Perspectives, H. Eyring and D. Henderson (eds.), Vol.3, Academic Press, New York (1978), p.1.

References (cont.)

72. K.N. Houk, N.G. Rondan, M.N. Paddon-Row, C.W. Jefford, P.T. Huy, P.D. Burrow and K.D. Jordan, J.Am.Chem.Soc. 105, 5563 (1983).

73. L.S. Cederbaum, W. Domcke and W. von Niessen, J.Phys.B: Atom. Molec.Phys. 10, 2963 (1977).

74. P.J. Bruna, S.D. Peyerimhoff and R.J. Buenker, Chem.Phys. Letters 39, 211 (1976).

75. L.S. Cederbaum and W. Domcke, Z.Phys. A277, 221 (1976).

76. B.K. Janousek and J.I. Brauman, in Gas-Phase Ion Chemistry, Vol.2, M.T. Bowers (ed.), Academic Press, New York (1979), p.53.

77. J.E. Bloor, R.E. Sherrod and F.A. Grimm, Chem.Phys.Letters 78, 351 (1981).

78. E.E. Koch, V. Saile and N. Schwentner, Chem.Phys.Letters 33, 322 (1975).

79. E.E. Koch and M. Skibowski, Chem.Phys.Letters 9, 429 (1971).

80. M.H. Palmer, W. Moyes, M. Spiers and J.N.A. Ridyard, J.Mol. Structure, 55, 243 (1979).

81. S. Nagakura, M. Kojima and Y. Maruyama, J.Mol.Spectrosc. 13, 174 (1964).

82. T. Shida and S. Iwata, J.Phys.Chem. 75,2591 (1971).

83. M. Zeitz, S.D. Peyerimhoff and R.J. Buenker, Chem.Phys.Letters 64, 243 (1979).

84. T.A. Carlson, W.B. Dress, F.A. Grimm and J.S. Haggerty, J.Electron Spectrosc. 10, 147 (1977).

85. M.S. Banna and D.A. Shirley, J.Electron Spectrosc. 8, 255 (1976).

86. A. Gardner, P.K. Mukherjee and D.P. Chong, J.Mol.Struct. THEOCHEM 108, 25 (1984).

87. G.J. Schulz, Rev.Mod.Phys. 45, 423 (1971).

88. J.A. Tossell, J.Phys.B: Atom.Molec.Phys. 18, 387 (1985).

89. J.E. Bloor, R.A. Paysen and R.E. Sherrod, Chem.Phys.Letters 60, 476 (1979).

K. Studies of 1s electrons

K.1. Calculation of 1s ionization (ESCA) energies in the HAM model

Ionization of one 1s electron is treated in the same way as ionization of one valence electron. As discussed in Sec.H.1. the ionization energy IP_{1s}^{α} is therefore the difference of two total energies

$$IP_{1s}^{\alpha} = E\left(q_{1s}^{\alpha} - 1\right) - E\left(q_{1s}^{\alpha}\right) \tag{K.1}$$

As an expression for the total energy E we use here the sum of the expressions (D.38) + (F.21) + (F.22).

The reason for our neglect of (F.20), (F.23) and (F.24), which were included in the total energy in Sec.F.3., is that the 1s electrons interact very little with the other electrons in the molecule. A common procedure is to include the 1s electrons in the charge of the nucleus, which then is changed from Z to $Z-2$. Calculations of valence electrons with such nuclear charges have usually given good results. We will therefore assume here that we can take $P_{1s\nu} = 0$. The expressions (F.20), (F.23) and (F.24) vanish then.

We assume further that before the ionization has taken place we have two 1s electrons in atom A. We have thus $q_{1s}^{\alpha} = 1$ and $q_{1s}^{\beta} = 1$. We assume also that before the ionization has taken place the valence orbitals are independent of spin: $\zeta_{\nu}^{\alpha} = \zeta_{\nu}^{\beta}$ and $\sigma_{1s\nu}^{\alpha\alpha} = \sigma_{1s\nu}^{\alpha\beta}$.

We obtain then directly the following expression for the ESCA energy of a molecule, in which atom A is ionized.

$$IP_{1s}^{\alpha} = \frac{1}{2}\left[Z_A - \sum_{\nu}^{\text{valence}} N_{\nu}\,\sigma_{\nu 1s} - 2\,\sigma_{1s1s}^{\alpha\beta}\right]^2 - \left[\sigma_{1s1s}^{\alpha\beta}\right]^2 - \tag{K.2}$$

$$-\sum_{\nu}^{\text{valence}} N_{\nu}\cdot\frac{1}{n_{\nu}}\cdot\left[\zeta_{\nu}\,\sigma_{1s\nu} + \frac{1}{2}\,\sigma_{1s\nu}^{2}\right] + \tag{K.3}$$

$$+\sum_{B}\left(Q_B + \frac{\mu_{tB}}{R_{AB}}\right)\cdot\frac{1}{R_{AB}} \tag{K.4}$$

In eqs. (K.2) and (K.3) the summation is over valence orbitals on atom A only. In eq. (K.4) we have replaced γ_{AB} and V_{AB} by R_{AB}^{-1} since the 1s orbital is small. Q_B is the atomic charge (see eq. (A.12)). ζ_{ν} is the orbital exponent before ionization.

The ESCA energy of an atom

Eqs.(K.2) and (K.3) describe the 1s ionization energy of an atom. We can easily estimate this energy for a carbon atom in the following way.

For carbon $1s^2 2s^2 2p^2$ we have $Z_A = 6$ and $\sum_{\nu}^{valence} N_\nu = 4$. We use the shielding efficiencies, proposed by Slater (see Sec.E.4.), and take also his orbital exponents $\zeta_{2s} = \zeta_{2p} = 1.625$. This gives

$$IP_{1s} = \left\{ \frac{1}{2} \cdot 5.4^2 - 0.3^2 - 4 \cdot \frac{1}{2} \cdot \left[1.625 \cdot 0.85 + \frac{1}{2} \cdot 0.85^2 \right] \right\} \cdot 27.2 =$$

$$= 299.3 \text{ eV}$$

to be compared with the experimental value, 296.49 eV, given in Sec.E.5. (page E.15.).

Use of better shielding efficiencies will of course give better agreement with experiment. The HAM/4 work by Åsbrink [1] (see Sec.E.10.) gave this energy as (2 388 991.0 + 4 858.4) cm^{-1}, which means 2 393 849.4 · 12 398.520 · 10^{-8} = 296.80 eV.

The ESCA energy of a molecule: the potential model

Also in a molecule we can treat eqs.(K.2-4) in an approximate way. We will study a carbon atom in the molecule.

We observe first that (K.2) will not be changed much upon molecule formation since $\sigma_{\nu 1s}$ is very small. A change of N_ν will therefore change (K.2) only little and we consider it therefore as a constant.

In (K.3) we take Slater's values for $\sigma_{1s\nu}$ and ζ_ν and obtain

$$(K.3) = -(Z_A - 2 - Q_A) \cdot \frac{1}{2} \cdot \left[1.625 \cdot 0.85 + \frac{1}{2} \cdot 0.85^2 \right] =$$

$$= Q_A \cdot 0.87 - (Z_A - 2) \cdot 0.87 \quad (\text{a.u.})$$

$$= Q_A \cdot 23.7 - \text{constant} \quad (\text{eV})$$

If we finally neglect the local dipole term we find

$$IP_{1s} = \text{constant} + 23.7\ Q_A + \sum_B Q_B / R_{AB} \qquad (K.5)$$

Since the charge Q_A on atom A is different for different molecules, we get a "chemical shift" which can be used in "Electron Spectroscopy for Chemical Analysis" [2,3].

The expression (K.5) is called the "potential model formula". It was first deduced in an intuitive, instructive way ([3] page 108). (For later work see ref. [4] page 343.)

Our crude estimate, 23.7 eV, for the coefficient in eq.(K.5) agrees well with the experimental value 23.5 eV ([4] page 345), however obtained by means of charges obtained from CNDO calculations.

The ESCA energy of a molecule: the HAM calculation

In the HAM/3 computer program [5] the ESCA energy is calculated by use of eq.(K.1). Two SCF calculations are therefore necessary.

First, the neutral molecule is calculated. The print-out gives then the total energy of the molecule in eV. The reference for the energy has been chosen so that the printed energy means "heat of formation".

Secondly, the ionized molecule is calculated. To remove one electron from the 1s orbital in for instance the atom with number 3 in the molecule, the codeword

ESCA 3.

is used. The print-out gives then the heat of formation.

Finally, the difference of the two energies gives the ESCA energy.

It is obvious that this ΔE_{SCF} calculation (see Sec.H.2.) takes into account all reorganization energy. The correlation energy change is handled by the shielding efficiencies.

K.2. Experimental methods in ESCA

The figure shows an ESCA instrument, completed 1972 in Uppsala by K. Siegbahn and coworkers. An X-ray line is always broad (≈ 0.8 eV) due to the short lifetime of the excited state. The breadth is therefore reduced by spectroscopical analysis.

The photoelectrons are analyzed by electrostatic deflection.

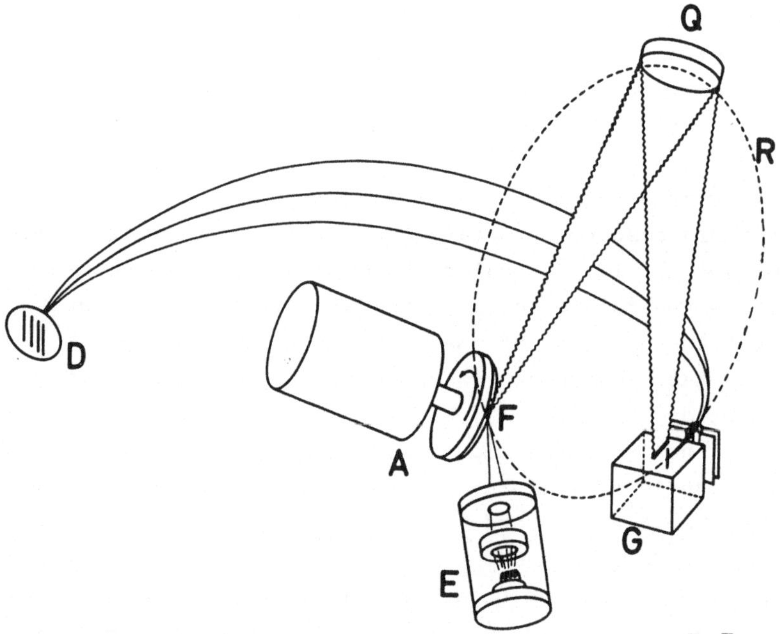

[6]

E = electron gun

A and F = rotating anode

Q = spherically bent quartz crystal

R = Rowland circle

G = chamber, where the gas is ionized

D = detector

K.3. ESCA energies: some results

The ionization of an 1s electron is an atomic and not a molecular process. In for instance N_2 the ionized electron comes from one of the nitrogens, and afterwards N_2^+ is therefore unsymmetric.

It has been shown, that the process in which $\frac{1}{2}$ electron goes away from each nitrogen, requires much more (7 eV) energy.

According to the potential model formula the ionization energy depends upon the situation of the atom in the molecule.

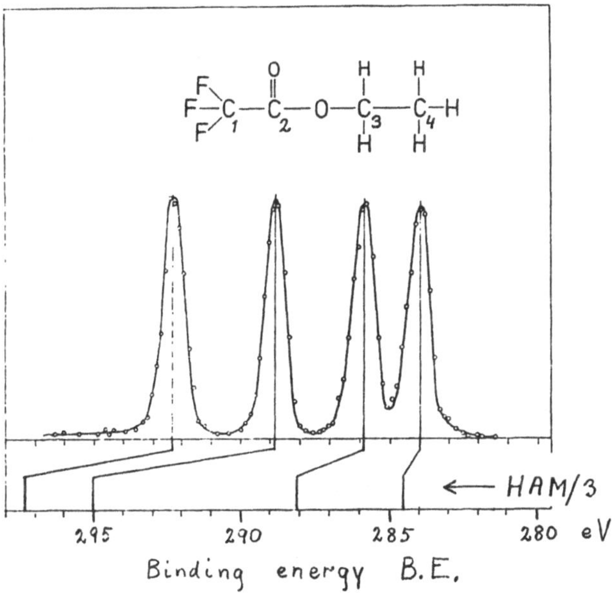

[6]

The HAM/3 calculation prints always the heat of formation, ΔH_f (see Sect. L.2.), for the molecule or the ion which is studied in the SCF calculation. Five calculations are necessary for this molecule:

1. No codeword gives for the molecule ΔH_f = -6.65 eV
2. Codeword ESCA 1. gives ΔH_f = +290.73 and B.E. = 297.38 eV
3. " " 2. = +288.39 = 295.04 eV
4. " " 3. = +281.44 = 288.09 eV
5. " " 4. = +277.93 = 284.58 eV

These results are shown in the figure. It is seen that the calculation exaggerates the chemical shift. One reason is certainly that the local dipoles on O and F are neglected in the HAM/3 program.

The ionization energies of 1s electrons from molecules in gas phase have recently been tabulated [7].

Calculation of ESCA energies using HAM/3 has been performed by Chong [8] for carbonsuboxide O = C = C = C = O

Table K.3.a. Vertical ionization potentials (in eV) of C_3O_2

MO	HAM/3	Experiment PES (12)	Ref. 14	XPS (13)	CNDO/2 (15)	INDO (11)	Xα (16)	*Ab initio* (13)
$2\pi_u$	10.9	10.7	10.75	10.8	12.42	9.57	13.5	11.1
$1\pi_g$	15.2	14.8	14.8	15.0*	20.09	16.2	16.9	17.7
$1\pi_u$	15.8	15.9	17.5	16.0*	24.93	17.6	17.4	18.7
$5\sigma_u$	17.3	17.0	15.8	17.3*		15.7	20.4	20.3
$6\sigma_g$	17.3	17.3	17.0	17.5*		17.1	20.6	20.6
$4\sigma_u$	20.7	19		21.9		23.9	24.0	26.1
$5\sigma_g$	25.5			25.6		33.6	25.8	30.7
$3\sigma_u$	35.8			35.5		47.1	32.4	41.8
$4\sigma_g$	36.0			35.5		46.7	32.7	41.8
C_2K	289.2			291.5			306.0	307.3
C_1K	292.2			294.9			309.4	312.3
OK	538.1			539.7			536.4	562.6

*Best fit.

and for some other molecules [9]

Table K.3.b. Core-electron binding energies (in eV) of some typical molecules

Molecule	HAM/3	Experiment c	Experiment Others
$\underline{C}O$	294.78 a	295.9	296.1 c,d
$\underline{C}O_2$	296.04 a	297.5	297.69 d
$O\underline{C}CCO$	292.17 b		294.9 e
$OC\underline{C}CO$	289.23 b		291.5 e
$\underline{C}_6H_6$	286.18 a		290.2 f
$\underline{C}_6H_5NO_2$	287.01-289.99		291.0-292.1 g
$\underline{N}_2$	406.74	409.9	409.93 d
$\underline{N}NO$	406.32	408.5	
$N\underline{N}O$	409.56	412.5	
$C_6H_5\underline{N}O_2$	409.43		411.5 g
$C\underline{O}$	539.32 a	542.1	
$C\underline{O}_2$	539.05 a	540.8	541.28 d
$C_3\underline{O}_2$	538.11 b		539.7 e
$H_2\underline{O}$	536.39	539.7	
$C_6H_5N\underline{O}_2$	538.02		538.3 g

a Åsbrink [59]. b Chong [19].
c Siegbahn et al [54]. d Johansson et al [60].
e Gelius et al [61]. f Gelius et al [62].
g Ohta et al [63].

K.4. Excitation of 1s electrons studied in electron impact

An 1s electron may be excited to an empty orbital in the molecule of π^* or σ^* or Rydberg type. The excitation energy is large, in carbon about 300 eV. The energy can be supplied using electron impact [10] or photon impact (see Sec.K.5.). In the first case the apparatus is shown in Sec.I.7.e.

Excitation to π^* in ethylene

The excitation of an 1s electron to π^* produces an intense and narrow peak in all molecules. In e.g. ethylene its energy

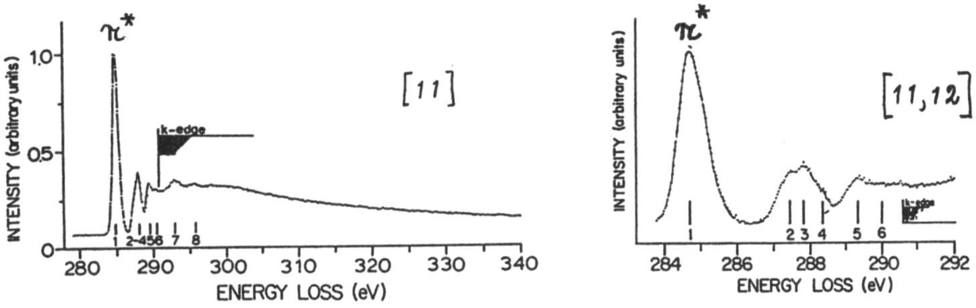

Excitation of carbon 1s in ethylene. To the right details.

is 284.7 eV. Since the energy required to excite the electron to infinity (K-edge) is 290.6 eV in ethylene, the experimental electron affinity of π^* in the ESCA-ionized molecule is +5.9 eV.

To calculate this we remove one 1s electron from the carbon atom C1 (codeword ESCA 1.) and calculate the electron affinity by adding 1/2 electron (codeword EA). The negative eigenvalue of orbital 7 gives then EA as +4.75 eV.

ONE 1S ELECTRON REMOVED FROM ATOM 1 ONE HALF ELECTRON DIFFUSELY ADDED.

			1	2	3	4	5	6	7	8	9	10	11	12
			-27.364	-21.410	-18.942	-16.477	-13.680	-12.479	-4.750	17.122	25.333	27.292	35.190	41.644
1	C	1	0.680	0.197	0.0	0.133	0.0	0.0	0.0	0.112	0.678	0.0	-0.094	0.0
2	C	1	-0.044	0.521	0.0	-0.517	0.0	0.0	0.0	0.285	-0.135	0.0	-0.600	0.0
3	C	1	0.0	0.0	-0.729	0.0	-0.250	0.0	0.0	0.0	0.0	-0.282	0.0	-0.572
4	C	1	0.0	0.0	0.0	0.0	0.0	0.795	0.607	0.0	0.0	0.0	0.0	0.0
5	C	2	0.412	-0.541	0.0	-0.098	0.0	0.0	0.0	0.637	-0.347	0.0	-0.036	0.0
6	C	2	0.206	0.083	0.0	0.677	0.0	0.0	0.0	-0.209	-0.404	0.0	-0.534	0.0
7	C	2	0.0	0.0	-0.300	0.0	0.707	0.0	0.0	0.0	0.0	-0.542	0.0	0.342
8	C	2	0.0	0.0	0.0	0.0	0.0	0.607	-0.795	0.0	0.0	0.0	0.0	0.0
9	H	3	0.365	0.286	-0.409	-0.144	-0.191	0.0	0.0	-0.184	-0.344	0.274	0.335	0.470
10	H	4	0.365	0.286	0.409	-0.144	0.191	0.0	0.0	-0.184	-0.344	-0.274	0.335	-0.470
11	H	5	0.168	-0.337	-0.147	-0.321	0.427	0.0	0.0	-0.441	0.031	0.488	-0.245	-0.239
12	H	6	0.168	-0.337	0.147	-0.321	-0.427	0.0	0.0	-0.441	0.031	-0.488	-0.245	0.239

Singlet-triplet split

The band at 284.7 eV shows the singlet excitation. The triplet can then be expected at a lower energy. Our EA calculation refers to the average of triplet and singlet. To calculate the observed singlet we must therefore subtract $K_{1s\pi^*}$ from our EA.

To calculate $K_{1s\pi^*}$ we assume that the 1s electron in atom A is excited to π^*. With eq.(I.23) we find

$$K_{1s\pi^*} \approx c_{\mu 1s}\, c_{\nu\pi^*}\, c_{\nu\pi^*}\, c_{\mu 1s}\, \left(1s_A\, z_A \middle| z_A\, 1s_A\right) = \tag{K.6}$$

$$= 1 \cdot \left(c_{\nu\pi^*}\right)^2 \cdot 1 \cdot \tfrac{1}{3}\, G^1(1s,2p) \tag{K.7}$$

where we have assumed that the 1s orbital is localized on A and that eq.(I.28) can be used.

Slater [13] has given $G^1(1s,2p)$ for nitrogen and oxygen. Since these two values are approximately equal, we will use the nitrogen value for C, N and O and find

$$K_{1s_A\pi^*} \approx \left(c_{\nu\pi^*}\right)^2 \cdot 0.8\ \text{eV} \tag{K.8}$$

For ethylene we have

$$K_{1s_A\pi^*} \approx (0.607)^2 \cdot 0.8 = 0.30\ \text{eV}$$

and the calculated EA for the singlet is 4.75 - 0.30 = 4.45 eV.

The difference between the experimental EA, +5.9 eV, and the calculated, +4.45 eV, is thus 1.45 eV. Since this difference is reasonably constant for several molecules (see below), we will ascribe it to a deficiency of our calculation. The charge in the ESCA-ionized molecule makes the orbital size smaller, which should influence all interactions. In spite of this, the HAM program calculates the overlaps with unchanged orbital size. This will give a too small EA.

It is finally clear that the EA calculation here cannot be more accurate that the EA calculation in Chap.J. We expect thus random errors of the same magnitude as in Table J.3.

Excitation to π^* in CO and N_2

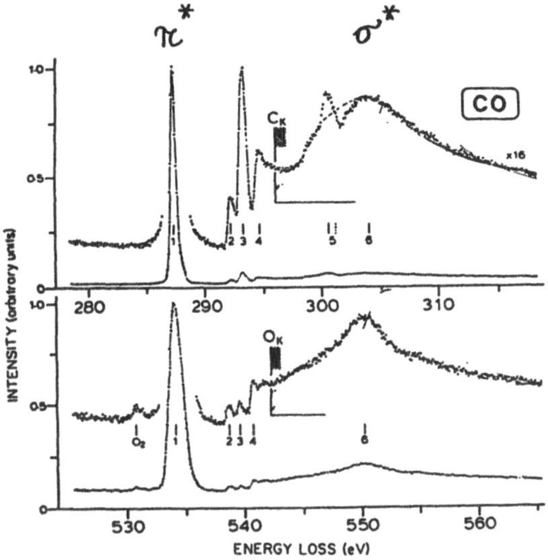

[14]

Excitation of carbon 1s and oxygen 1s, respectively, in CO.

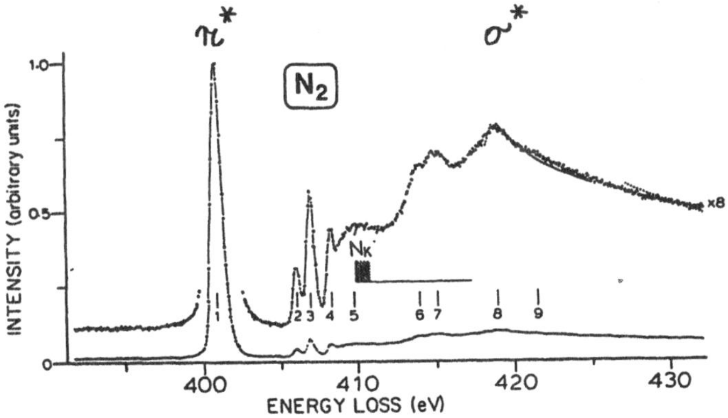

[14]

Excitation of nitrogen 1s in N_2

In CO and N_2 also the triplets have been observed by use of impact of electrons with lower velocity [15, 16] . This enables us to check eq. (K.8):

$\underline{C}O$: $K_{1s_C \pi^*} = (0.786)^2 \cdot 0.8 = 0.49$ eV. Exp: split/2 = 0.73 eV.

N_2: $(0.626)^2 \cdot 0.8 = 0.31$ eV. Exp: 0.41 eV.

The calculated excitation energies differ from the experimental values by about 1 eV in both cases. A recent ab-initio study [21] achieves a similar agreement.

Excitation to π^* in benzene

In benzene there are several π^* orbitals with different energies. We expect therefore several bands in the core-excitation spectrum.

ONE 1S ELECTRON REMOVED FROM ATOM 1 ONE HALF ELECTRON DIFFUSELY ADDED.

	12	13	14	15	16	17	18	19	20	21	22	23
	-12.484	-12.427	-10.278	-9.838	-4.503	-4.273	-0.328	14.544	15.407	16.581	17.729	18.610
1 C 1	-0.025	0.0	0.0	0.0	0.0	0.0	0.0	0.015	0.0	0.197	0.172	0.0
2 C 1	-0.000	-0.183	0.0	0.0	0.0	0.0	0.0	0.0	0.193	0.0	0.0	0.432
3 C 1	0.127	-0.000	0.0	0.0	0.0	0.0	0.0	0.201	0.0	-0.063	-0.125	0.0
4 C 1	0.0	0.0	-0.364	-0.000	-0.490	0.000	0.311	0.0	0.0	0.0	0.0	0.0
5 C 2	0.065	0.016	0.0	0.0	0.0	0.0	0.0	0.236	-0.390	-0.180	0.138	-0.227
6 C 2	-0.027	0.189	0.0	0.0	0.0	0.0	0.0	0.127	0.058	0.080	0.156	0.185
7 C 2	-0.376	0.061	0.0	0.0	0.0	0.0	0.0	0.140	0.216	0.124	-0.410	-0.111
8 C 2	0.0	0.0	-0.007	0.539	0.401	0.458	-0.390	0.0	0.0	0.0	0.0	0.0
9 C 3	-0.026	0.039	0.0	0.0	0.0	0.0	0.0	0.212	0.403	-0.125	-0.328	-0.157
10 C 3	0.069	-0.451	0.0	0.0	0.0	0.0	0.0	0.157	-0.106	0.317	0.055	0.253
11 C 3	0.384	-0.107	0.0	0.0	0.0	0.0	0.0	-0.135	0.219	0.327	-0.308	0.237
12 C 3	0.0	0.0	0.467	0.458	0.223	-0.539	0.439	0.0	0.0	0.0	0.0	0.0
13 C 4	-0.041	0.0	0.0	0.0	0.0	0.0	0.0	0.103	0.0	0.572	0.251	0.0
14 C 4	0.000	0.536	0.0	0.0	0.0	0.0	0.0	0.0	-0.283	0.0	0.0	0.564
15 C 4	-0.452	0.000	0.0	0.0	0.0	0.0	0.0	-0.255	0.0	0.206	0.015	0.0
16 C 4	0.0	0.0	0.656	0.000	-0.582	0.000	-0.462	0.0	0.0	0.0	0.0	0.0
17 C 5	-0.027	-0.039	0.0	0.0	0.0	0.0	0.0	0.212	-0.403	-0.125	-0.328	0.157
18 C 5	-0.070	-0.451	0.0	0.0	0.0	0.0	0.0	-0.157	-0.106	-0.317	-0.055	0.253
19 C 5	0.384	0.107	0.0	0.0	0.0	0.0	0.0	-0.135	-0.219	0.327	-0.308	-0.237
20 C 5	0.0	0.0	0.468	-0.458	0.224	0.538	0.439	0.0	0.0	0.0	0.0	0.0
21 C 6	0.065	-0.017	0.0	0.0	0.0	0.0	0.0	0.236	0.390	-0.180	0.138	0.227
22 C 6	0.027	0.189	0.0	0.0	0.0	0.0	0.0	-0.127	0.058	-0.079	-0.156	0.185
23 C 6	-0.376	-0.060	0.0	0.0	0.0	0.0	0.0	0.140	-0.216	0.124	-0.410	0.111
24 C 6	0.0	0.0	-0.007	-0.539	0.401	-0.459	-0.390	0.0	0.0	0.0	0.0	0.0
25 H 7	0.105	0.0	0.0	0.0	0.0	0.0	0.0	-0.205	0.0	-0.116	0.072	0.0
26 H 8	-0.128	0.160	0.0	0.0	0.0	0.0	0.0	-0.319	0.110	-0.030	-0.013	-0.034
27 H 9	-0.118	-0.240	0.0	0.0	0.0	0.0	0.0	-0.312	-0.075	0.012	0.018	-0.063
28 H 10	0.327	-0.000	0.0	0.0	0.0	0.0	0.0	-0.292	0.0	-0.136	-0.229	0.0
29 H 11	-0.118	0.240	0.0	0.0	0.0	0.0	0.0	-0.312	0.075	0.012	0.018	0.063
30 H 12	-0.128	-0.160	0.0	0.0	0.0	0.0	0.0	-0.319	-0.110	-0.030	-0.013	0.034
					EA = +4.503	no intensity	EA = +0.328	σ^*_{CH}	σ^*_{CC}	σ^*_{CC}	σ^*_{CC}	σ^*_{CC}

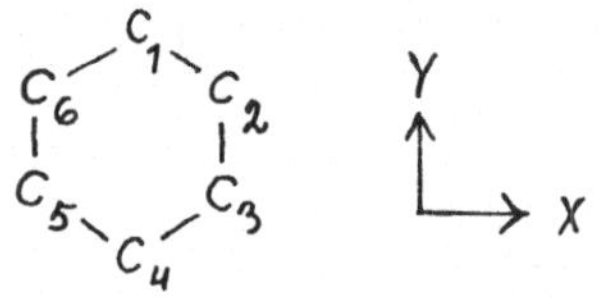

We observe first that ESCA ionization of carbon C1 removes the benzene symmetry. Orbitals 16 and 17 have therefore different energies.

The intensities are proportional to $(c_{\nu a})^2$ according to eqs.(I.37) and (I.35). Orbital 17 gives zero intensity and the intensity ratio of band 16 and band 18 is $(0.490)^2/(0.311)^2 = 2.5$.

Due to the small coefficients in the printout, the exchange integrals $K_{1s\pi^*}$ are small: 0.19 eV for band 16 and 0.08 eV for band 18. Often in a large molecule the exchange integrals may be neglected.

In the observed spectrum two bands with a slightly larger intensity ratio can be observed. The weaker band was previously interpreted as Rydberg [11].

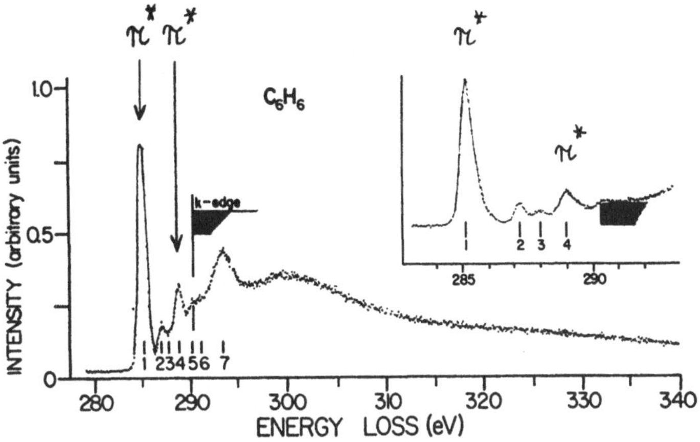

[11]

Excitation of carbon 1s in benzene.

Also in butadiene two π^* bands can be seen [17].

Comparison of results

In Table K.4.a. the results for different molecules are collected.

Table K.4.a. Electron affinities of ESCA-ionized molecules. The ESCA-ionized atom is underlined.

Molecule	Orbital	EA (exp.)	EA - K (HAM)	"Error"	Ref.
CO	π^*	+8.8	+8.3	+0.5	[14]
CO	π^*	+8.3	+7.0	+1.3	[14]
OCO	π^*	+6.8	+6.2	+0.6	[18]
OCO	π^*	+5.4	+5.1	+0.3	[18]
NN	π^*	+8.9	+7.8	+1.1	[14]
ethylene	π^*	+5.9	+4.5	+1.4	[11]
benzene	π^* orb.16	+5.1	+4.3	+0.8	[11]
	π^* orb.17		zero intensity		
	π^* orb.18	+1.4	+0.2	+1.2	

The reasons for the "error" have been discussed above.

It is remarkable that π^* can be observed in CO and CO_2 and other molecules, although in the UV spectra these transitions are hidden by strong Rydbergs and have never been identified.

Excitation to σ^*

Broad diffuse bands can be seen in all energy-loss spectra above the K-edge. They will be interpreted as due to excitation to σ^*.

Since no reliable calculations of σ^* energies have been performed, we will simply compare the energy difference between σ^* and π^* here with the corresponding energy difference in Chap.J. (Table J.3.).

Table K.4.b. Electron affinities of ESCA-ionized molecules.

Molecule	Orbital	EA (exp.)	$\pi^*\sigma^*$-difference here	Table J.3.
$\underline{C}O$	σ^*	-7.8	16.6	17.7
$C\underline{O}$	σ^*	-8.5	16.8	17.7
$O\underline{C}O$	σ_g^*		zero intensity	
	σ_u^*	-14.5	21.3	26
$\underline{O}CO$	$\sim\sigma_g^*$	0.0	5.4	7.0
	$\sim\sigma_u^*$	-17	22.4	26
$\underline{N}N$	σ^*	-9.0	17.9	19.8
ethylene	σ_{CH}^* orb.8	+3	2.9	3.2
	σ_{CH}^* orb.9	low	intensity	5.2
	σ_{CH}^* orb.10	-2.0	7.9	7.2
	σ_{CH}^* orb.12	-4.6	10.5	9.2
benzene	σ_{CH}^* orb.19	-3.2		
acetylene	σ_{CH}^*	+2.1		
	σ_{CH}^*	-4.5		

The similarity of the values in the last two columns in Table K.4.b. proves that the diffuse bands are of σ^* type.

In the ethylene spectrum above, the bands, denoted as "7" and "8", are above the ionization limit and therefore cannot be of Rydberg type. They correspond probably to the orbitals 10 and 12, which are of σ_{CH}^* type. Orbital 9 has a smaller 2p character on carbon C1. We expect therefore that this band has a low intensity and cannot be seen. On the other hand, orbital 8 should be observed. It could be band "5" which has correct energy.

In benzene there is a strong band, marked "7", which probably corresponds to the σ_{CH}^* orbital 19. The higher orbitals are mainly σ^*_{cc} according to the print-out and have therefore probably higher energies. We would expect that orbital 19 should appear in EA studies using the transmission method at about 9 eV, but it has not yet been observed.

In O<u>C</u>O no transition to σ^*_g can be seen. The explanation is

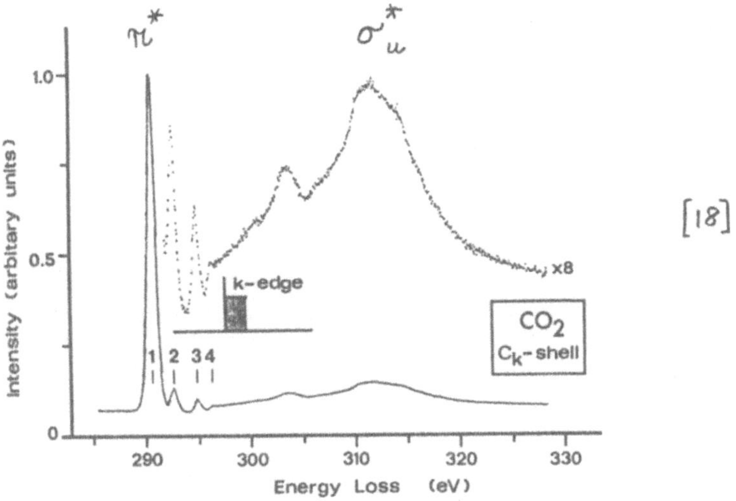

Excitation of carbon 1s in CO_2

that according to the printout this orbital has no p character on the carbon atom, and therefore the intensity is zero.

ONE 1S ELECTRON REMOVED FROM ATOM 1   ONE HALF ELECTRON DIFFUSELY ADDED.

	1	2	3	4	5	6	7	8	9	10	11	12
	-40.723	-39.469	-20.701	-20.172	-20.172	-19.154	-14.391	-14.391	-6.546	-6.546	10.798	44.424
1 C 1	-0.654	0.0	-0.317	0.0	0.0	0.0	0.0	0.0	0.0	0.0	0.687	0.0
2 C 1	0.0	-0.627	0.0	0.0	0.0	-0.252	0.0	0.0	0.0	0.0	0.0	0.738
3 C 1	0.0	0.0	0.0	0.0	0.719	0.0	0.0	0.0	0.0	-0.695	0.0	0.0
4 C 1	0.0	0.0	0.0	0.719	0.0	0.0	0.0	0.0	0.695	0.0	0.0	0.0
5 O 2	-0.503	-0.529	0.401	0.0	0.0	0.326	0.0	0.0	0.0	0.0	-0.294	-0.338
6 O 2	0.182	0.156	0.538	0.0	0.0	0.602	0.0	0.0	0.0	0.0	0.421	0.337
7 O 2	0.0	0.0	0.0	0.0	0.492	0.0	-0.707	0.0	0.0	0.508	0.0	0.0
8 O 2	0.0	0.0	0.0	0.492	0.0	0.0	0.0	0.707	-0.508	0.0	0.0	0.0
9 O 3	-0.503	0.529	0.400	0.0	0.0	-0.326	0.0	0.0	0.0	0.0	-0.294	0.338
10 O 3	-0.182	0.156	-0.538	0.0	0.0	0.602	0.0	0.0	0.0	0.0	-0.421	0.337
11 O 3	0.0	0.0	0.0	0.0	0.492	0.0	0.707	0.0	0.0	0.508	0.0	0.0
12 O 3	0.0	0.0	0.0	0.492	0.0	0.0	0.0	-0.707	-0.508	0.0	0.0	0.0
										EA = + 6.54	σ^*_g no intensity	σ^*_u

$O_3{=}C_1{=}O_2 \longrightarrow X$

K.5. Excitation of 1s electrons studied spectroscopically

The UV absorption of e.g. N_2 between 400 eV and 450 eV [19] results in a curve, which is similar to the energy loss spectrum in Sect. K.4. The light is obtained from a synchrotron and is measured by counting the photons.

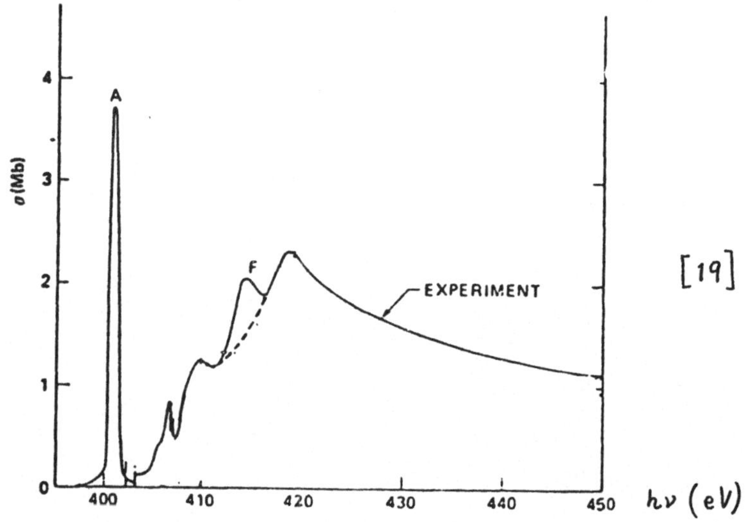

Excitation of nitrogen 1s in N_2 studied in UV absorption. The band F is due to double excitation.

These phenomena can also be observed in emission [20] if the gas is bombarded with high-energy electrons (10000 eV). When the 1s hole is filled with an electron from the π orbital, the emitted photons have energy 393.34 eV. Since a few electrons had been excited to π^* during the bombardment, one of them could fill the 1s hole. The emitted photon energy is then 400.81 eV, i.e. the same as above and in Sect. K.4. No σ^* bands have been observed using this technique, but if this could be achieved, it would give increased credibility to the interpretation of such bands as due to σ^*.

References

1. L. Åsbrink, Phys.Scripta 28, 394 (1983); see also: L. Åsbrink, TRITA-FYS-1001, Physics Department, The Royal Institute of Technology, Stockholm (1980).
2. K. Siegbahn, C. Nordling, A. Fahlman, R. Nordberg, K. Hamrin, J. Hedman, G. Johansson, T. Bergmark, S.-E. Karlsson, I. Lindgren, B. Lindberg, ESCA - Atomic, molecular and solid state structure studied by means of electron spectroscopy, Almqvist and Wixell, Uppsala (1967).
3. K. Siegbahn, C. Nordling, G. Johansson, J. Hedman, P.-F. Hedén, K. Hamrin, U. Gelius, T. Bergmark, L.-O. Werme, R. Manne and Y. Baer, ESCA applied to free molecules, North-Holland, Amsterdam (1969).
4. H. Siegbahn and L. Karlsson, in Handbuch der Physik, Vol. XXXI: Corpuscles and Radiation in Matter I, W. Mehlhorn (ed.), Springer, Berlin (1982), p.215.
5. L. Åsbrink, C. Fridh and E. Lindholm, QCPE No.393 (1980).
6. U. Gelius, E. Basilier, S. Svensson, T. Bergmark and K. Siegbahn, J. Electron Spectrosc. 2, 405 (1974).
7. A.A. Bakke, H.W. Chen and W.L. Jolly, J. Electron Spectrosc. 20, 333 (1980).
8. D.P. Chong, Can.J.Chem. 58, 1687 (1980).
9. D.P. Chong, J.Mol.Sci. 2, 55 (1982).
10. G.R. Wight and C.E. Brion, J.Electron Spectrosc. 4, 347 (1974).
11. A.P. Hitchcock and C.E. Brion, J.Electron Spectrosc. 10, 317 (1977).
12. M. Tronc, G.C. King and F.H. Read, J.Phys.B: Atom.Molec.Phys. 12, 137 (1979).
13. J.C. Slater, Quantum Theory of Atomic Structure, Vol.I, McGraw-Hill, New York (1960), Table 15-10.
14. A.P. Hitchcock and C.E. Brion, J.Electron Spectrosc. 18, 1 (1980).
15. D.A. Shaw, G.C. King, D. Cvejanovic and F.H. Read, J.Phys.B: At.Mol.Phys. 17, 2091 (1984).

16. D.A. Shaw, G.C. King, F.H. Read and D. Cvejanovic, J.Phys.B: At.Mol.Phys. 15, 1785 (1982).
17. A.P. Hitchcock, S. Beaulieu, T. Steel, J. Stöhr and F. Sette, J.Chem.Phys. 80, 3927 (1984).
18. G.R. Wight and C.E. Brion, J.Electron Spectrosc. 3, 191 (1974).
19. H. Petersen, A. Bianconi, F.C. Brown and R.S. Bachrach, Chem.Phys.Letters 58, 263 (1978).
20. J. Nordgren, H. Ågren, L. Selander, C. Nordling and K. Siegbahn, Phys.Scripta 16, 280 (1977).
21. A. Barth and J. Schirmer, J.Phys.B: Atom.Molec.Phys. 18, 867 (1985).

L. Shake up in PES and EA.

L.1. Shake up in PES

In the photoelectron spectrum of ethylene, shown in Sec.H.5., the strong band at 27.2 eV is a shake-up band. It is due to simultaneous ionization af one $2b_{1u}$ electron at 19.0 eV and excitation of another electron from π $(1b_{3u})$ to π^* $(1b_{2g})$. Since two-electron transitions are forbidden, the shake-up configuration has intensity = 0. However, the shake-up configuration has the same symmetry as the PES band at 23.5 eV. Interaction will therefore take place, and the shake-up configuration will steal intensity from the PES band.

The shake-up wavefunction can be illustrated as a product of three orbitals.

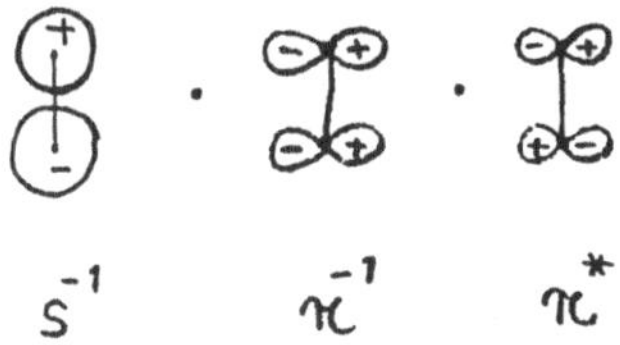

gives the product

which has the same symmetry as the PES band at 23.5 eV:

L.2. Calculation of the PES shake-up energy [1,2]

We assume that in the shake-up the ionization takes place from orbital ψ_m and that simultaneously another electron is excited from orbital ψ_i to orbital ψ_a.

If we now observe the spins we have several cases.

In the first case we have the energy of the shake-up configuration

$$\text{sh.u. energy} = E\left(q_m^\alpha - 1,\ q_i^\alpha - 1,\ q_a^\alpha + 1\right) - E\left(q_m^\alpha,\ q_i^\alpha,\ q_a^\alpha\right) \qquad (L.1)$$

Use of eqs.(B.10), (B.11) and (G.31) gives

$$
\begin{aligned}
(L.1) = {} & H^{\alpha}_{aa} + \sum_k q_k J_{ak} - \sum_k q^{\alpha}_k K_{ak} - \\
& - H^{\alpha}_{ii} - \sum_k q_k J_{ik} + \sum_k q^{\alpha}_k K_{ik} - \\
& - H^{\alpha}_{mm} - \sum_k q_k J_{mk} + \sum_k q^{\alpha}_k K_{mk} - \\
& - J_{ia} + K_{ia} - J_{ma} + K_{ma} + J_{mi} - K_{mi} + q^{\alpha}_a L_{aa} + \left[1-q^{\alpha}_i\right] L_{ii} + \left[1-q^{\alpha}_m\right] L_{mm}
\end{aligned}
\qquad (L.2)
$$

In this general expression the terms can be rearranged in different ways, giving different formulations for the excitation energy.

One such way is the following:

$$
\begin{aligned}
(L.1) = {} & H^{\alpha}_{aa} + \sum_{\substack{k \\ \neq i,m}} q_k J_{ak} + \left(q_i - \tfrac{1}{2}\right) J_{ai} + \left(q_m - \tfrac{1}{2}\right) J_{am} - \sum_{\substack{k \\ \neq i,m}} q^{\alpha}_k K_{ak} - \left(q^{\alpha}_i - \tfrac{1}{2}\right) K_{ai} - \left(q^{\alpha}_m - \tfrac{1}{2}\right) K_{am} \\
& - H^{\alpha}_{ii} - \sum_{\substack{k \\ \neq a,m}} q_k J_{ik} - \left(q_a + \tfrac{1}{2}\right) J_{ia} - \left(q_m - \tfrac{1}{2}\right) J_{im} + \sum_{\substack{k \\ \neq a,m}} q^{\alpha}_k K_{ik} + \left(q^{\alpha}_a + \tfrac{1}{2}\right) K_{ia} + \left(q^{\alpha}_m - \tfrac{1}{2}\right) K_{im} \\
& - H^{\alpha}_{mm} - \sum_{\substack{k \\ \neq i,a}} q_k J_{mk} - \left(q_a + \tfrac{1}{2}\right) J_{ma} - \left(q_i - \tfrac{1}{2}\right) J_{mi} + \sum_{\substack{k \\ \neq i,a}} q^{\alpha}_k K_{mk} + \left(q^{\alpha}_a + \tfrac{1}{2}\right) K_{ma} + \left(q^{\alpha}_i - \tfrac{1}{2}\right) K_{mi} \\
& + \left[\left(q^{\alpha}_a + \tfrac{1}{2}\right) - \tfrac{1}{2}\right] L_{aa} - \left[\left(q^{\alpha}_i - \tfrac{1}{2}\right) - \tfrac{1}{2}\right] L_{ii} - \left[\left(q^{\alpha}_m - \tfrac{1}{2}\right) - \tfrac{1}{2}\right] L_{mm}
\end{aligned}
\qquad (L.3)
$$

which by use of eqs. (B.13) and (G.33) simply is equivalent to the formulation

$$
(L.1) = {}^{t}\varepsilon^{\alpha}_{a} - {}^{t}\varepsilon^{\alpha}_{i} - {}^{t}\varepsilon^{\alpha}_{m} \qquad (L.4)
$$

where in the transition state we have removed one half electron from ψ^{α}_{m} and excited another half electron from ψ^{α}_{i} to ψ^{α}_{a}.

Another formulation is obtained from the following:

$$
\begin{aligned}
(L.1) = {} & H_{aa}^{\alpha} + \sum_{\substack{k\\ \neq m}} q_k J_{ak} + \left(q_m - \tfrac{1}{2}\right) J_{am} - \sum_{\substack{k\\ \neq m}} q_k^{\alpha} K_{ak} - \left(q_m^{\alpha} - \tfrac{1}{2}\right) K_{am} \\
& - H_{ii}^{\alpha} - \sum_{\substack{k\\ \neq m}} q_k J_{ik} - \left(q_m - \tfrac{1}{2}\right) J_{im} + \sum_{\substack{k\\ \neq m}} q_k^{\alpha} K_{ik} - \left(q_m^{\alpha} - \tfrac{1}{2}\right) K_{im} \\
& - H_{mm}^{\alpha} - \sum_{k} q_k J_{mk} + \sum_{k} q_k^{\alpha} K_{mk} \qquad (L.5) \\
& + \left[q_a^{\alpha} - \tfrac{1}{2} \right] L_{aa} - \left[q_i^{\alpha} - \tfrac{1}{2} \right] L_{ii} - \left[\left(q_m^{\alpha} - \tfrac{1}{2}\right) - \tfrac{1}{2} \right] L_{mm} \\
& + \tfrac{1}{2} L_{aa} + \tfrac{1}{2} L_{ii} - J_{ia} - \tfrac{1}{2} J_{am} + \tfrac{1}{2} J_{im} + K_{ia} + \tfrac{1}{2} K_{ma} - \tfrac{1}{2} K_{im}
\end{aligned}
$$

which means

$$
\begin{aligned}
(L.1) = {} & \varepsilon_a^{\alpha} - \varepsilon_i^{\alpha} - {}^{t}\varepsilon_m^{\alpha} + \qquad (L.6) \\
& + \left(\tfrac{1}{2} L_{aa} + \tfrac{1}{2} L_{ii} - J_{ia} + K_{ia} - \tfrac{1}{2} J_{am} + \tfrac{1}{2} K_{am} + \tfrac{1}{2} J_{im} - \tfrac{1}{2} K_{im} \right)
\end{aligned}
$$

Here, the transition state means that one half electron has been ionized from ψ_m^{α}, but in ψ_i^{α} and ψ_a^{α} the occupancies are the same as in the ground state. Since the sum of the terms in the parenthesis in eq. ($L.6$) is usually small, we can often use the formula eq. ($L.4$) for both kinds of transition states.

In the second case we calculate

$$
\text{sh.u. energy} = E\left(q_m^{\alpha}-1,\, q_i^{\beta}-1,\, q_a^{\beta}+1\right) - E\left(q_m^{\alpha},\, q_i^{\beta},\, q_a^{\beta}\right) \qquad (L.7)
$$

which gives

$$
(L.7) = {}^{t}\varepsilon_a^{\beta} - {}^{t}\varepsilon_i^{\beta} - {}^{t}\varepsilon_m^{\alpha} - K_{ma} + K_{mi} \qquad (L.8)
$$

In the third case we calculate

$$\text{sh.u. energy} = E\left(q_m^\beta - 1,\; q_i^\alpha - 1,\; q_a^\beta + 1\right) - E\left(q_m^\beta,\; q_i^\alpha,\; q_a^\beta\right) \quad (L.9)$$

which gives

$$(L.9) = {}^t\varepsilon_a^\beta - {}^t\varepsilon_i^\alpha - {}^t\varepsilon_m^\beta \qquad - K_{ia} + K_{im} \quad (L.10)$$

We remember now that the three shake-up energies were basically obtained by use of eq. (B.3) and understand therefore that the off-diagonal matrix elements can be obtained by use of the same formula

$$H_{kl} = \int \Psi_k^* \mathcal{H} \Psi_l \, d\tau \quad (L.11)$$

We denote the three shake-up configurations as D_1 , D_2 and D_3 and obtain

$$H_{12} = \int D_1^* \mathcal{H} D_2 \, d\tau = - K_{ia} \quad (L.12)$$

$$H_{13} = K_{am} \quad (L.13)$$

$$H_{23} = K_{im} \quad (L.14)$$

The configurations D_1 , D_2 and D_3 do not describe the total spin of the molecule properly (see [3] page 262). It is therefore necessary to form three new wavefunctions [4 − 6] .

$$\Psi_{C\alpha} = \frac{1}{\sqrt{2}}\left(D_1 - D_2\right) \quad (L.15)$$

$$\Psi_{C\beta} = \frac{1}{\sqrt{6}}\left(D_1 + D_2 + 2D_3\right) \quad (L.16)$$

$$\Psi_{CQ} = \frac{1}{\sqrt{3}}\left(D_1 + D_2 - D_3\right) \quad (L.17)$$

In $C\alpha$ and $C\beta$ the resulting spin is = 1/2 (doublet) but in CQ it is = 3/2 (quartet).

The total energies of the new wavefunctions are again obtained from eq.(L.11), which gives

$$H_{\alpha\alpha} = \frac{1}{2}\left[H_{11} + H_{22} - 2H_{12} \right] \qquad (L.18)$$

$$H_{\beta\beta} = \frac{1}{6}\left[H_{11} + H_{22} + 4H_{33} + 2H_{12} + 4H_{13} + 4H_{23} \right] \qquad (L.19)$$

$$H_{QQ} = \frac{1}{3}\left[H_{11} + H_{22} + H_{33} + 2H_{12} - 2H_{13} - 2H_{23} \right] \qquad (L.20)$$

which gives the shake-up energies E^{o}_{α} , E^{o}_{β} and E^{o}_{Q} as

$$E^{o}_{\alpha} = {}^{t}\varepsilon_{a} - {}^{t}\varepsilon_{i} - {}^{t}\varepsilon_{m} + K_{ia} - \frac{1}{2}K_{am} + \frac{1}{2}K_{im} \qquad (L.21)$$

$$E^{o}_{\beta} = {}^{t}\varepsilon_{a} - {}^{t}\varepsilon_{i} - {}^{t}\varepsilon_{m} - K_{ia} + \frac{1}{2}K_{am} + \frac{3}{2}K_{im} \qquad (L.22)$$

$$E^{o}_{Q} = {}^{t}\varepsilon_{a} - {}^{t}\varepsilon_{i} - {}^{t}\varepsilon_{m} - K_{ia} - K_{am} \qquad (L.23)$$

where we have assumed that $\varepsilon^{\alpha} = \varepsilon^{\beta}$.

It may be remarked that these formulas are easy to handle since they do not contain any differences between large numbers of the form $J_{ia} - J_{im}$. The reason for this is the introduction of the transition state.

The configurational energies.

The energy expressions eqs.(L.21-23) constitute the most important part of every study of valence electron shake up, since if incorrect energies are obtained here, the energies of the shake-up bands will be incorrect also. The reason for this is that the CI, which will be discussed below, will influence the energies very little. The accuracy of the energies is especially important since a small energy error usually will introduce a large intensity error. We must therefore study these expressions in detail.

The eigenvalues in these expressions are obtained from a SCF calculation of a transition state in which one half electron has been removed from ψ_m and one half excited from ψ_i to ψ_a.

The eigenvalues cannot be connected with experimental values directly, but if we remember, that an excitation energy is approximately independent of the charge of the molecule (see e.g. the ethylene computer print-outs in Sec.I.8.) and that the sum of the terms in parenthesis in eq.(L.6) is small, we can compare with ionization and excitation energies. We find from eqs.(L.21 - 23), denoting the singlet and triplet excitation energies as S_{ia} and T_{ia} , respectively

$$E_{\alpha}^{o} = IP_m + S_{ia} - \frac{1}{2} K_{am} + \frac{1}{2} K_{im} \qquad (L.24)$$

$$E_{\beta}^{o} = IP_m + T_{ia} + \frac{1}{2} K_{am} + \frac{3}{2} K_{im} \qquad (L.25)$$

$$E_{Q}^{o} = IP_m + T_{ia} - K_{am} \qquad (L.26)$$

A theoretically better way would be to perform two HAM calculations: the first with $q_m - \frac{1}{2}$ only and the second with $q_m - \frac{1}{2}$, $q_i - \frac{1}{2}$ and $q_a + \frac{1}{2}$. This gives the necessary correction to IP_m in eqs.(L.24 - 26). In the same way the corrections to S_{ia} and T_{ia} can be obtained. It is evident from Sec.L.3. that these corrections are small for ethylene.

The interaction matrix elements.

Since two-electron processes are forbidden, the shake-up configurations $C\alpha$, $C\beta$ and CQ can never be observed directly in a PES. The shake-up configuration can, however, acquire intensity by interaction with a primary hole configuration A in which one electron has been ionized from ψ_h (with ionization energy I_h). The matrix elements are [1,6,7,8,9]

$$G_\alpha = \frac{\sqrt{2}}{2}\left[2(hm|ia) - (hi|ma)\right] \quad \text{for interaction } C\alpha - A \quad (L.27)$$

$$G_\beta = -\frac{\sqrt{6}}{2}(hi|ma) \quad \text{for interaction } C\beta - A \quad (L.28)$$

$$G_{\alpha\beta} = \frac{\sqrt{3}}{2}\left[K_{am} - K_{im}\right] \quad \text{for interaction } C\alpha - C\beta \quad (L.29)$$

but for the interaction between CQ (quartet) and A (doublet) we have $G = 0$ and no shake ups of CQ type can therefore be observed.

A treatment similar to that in Sec.G.8. or Sec.I.2. or Sec.I.9. gives then the secular determinant

$$\begin{vmatrix} I_h^\circ - E & G_\alpha & G_\beta \\ G_\alpha & E_\alpha^\circ - E & G_{\alpha\beta} \\ G_\beta & G_{\alpha\beta} & E_\beta^\circ - E \end{vmatrix} = 0 \quad (L.30)$$

Solving the matrix gives first of all three energy values: I_h , E_α and E_β, which are slightly different from I_h° , E_α° and E_β°.

Further, the wavefunction of the primary ionization gives a small contribution to the wavefunction of the shake up. The square of the coefficient indicates directly the intensity of the shake-up band.

L.3. Shake ups in PES: some results

Shake up in ethylene [1]

The orbitals in ethylene have been presented in Sec.A.4. and discussed in Sec.H.5. They are given again in the table below

1	2	3	4	5	6	7
$2a_g$	$2b_{1u}$	$1b_{2u}$	$3a_g$	$1b_{3g}$	$1b_{3u}$	$1b_{2g}$
24.3 eV	19.0 eV	15.9 eV	14.6 eV	12.8 eV	10.4 eV	4.5 eV
s + s	s - s		σ		π	π^*
h	m				i	a

The energy values in our calculation should be transition state eigenvalues. Since the HAM/3 calculation is not perfect, we have instead taken the corresponding experimental ionization energies except for the first orbital, for which we have taken the HAM value.

The matrix elements are (see Sec.I.2.)

$$g_\alpha = \frac{\sqrt{2}}{2}\left[2(hm|ia) - (hi|ma)\right] = \frac{\sqrt{2}}{2}\left[2 \cdot 1.236 - 0.533\right] = 1.372 \text{ eV}$$

$$g_\beta = -\frac{\sqrt{6}}{2} \cdot (hi|ma) = -\frac{\sqrt{6}}{2} \cdot 0.533 = -0.656 \text{ eV}$$

Further

$$E^\circ_\alpha = 19.0 + 7.6 - \frac{1}{2} \cdot 0.45 + \frac{1}{2} \cdot 0.45 = 26.6 \text{ eV}$$

$$E^\circ_\beta = 19.0 + 4.2 + \frac{1}{2} \cdot 0.45 + \frac{3}{2} \cdot 0.45 = 24.1 \text{ eV}$$

This gives

$$\begin{vmatrix} 24.3 - E & 1.372 & -0.656 \\ 1.372 & 26.6 - E & 0 \\ -0.656 & 0 & 24.1 - E \end{vmatrix} = 0$$

Solving the matrix gives the following results:

a) The primary hole band is displaced to 23.2 eV and is supposed to have intensity 100 %.

b) $C\alpha$ is a little displaced to 27.3 eV and has intensity 33 %. This is in excellent agreement with the experimental results in Sec.H.5.

c) $C\beta$ is displaced to 24.5 eV and has a high intensity (42 %). This means that $C\beta$ is part of the strong band in the PES at this energy and can therefore never be observed.

The displacement of the primary hole band is important for the following reason.

In the work to determine the parameters in HAM the photoelectron spectra of many molecules, amongst them ethylene, were used. It appeared during this work to be impossible to achieve agreement on the ionization energy 23.5 eV in ethylene. The photoelectron spectra of the other molecules determined the results of the least-squares fit, since they were so many, and therefore a calculated ionization energy about 24.3 eV was always obtained. The present study proves now that the calculated value can be considered as correct although it deviates from experiment by 0.8 eV.

The very weak shake-up band at 31.4 eV (see Sec.H.5.) is handled in the same way. We have

$$E^{\circ}_{\alpha} = 23.5 + 7.6 - \frac{1}{2}\cdot 0.77 + \frac{1}{2}\cdot 0.77 = 31.1 \text{ eV}$$

The interaction with the band at 19.0 eV gives $S_{\alpha} = 1.372$ eV. We obtain then $C\alpha$ at 31.2 eV with intensity 1 % in excellent agreement with experiment. ($C\beta$ is calculated at 29.2 eV with very low intensity, 0.2 %).

Study of the transition state

We intend to show that the transition state used in eq.(L.4) and the approximate transition state, used in eq.(L.6), give about the same values for ${}^{t}\varepsilon_{a} - {}^{t}\varepsilon_{i} - {}^{t}\varepsilon_{m}$.

With one half electron ionized from ψ_m and one half electron excited from ψ_i to ψ_a we have

$${}^{t}\varepsilon_{a} - {}^{t}\varepsilon_{i} - {}^{t}\varepsilon_{m} = 26{,}350 \text{ eV.}$$

	1	2	3	4	5	6	7	8	9	10	11	12
	-24.609	-20.096	-16.350	-14.977	-13.274	-10.567	-4.313	15.243	19.401	25.466	29.874	36.439
$q_j \rightarrow$	2.000	1.500	2.000	2.000	2.000	1.500	0.500	0.0	0.0	0.0	0.0	0.0
1 C 1	0.467	0.336	0.000	0.084	-0.000	-0.000	0.000	0.675	0.834	0.000	0.004	0.000
2 C 1	-0.053	0.210	0.000	-0.521	0.000	0.000	0.000	0.415	-0.382	-0.000	1.004	0.000
3 C 1	-0.000	0.000	-0.419	-0.000	-0.481	-0.000	0.000	0.000	-0.000	0.688	0.000	-0.858
4 C 1	-0.000	-0.000	-0.000	0.000	-0.000	0.642	0.797	0.000	0.000	-0.000	0.000	-0.000
5 C 2	0.467	-0.336	-0.000	0.084	0.000	-0.000	0.000	0.675	-0.834	-0.000	-0.004	-0.000
6 C 2	0.053	0.210	0.000	0.521	-0.000	-0.000	-0.000	-0.415	-0.382	-0.000	1.004	0.000
7 C 2	0.000	0.000	-0.419	0.000	0.481	0.000	-0.000	0.000	-0.000	0.688	0.000	0.858
8 C 2	0.000	0.000	-0.000	0.000	-0.000	0.642	-0.797	-0.000	-0.000	-0.000	-0.000	0.000
9 H 3	0.177	0.258	-0.251	-0.181	-0.302	-0.000	0.000	-0.463	-0.288	-0.634	-0.476	0.651
10 H 4	0.177	0.258	0.251	-0.181	0.302	0.000	0.000	-0.463	-0.288	0.634	-0.476	-0.651
11 H 5	0.177	-0.258	-0.251	-0.181	0.302	0.000	-0.000	-0.463	0.288	-0.634	0.476	-0.651
12 H 6	0.177	-0.258	0.251	-0.181	-0.302	-0.000	0.000	-0.463	0.288	0.634	0.476	0.651

With one half electron ionized from ψ_m we have

$${}^t\varepsilon_a - {}^t\varepsilon_i - {}^t\varepsilon_m \approx 26,185 \text{ eV}.$$

	1	2	3	4	5	6	7	8	9	10	11	12
	-24.338	-19.907	-16.225	-14.812	-13.154	-10.355	-4.077	15.467	19.735	25.590	30.122	36.584
$q_j \rightarrow$	2.000	1.500	2.000	2.000	2.000	2.000	0.0	0.0	0.0	0.0	0.0	0.0
1 C 1	0.464	0.333	-0.000	0.087	-0.000	-0.000	0.000	0.677	0.836	0.000	-0.001	0.000
2 C 1	-0.052	0.210	0.000	-0.521	-0.000	0.000	-0.000	0.415	-0.375	-0.000	1.007	0.000
3 C 1	-0.000	0.000	0.417	0.000	0.478	-0.000	0.000	0.000	-0.000	0.689	0.000	-0.859
4 C 1	-0.000	0.000	-0.000	-0.000	0.000	0.642	0.797	-0.000	0.000	-0.000	-0.000	-0.000
5 C 2	0.464	-0.333	0.000	0.087	0.000	0.000	0.000	0.677	-0.836	-0.000	0.001	-0.000
6 C 2	0.052	0.210	-0.000	0.521	0.000	0.000	0.000	-0.415	-0.375	-0.000	1.007	0.000
7 C 2	0.000	0.000	0.417	0.000	-0.478	0.000	-0.000	0.000	-0.000	0.689	0.000	0.859
8 C 2	0.000	0.000	-0.000	0.000	0.000	0.642	-0.797	0.000	0.000	0.000	-0.000	0.000
9 H 3	0.180	0.260	0.252	-0.181	0.304	-0.000	-0.000	-0.462	-0.290	-0.634	-0.474	0.650
10 H 4	0.180	0.260	-0.252	-0.181	-0.304	0.000	-0.000	-0.462	-0.290	0.634	-0.474	-0.650
11 H 5	0.180	-0.260	0.252	-0.181	-0.304	0.000	-0.000	-0.462	0.290	-0.634	0.474	-0.650
12 H 6	0.180	-0.260	-0.252	-0.181	0.304	-0.000	-0.000	-0.462	0.290	0.634	0.474	0.650

With one half electron diffusely ionized we have

$${}^t\varepsilon_a - {}^t\varepsilon_i - {}^t\varepsilon_m \approx 26.117 \text{ eV}.$$

ONE HALF ELECTRON DIFFUSELY REMOVED. FILLED ORBITALS GIVE IONIZATION ENERGIES

	1	2	3	4	5	6	7	8	9	10	11	12
	-24.292	-19.848	-16.256	-14.883	-13.173	-10.538	-4.269	15.420	19.620	25.496	29.969	36.434
1 C 1	0.466	0.334	-0.000	0.081	0.000	0.000	-0.000	0.676	0.835	0.000	0.003	0.000
2 C 1	-0.055	0.213	0.000	-0.522	0.000	-0.000	-0.000	0.413	-0.381	-0.000	1.004	0.000
3 C 1	-0.000	0.000	-0.420	-0.000	-0.482	0.000	-0.000	-0.000	-0.000	0.687	0.000	-0.857
4 C 1	-0.000	-0.000	-0.000	0.000	-0.000	-0.642	-0.797	0.000	0.000	-0.000	0.000	-0.000
5 C 2	0.466	-0.334	-0.000	0.081	-0.000	0.000	-0.000	0.676	-0.835	-0.000	-0.003	-0.000
6 C 2	0.055	0.213	0.000	0.522	-0.000	0.000	0.000	-0.413	-0.381	-0.000	1.004	0.000
7 C 2	0.000	0.0	-0.420	0.000	0.482	-0.000	-0.000	0.000	-0.000	0.687	0.000	0.857
8 C 2	0.000	0.000	-0.000	0.000	-0.000	-0.642	0.797	-0.000	-0.000	-0.000	0.000	-0.000
9 H 3	0.177	0.257	-0.249	-0.180	-0.301	0.000	-0.000	-0.464	-0.288	-0.635	-0.476	0.652
10 H 4	0.177	0.257	0.249	-0.180	0.301	-0.000	-0.000	-0.464	-0.288	0.635	-0.476	-0.652
11 H 5	0.177	-0.257	-0.249	-0.180	0.301	-0.000	0.000	-0.464	0.288	-0.635	0.476	-0.652
12 H 6	0.177	-0.257	0.249	-0.180	-0.301	0.000	-0.000	-0.464	0.288	0.635	0.476	0.652

The results from these three calculations show thus that any of them can be used in studies of shake-up energies.

Shake up in cyanogen [2]

In cyanogen there is a strong shake-up band at 23.7 eV. The spectrum and the orbitals are presented in Sec.H.5. Since the HAM/3 calculation is not perfect, also the experimental ionization energies are given there. The energies of the two π^* orbitals are discussed in Sec.I.10.

The shake-up process in cyanogen is defined from the indices in the table below:

3	4	5	6	7	8	9	10	11	12	13
σ_g	y	x	σ_u	σ_g	y	x	X	Y	Y	X
22.8	15.6	15.6	14.9	14.5	13.4	13.4	7.04	7.04	1.31	1.31
h			m			i	a			

We have

$$S_{ia} = 13.4 - 7.04 + K_{ia} = 7.65 \text{ eV}$$

$$T_{ia} = 13.4 - 7.04 - K_{ia} = 5.07 \text{ eV}$$

which gives

$$E^0_\alpha = 14.9 + 7.65 - \tfrac{1}{2}\cdot 0.44 + \tfrac{1}{2}\cdot 0.44 = 22.55 \text{ eV}$$

$$E^0_\beta = 14.9 + 5.07 + \tfrac{1}{2}\cdot 0.44 + \tfrac{3}{2}\cdot 0.44 = 20.85 \text{ eV}$$

Further

$$g_\alpha = \frac{\sqrt{2}}{2}\cdot\left[2\cdot(-0.51) - (-0.23)\right] = -0.56 \text{ eV}$$

$$g_\beta = -\frac{\sqrt{6}}{2}\cdot(-0.23) = +0.28 \text{ eV}$$

These numbers are sufficient to handle the interaction in the xX -plane.

Simultaneously, however, we have similar interactions in the yY -plane. The two excitations interact mutually in the way studied in Sec.I.10. and the matrix elements are given by eqs.

(I.48) and (I.41). This give

$$\varsigma_{9,10,8,11} = 2\,(9,10|8,11) - (9,8|10,11) = 2\cdot(-0.91) - (-0.19) = -1.63\ \mathrm{eV}$$
$$= \qquad - (9,8|10,11) = +0.19\ \mathrm{eV}$$

for singlet and triplet excitation, respectively. The interaction with the primary hole is the same as above.

The secular determinant is therefore

$$\begin{vmatrix} 22.8-E & -0.56 & 0.28 & 0.56 & -0.28 \\ -0.56 & 22.55-E & 0 & -1.63 & 0 \\ 0.28 & 0 & 20.85-E & 0 & 0.19 \\ 0.56 & -1.63 & 0 & 22.55-E & 0 \\ -0.28 & 0 & 0.19 & 0 & 20.85-E \end{vmatrix} = 0 \qquad (L.31)$$

Solving the matrix gives the following results.

a) The primary hole is displaced a little to 22.5 eV, which indicates, that we should have used 23.2 eV instead for its one-determinant value.

b) $C\alpha$ is obtained at 24.5 eV with intensity 22 %. It corresponds to the observed shake-up band at 23.7 eV. Another $C\alpha$ without intensity is obtained at 20.9 eV

c) $C\beta$ is obtained at 20.6 eV with intensity 4 %. It is too weak to be observed. Another $C\beta$ without intensity is obtained at 21.0 eV.

The calculated energy of the shake up is thus 0.8 eV too high. The explanation is probably that our calculation of matrix elements and our choice of energy for π^* give somewhat inaccurate results. The calculated intensity is also too small, but it is clear that if the energy is slightly improved, this will be accompanied by a much increased intensity.

A mathematical note

The secular determinant for the linear molecule cyanogen looks very complex. An approximate solution can, however, be obtained in the following way.

Let us, to begin with, neglect the interaction with the primary hole (at 22.8 eV in the example above). This means that the left column and the top line are omitted from the determinant.

In the remaining 4 x 4 matrix we have now two singlet excitations at 22.55 eV and two triplet excitations at 20.85 eV.

The two singlet excitations interact with a matrix element - 1.63 eV. The interaction produces two states:

a) one at 22.55 + 1.63 = 24.18 eV, which corresponds to $^1\Sigma^+$ in the $\pi\pi^*$ excitation of a linear molecule, and

b) one at 22.55 - 1.63 = 20.92 eV, which in the same way corresponds to $^1\Delta^+$.

The two triplet excitations interact with a matrix element 0.19 eV. This produces two states:

c) one at 20.85 + 0.19 = 21.04 eV, which corresponds to $^3\Delta^+$, and

d) one at 20.85 - 0.19 = 20.66 eV, which corresponds to $^3\Sigma^+$.

It is obvious that a Δ excitation cannot interact with a Σ primary ionization, and this explains why we get two bands without intensity.

The secular determinant, eq.(L.31), can also be treated in the following way. Since a determinant is not changed if we add one column to another column or one row to another row, we find easily

$$(L.31) = (22.55 - 1.63 - E)(20.85 + 0.19 - E)\cdot\begin{vmatrix} 22.8 - E & 0.56 & -0.28 \\ 2\cdot 0.56 & 22.55 + 1.63 - E & 0 \\ 2\cdot(-0.28) & 0 & 20.85 - 0.19 - E \end{vmatrix}$$

$$\underset{C\alpha(^1\Delta^+)}{\uparrow} \qquad \underset{C\beta(^3\Delta^+)}{\uparrow} \qquad \underset{C\alpha(^1\Sigma^+)}{\uparrow} \qquad \underset{C\beta(^3\Sigma^+)}{\uparrow}$$

The interaction in the determinant takes place between the primary hole and the shake-up configurations, formed from the $^1\Sigma^+$ and $^3\Sigma^+$ excitations, respectively.

It is therefore clear that a study of the shake-up bands should be predeeded by a study of the excitation bands, and if there are any irregularities in the excitations, they will appear also in the shake ups.

L.4. Discussion of calculations of shake up in PES

The shake-up process means ionization of one electron and excitation of another.

It is obvious that a condition for a successful calculation of shake ups is that the theoretical method is able to calculate ionizations and excitations separately. We have demonstrated in Chaps. H. and I. that the HAM/3 program gives reasonable results for these two processes for a large number of molecules. We expect therefore that the method will give reasonable results also for shake-up calculations.

Unfortunately, this is not easy to prove in an unequivocal way by comparison with experiment. Shake-up bands have been observed in only few molecules, and when they occur, they have often been interpreted in different ways.

An important point in a discussion of these problems concerns the energy difference between $C\alpha$ and $C\beta$. According to eqs. (L.21) and (L.22) this energy difference should be smaller than $2K_{ia}$. Since K_{ia} seldom is larger that 1.5 eV, the energy difference should be of the order of 2.5 eV.

In a linear molecule we have to add the matrix elements G_{xXyY} for singlet and triplet interactions, which gives an energy difference in this case of about 4.5 eV.

Ab-initio calculations

According to Sec.I.6. ab-initio calculations usually give much too high singlet $\pi\pi^*$ excitation energies, whereas the corresponding triplet energies are in good agreement with experiment, and according to Sec.H.2. the calculated ionization energies are usually too high also. In both cases the results are caused by the difficulties to calculate the correlation energies in CI work.

Since the shake-up process is much more complicated than the ionization process or the excitation process separately, being the sum of these two processes, we expect more severe difficulties here. We expect therefore that ab-initio calculations of $C\beta$ energies will give too high results and of $C\alpha$ energies much too high.

A comparison with the HAM results and the experimental results will now be performed for the two molecules, studied above.

Ethylene

Ethylene is the first example of our discussion.

Martin and Davidson [10] used CI and studied $2b_{1u}^{-1} \cdot \pi\pi^*$. $C\beta$ was calculated at 27.8 eV with intensity 30 % and $C\alpha$ at 31.1 eV with intensity 5 %. There is thus good agreement with experiment although their shake-up energies are much higher than ours. Cederbaum et al. [11] and Baker [12] obtained results which are very similar and approximately equal to the results by Martin and Davidson.

Recent work by Nakatsuji [13] is different. His CI study is very all round. The PES of ethylene is first calculated with very good ionization energies. The excitation study gives reasonable results with the singlet $\pi\pi^*$ energy about 0.5 eV too high. But the shake ups cause problems. Only one transition of type $2b_{1u}^{-1}\pi\pi^*$ is found at 30.15 eV with low intensity. If this is $C\beta$ it means that the triplet excitation energy is about 11 eV, which seems too high and which explains the low intensity. The result is therefore that no strong shake ups are obtained in pronounced disagreement with the experimental results.

Cyanogen

Cyanogen has been studied by Cederbaum et al. [14]. A TDA calculation gave $C\beta$ at 25.49 eV with intensity 4 % and $C\alpha$ at 35.56 eV with intensity 13 %. We observe first that these shake-up energies are much higher that our values. The explanation for the low intensity of $C\beta$ is, of course, that it is too high above the primary hole band at 22.8 eV.

Since the authors thus could not find any simple explanation for a shake-up band at 23.7 eV, they studied vibrational interaction between a weak shake-up band of Π type at this energy and the primary band of Σ type at 22.8 eV. This explanation requires however that the energy of the Π shake up is as high as 22.8 eV, which can be doubted.

Shake up in acetylene

The PES of acetylene, studied in Uppsala with 1486 eV photons, shows two shake-up bands at 28.0 eV and 26.7 eV [15]. Very large theoretical studies were then made to interpret the two bands, but although several ab-initio procedures were tried, giving different results, it appeared to be impossible to find an interpretation of more that one band [15]. The measurements were therefore repeated with 50 eV and 70 eV photons [16]. Exactly the same two bands were then obtained.

If one tries to understand the calculated results in ref. [15], one will find that the T_{ia} used in two of the tables in ref. [15] is about 10.1 eV and 7.3 eV, respectively, and the S_{ia} about 15.6 eV and 14.2 eV (neglecting displacements and matrix elements). These values are too large compared with what can be deduced from UV spectra of acetylene [17], but not unexpected large if one compares with recent calculations of $\pi\pi^*$ excitation (see Sec.I.6.).

On the contrary, a HAM calculation gives directly two shake-up bands at approximately correct energies. Unfortunately, a third shake-up band appears near the primary hole band at 23.5 eV. We must therefore discuss the molecule in detail before we start the calculations.

It was pointed out in the mathematical note in Sec.L.3. that irregularities in the excitations will appear also in the shake ups. This concerns especially the $^3\Sigma_u^+$ excitation of acetylene, which is very irregular due to the presence of strong HCC bending vibrations.

Demoulin [18] has drawn theoretical potential curves for many excited states, which seem to be in very good agreement with recent electron impact energy loss measurements [19]. The calculated curve for $^3\Sigma_u^+$ exhibits a minimum at the HCC angle 135°, about 1 eV deep.

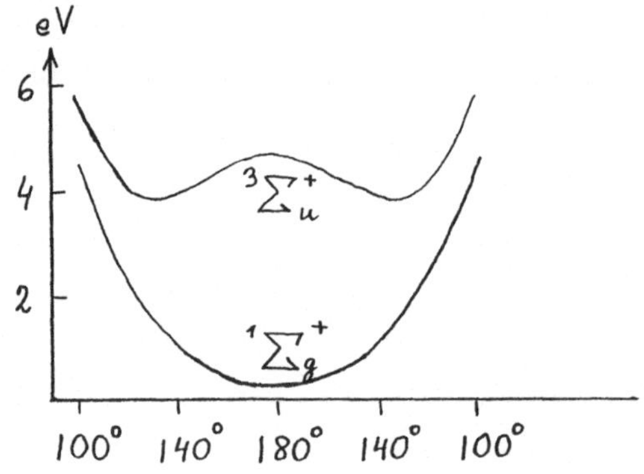

[18]

Bending vibrations will thus appear with vibrational energy about 0.09 eV [20]. The progression has been observed by Wilden, Hicks and Comer [19] with 14 peaks.

When therefore the $^3\Sigma^+$ state is formed in a vertical transition, it will have a vibrational quantum number of the order of 14, and the same will be the case for the shake-up configuration. When this shake-up configuration interacts with the primary hole configuration, in which the vibrational quantum number presumably is small, the strength of the interaction will be reduced, since the vibrational wavefunctions enter the interaction matrix elements.

Orbitals 1 and 2 in acetylene are both of s type and have both high IP's. Shake-up configurations can therefore acquire intensity in two ways, which must be treated separately.

The shake-up process of type $^2\Sigma_g^+$ is defined from the indices in the table

1	2	3	4	5	6	7
σ_g	σ_u	σ_g	y	x	X	Y
23.5	18.7	16.7	11.4	11.4	4.7	4.7
h	m			i	a	

The ionization energy of π^*, 4.7 eV, corresponds to a calculated electron affinity -2.5 eV in good agreement with the experimental values in Table J.3.

To calculate the singlet and triplet excitation energies we will use K_{ia} = 1.20 eV instead of the value 1.48 eV obtained from the HAM/3 calculation. This gives

$$S_{ia} = 11.4 - 4.7 + 1.2 = 7.9 \text{ eV}$$

$$T_{ia} = 11.4 - 4.7 - 1.2 = 5.5 \text{ eV}$$

and

$$E^0_\alpha = 18.7 + 7.9 - \frac{1}{2}\cdot 0.46 + \frac{1}{2}\cdot 0.46 = 26.60 \text{ eV}$$

$$E^0_\beta = 18.7 + 5.5 + \frac{1}{2}\cdot 0.46 + \frac{3}{2}\cdot 0.46 = 25.12 \text{ eV}$$

Further

$$G_{\alpha} = \frac{\sqrt{2}}{2}\left[2\cdot(-0.91) - (-0.43)\right] = -\ 0.98\ \text{eV}$$

$$G_{\beta} = -\frac{\sqrt{6}}{2}\cdot(-0.43) = +\ 0.53\ \text{eV}$$

Due to the vibrations in ${}^3\Sigma_u^+$ we reduce G_{β} to half this value.

These numbers handle the interaction in the xX plane. In the yY plane we have similar interaction. The two excitations interact mutually in the way studied in Sec.I.10. with matrix elements

$$G_{5,6,4,7} = 2\cdot(-0.91) - (-0.28) = -\ 1.54\ \text{eV}$$

$$= \qquad -(-0.28) = +\ 0.28\ \text{eV}$$

for singlet and triplet excitation, respectively.

The secular determinant is therefore

$$\begin{vmatrix} 23.90-E & -0.98 & 0.25 & 0.98 & -0.25 \\ -0.98 & 26.60-E & 0 & -1.54 & 0 \\ 0.25 & 0 & 25.12-E & 0 & 0.28 \\ 0.98 & -1.54 & 0 & 26.60-E & 0 \\ -0.25 & 0 & 0.28 & 0 & 25.12-E \end{vmatrix} = 0$$

Solving the matrix gives the following results for ${}^2\Sigma_g^+$.

a) The primary hole band is displaced to 23.4 eV.

b) $C\alpha$ is obtained at 28.5 eV with intensity 9.4 % in reasonable agreement with experiment. Another $C\alpha$ without intensity is obtained at 25.1 eV.

c) $C\beta$ is obtained at 24.9 eV with intensity 5.2 %. This is the "third shake up" mentioned above. Another $C\beta$ without intensity is obtained at 25.4 eV.

The shake-up process of type $^2\Sigma_u^+$ is defined from the indices below

1	2	3	4	5	6	7
σ_g	σ_u	σ_g	y	x	X	Y
23.5	18.7	16.7	11.4	11.4	4.7	4.7
	h	m		i	a	

We obtain

$$E_\alpha^0 = 16.7 + 7.9 - \frac{1}{2}\cdot 0.21 + \frac{1}{2}\cdot 0.21 = 24.60 \text{ eV}$$

$$E_\beta^0 = 16.7 + 5.5 + \frac{1}{2}\cdot 0.21 + \frac{3}{2}\cdot 0.21 = 22.62 \text{ eV}$$

The secular determinant is then

$$\begin{vmatrix} 18.70-E & 1.26 & -0.10 & -1.26 & 0.10 \\ 1.26 & 24.60-E & 0 & -1.54 & 0 \\ -0.10 & 0 & 22.62-E & 0 & 0.28 \\ -1.26 & -1.54 & 0 & 24.60-E & 0 \\ 0.10 & 0 & 0.28 & 0 & 22.62-E \end{vmatrix} = 0$$

Solving the matrix gives the following results for $^2\Sigma_u^+$.

a) The primary hole is displaced to 18.3 eV.

b) $C\alpha$ is obtained at 26.5 eV with intensity 5.2 % in reasonable agreement with experiment. Another $C\alpha$ without intensity is obtained at 23.1 eV.

c) $C\beta$ is obtained at 22.3 eV with intensity 0.1 %. It is too weak to be observed. Another $C\beta$ without intensity is obtained at 22.9 eV.

Higher shake ups are easy to handle since for them all energies and matrix elements follow from the calculations above. We expect

$$E^{o}_{\alpha} = 23.5 + 7.4 - \frac{1}{2}\cdot 0.94 + \frac{1}{2}\cdot 0.94 = 30.90 \text{ eV}$$

$$E^{o}_{\beta} = 23.5 + 5.0 + \frac{1}{2}\cdot 0.94 + \frac{3}{2}\cdot 0.94 = 30.38 \text{ eV}$$

Interaction with the hole at 18.7 eV gives then

$$\begin{vmatrix} 18.70-E & -0.98 & 0.25 & 0.98 & -0.25 \\ -0.98 & 30.90-E & 0 & -1.54 & 0 \\ 0.25 & 0 & 30.38-E & 0 & 0.28 \\ 0.98 & -1.54 & 0 & 30.90-E & 0 \\ -0.25 & 0 & 0.28 & 0 & 30.38-E \end{vmatrix} = 0$$

Solving the matrix gives $C\alpha$ at 32.6 eV with intensity 1.0 % and $C\beta$ at 30.1 eV with intensity 0.1 %.

Acetylene: comparison with experiment

Comparison with experiment [15,16] gives the following result.

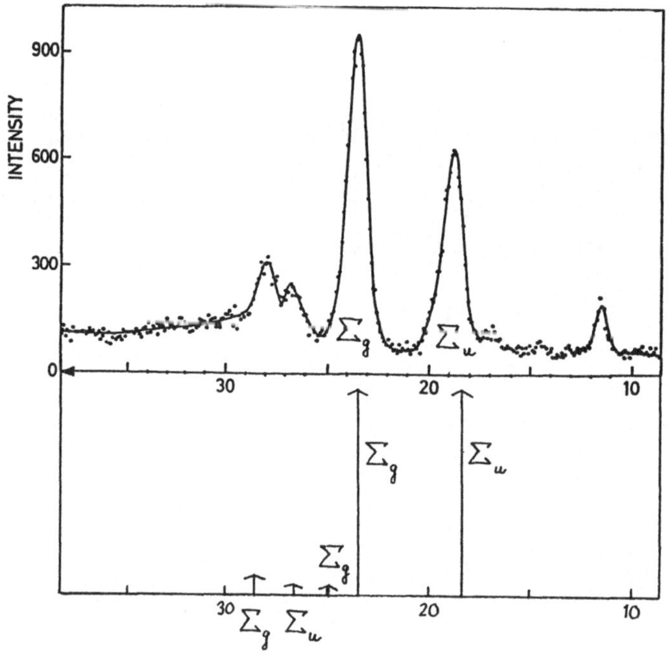

[15]

The two shake-up bands are in reasonable agreement with experiment. Both are denoted as $C\alpha$. One is $^2\Sigma_g$ and the other $^2\Sigma_u$.

The "third shake up" is seen at 24.9 eV. Due to its low intensity it cannot be observed. If the reduction of S_β had not been made, its intensity would have been 17 % .

The intensities of the shake ups are given in the following table [16]

Band	23.5 eV	26.7 eV	28.0 eV
1486.6 eV photons	100	11	19
70 eV photons	100	6.8	8.1
50 eV photons	100	6.6	9.4
HAM	100	5.2	9.4

Further, our calculated band at 32.6 eV corresponds possibly to part of a weak band, observed between 30 and 32 eV [16].

Finally, we compare our results with the calculations, performed by Müller et al. [15] . As mentioned above, they use several different ab-initio methods, which give the following energies for $C\alpha$ and $C\beta$.

Species	$^2\Sigma_g$				$^2\Sigma_u$		
Method	POL-CI	SE-CI	CASSCF	HAM/3	POL-CI	CASSCF	HAM/3
$C\beta$	28.3	28.8	28.4	24.9	24.0	24.2	22.3
$C\alpha$	?	34.3	?	28.5	30.9	?	26.5

Besides, they found in the CASSCF calculation two shake ups without intensity. Since these probably correspond to excitation of Δ type (see the mathematical note above), we have omitted them from the table. It can be seen that the energies, obtained by Müller et al., are much higher than the HAM/3 energies.

The result of the work by Müller et al. is that only <u>one of the two shake ups</u> can be explained.

L.5. Shake up in EA [2]

In the electron affinity spectrum of cyanogen, shown in Sec.J.3., the band at -9.1 eV is a shake-up band. It is due to simultaneous attachment of one electron to the π^* orbital $2\pi_u$ and excitation of another electron from π $(1\pi_g)$ to π^* $(2\pi_u)$. Being a two-electron process, the shake-up configuration cannot be formed when an extra electron is attached. It can, however, interact with the primary attachment to the higher π^* $(2\pi_g)$ and can steal intensity from this band.

We reproduce the print-out:

CYANOGEN CHARGE=-0.500

9 DOUBLY OCCUPIED MOLECULAR ORBITALS

TRANSITION STATE ONE HALF ELECTRON DIFFUSELY ADDED. EMPTY ORBITALS GIVE ELECTRON AFFINITIES

		3	4	5	6	7	8	9	10	11	12	13	14
		-14.419	-7.716	-7.716	-7.646	-7.220	-5.905	-5.905	-0.089	-0.089	4.713	4.713	11.844
1 N	1	0.343	0.0	0.0	0.382	-0.349	0.0	0.0	0.0	0.0	0.0	0.0	-0.214
2 N	1	0.0	0.0	-0.378	0.0	0.0	0.551	0.0	0.0	-0.612	0.470	0.0	0.0
3 N	1	0.0	-0.378	0.0	0.0	0.0	0.0	-0.551	-0.612	0.0	0.0	0.470	0.0
4 N	1	0.012	0.0	0.0	0.507	-0.526	0.0	0.0	0.0	0.0	0.0	0.0	0.440
5 C	2	-0.267	0.0	0.0	-0.192	0.057	0.0	0.0	0.0	0.0	0.0	0.0	0.754
6 C	2	0.0	0.0	-0.486	0.0	0.0	0.381	0.0	0.0	0.442	-0.714	0.0	0.0
7 C	2	0.0	-0.486	0.0	0.0	0.0	0.0	-0.381	0.442	0.0	0.0	-0.714	0.0
8 C	2	0.372	0.0	0.0	-0.107	0.239	0.0	0.0	0.0	0.0	0.0	0.0	-0.128
9 C	3	-0.267	0.0	0.0	0.192	0.057	0.0	0.0	0.0	0.0	0.0	0.0	-0.754
10 C	3	0.0	0.0	-0.486	0.0	0.0	-0.381	0.0	0.0	0.442	0.714	0.0	0.0
11 C	3	0.0	-0.486	0.0	0.0	0.0	0.0	0.381	0.442	0.0	0.0	0.714	0.0
12 C	3	-0.372	0.0	0.0	-0.107	-0.239	0.0	0.0	0.0	0.0	0.0	0.0	-0.128
13 N	4	0.343	0.0	0.0	-0.382	-0.349	0.0	0.0	0.0	0.0	0.0	0.0	0.214
14 N	4	0.0	0.0	-0.378	0.0	0.0	-0.551	0.0	0.0	-0.612	-0.470	0.0	0.0
15 N	4	0.0	-0.378	0.0	0.0	0.0	0.0	0.551	-0.612	0.0	0.0	-0.470	0.0
16 N	4	-0.012	0.0	0.0	0.507	0.526	0.0	0.0	0.0	0.0	0.0	0.0	0.440
							x	y	Y	X	X	Y	
							$1\pi_g$		$2\pi_u$		$2\pi_g$		
Case 1, 2 and 3:								i	a	m	h		
Case 4:							j			mm	h		

We have to discuss four cases.

In the first case we have the total energy of the shake-up configuration

$$\text{sh.u. energy} = E\left(q_m^\alpha+1,\; q_i^\alpha-1,\; q_a^\alpha+1\right) - E\left(q_m^\alpha, q_i^\alpha, q_a^\alpha\right) \qquad (L.32)$$

We observe that this definition differs from that in eq.(J.1) since we here study the total energy. Use of eqs.(B.10), (B.11) and (G.31) gives

$$\begin{aligned}(L.31) = {} & H_{aa}^\alpha + \sum_k q_k J_{ak} - \sum_k q_k^\alpha K_{ak} - \\ & - H_{ii}^\alpha - \sum_k q_k J_{ik} + \sum_k q_k^\alpha K_{ik} + \\ & + H_{mm}^\alpha + \sum_k q_k J_{mk} - \sum_k q_k^\alpha K_{mk} - \\ & - J_{ia} + K_{ia} + J_{ma} - K_{ma} - J_{mi} + K_{mi} + q_a^\alpha L_{aa} + \left[1-q_i^\alpha\right] L_{ii} + q_m^\alpha L_{mm}\end{aligned}$$

In this general expression the terms can be rearranged as above giving

$$(L.32) = {}^t\varepsilon_a^\alpha - {}^t\varepsilon_i^\alpha + {}^t\varepsilon_m^\alpha \qquad (L.33)$$

where in the transition state we have added one half electron to ψ_m^α and excited another half electron from ψ_i^α to ψ_a^α.

In the second case we calculate

$$\text{sh.u. energy} = E\left(q_m^\alpha+1,\; q_i^\beta-1,\; q_a^\beta+1\right) - E\left(q_m^\alpha, q_i^\beta, q_a^\beta\right) \qquad (L.34)$$

which gives

$$(L.34) = {}^t\varepsilon_a^\beta - {}^t\varepsilon_i^\beta + {}^t\varepsilon_m^\alpha + K_{ma} - K_{mi} \qquad (L.35)$$

In the third case we calculate

$$\text{sh.u. energy} = E\left(q_m^\beta+1,\; q_i^\beta-1,\; q_a^\alpha+1\right) - E\left(q_m^\beta, q_i^\alpha, q_a^\beta\right) \qquad (L.36)$$

which gives

$$(L.36) = {}^t\varepsilon_a^\alpha - {}^t\varepsilon_i^\beta + {}^t\varepsilon_m^\beta - K_{ia} + K_{ma} \qquad (L.37)$$

In the same way as above (eqs. (L.21–23)) we obtain

$$E^{o}_{\alpha} = {}^{t}\varepsilon_{a} - {}^{t}\varepsilon_{i} + {}^{t}\varepsilon_{m} + K_{ia} + \frac{1}{2}K_{ma} - \frac{1}{2}K_{mi} \qquad (L.38)$$

$$E^{o}_{\beta} = {}^{t}\varepsilon_{a} - {}^{t}\varepsilon_{i} + {}^{t}\varepsilon_{m} - K_{ia} + \frac{3}{2}K_{ma} + \frac{1}{2}K_{mi} \qquad (L.39)$$

$$E^{o}_{Q} = {}^{t}\varepsilon_{a} - {}^{t}\varepsilon_{i} + {}^{t}\varepsilon_{m} - K_{ia} - K_{mi} \qquad (L.40)$$

or, since the quartet lacks interest,

$$E^{o}_{\alpha} = -EA_{m} + S_{ia} + \frac{1}{2}K_{ma} - \frac{1}{2}K_{mi} \qquad (L.41)$$

$$E^{o}_{\beta} = -EA_{m} + T_{ia} + \frac{3}{2}K_{ma} + \frac{1}{2}K_{mi} \qquad (L.42)$$

The shake-up configurations can interact with a primary attachment configuration B in which one electron has been attached to ψ_h . The matrix elements are

$$\mathcal{G}_{B\alpha} = \frac{\sqrt{2}}{2}\left[2(hm|ai) - (hi|ma)\right] \quad \text{for interaction } C\alpha - B \qquad (L.43)$$

$$\mathcal{G}_{B\beta} = \frac{\sqrt{6}}{2}\cdot(hi|ma) \quad \text{for interaction } C\beta - B \qquad (L.44)$$

$$\mathcal{G}_{\alpha\beta} = \frac{\sqrt{3}}{2}\cdot\left[K_{ma} - K_{mi}\right] \quad \text{for interaction } C\alpha - C\beta \qquad (L.45)$$

but the quartet may be neglected.

The transition state used in eq.($L.33$) implies that we have added two half electrons to ψ_m^α and ψ_a^α , respectively, and removed one half from ψ_i^α . It would be of interest to see whether the diffuse calculation above is a reasonable approximation.

The exact calculation is shown below:

	3	4	5	6	7	8	9	10	11	12	13	14
	-14.860	-8.110	-8.048	-7.962	-7.648	-6.288	-6.180	-0.439	-0.342	4.448	4.518	11.464
$q_j \rightarrow$	2.000	2.000	2.000	2.000	2.000	1.500	2.000	0.500	0.500	0.0	0.0	0.0
1 N 1	0.345	-0.378	0.0	0.0	-0.343	0.0	0.0	0.0	0.0	0.0	0.0	-0.214
2 N 1	0.0	0.0	-0.014	0.382	0.0	0.024	-0.553	0.036	-0.608	0.038	-0.465	0.0
3 N 1	0.0	0.0	0.385	0.014	0.0	-0.554	-0.024	-0.607	-0.036	-0.464	-0.038	0.0
4 N 1	0.017	-0.506	0.0	0.0	-0.526	0.0	0.0	0.0	0.0	0.0	0.0	0.442
5 C 2	-0.276	0.196	0.0	0.0	0.056	0.0	0.0	0.0	0.0	0.0	0.0	0.753
6 C 2	0.0	0.0	-0.018	0.482	0.0	0.016	-0.377	-0.027	0.445	-0.059	0.713	0.0
7 C 2	0.0	0.0	0.480	0.018	0.0	-0.375	-0.016	0.447	0.026	0.715	0.059	0.0
8 C 2	0.366	0.109	0.0	0.0	0.244	0.0	0.0	0.0	0.0	0.0	0.0	-0.126
9 C 3	-0.276	-0.196	0.0	0.0	0.056	0.0	0.0	0.0	0.0	0.0	0.0	-0.753
10 C 3	0.0	0.0	-0.018	0.482	0.0	-0.016	0.377	-0.027	0.445	0.059	-0.713	0.0
11 C 3	0.0	0.0	0.480	0.018	0.0	0.375	0.016	0.448	0.026	-0.715	-0.059	0.0
12 C 3	-0.366	0.109	0.0	0.0	-0.244	0.0	0.0	0.0	0.0	0.0	0.0	-0.126
13 N 4	0.345	0.378	0.0	0.0	-0.343	0.0	0.0	0.0	0.0	0.0	0.0	0.214
14 N 4	0.0	0.0	-0.014	0.382	0.0	-0.024	0.553	0.036	-0.608	-0.038	0.465	0.0
15 N 4	0.0	0.0	0.385	0.014	0.0	0.554	0.024	-0.607	-0.036	0.464	0.038	0,0
16 N 4	-0.017	-0.506	0.0	0.0	0.526	0.0	0.0	0.0	0.0	0.0	0.0	0.442

We find ($L.33$) = -0.439 - 0.342 + 6.288 = 5.507 in reasonable agreement with the result from the "diffuse" calculation above which gives ($L.33$) = -0.089 - 0.089 + 5.905 ≈ 5.727.

In the fourth case, denoted as A , we calculate

$$\text{sh.u. energy} = E\left(q_m^\alpha + 1,\, q_m^\beta + 1,\, q_j^\beta - 1\right) - E\left(q_m^\alpha,\, q_m^\beta,\, q_j^\beta\right) \qquad (L.46)$$

Use of eqs.($B.10$), ($B.11$) and ($S.31$) gives

$$(L.46) = H_{mm}^\alpha + \sum_k q_k J_{mk} - \sum_k q_k^\alpha K_{mk} +$$

$$+ H_{mm}^\beta + \sum_k q_k J_{mk} - \sum_k q_k^\beta K_{mk} -$$

$$- H_{jj}^\beta - \sum_k q_k J_{jk} + \sum_k q_k^\beta K_{jk} +$$

$$+ J_{mm} - 2J_{jm} + K_{jm} + q_m^\alpha L_{mm} + q_m^\beta L_{mm} + \left[1 - q_j^\beta\right] L_{jj}$$

In this general expression the terms can be rearranged, giving a different formulation of eq.(L.46), namely

$$= H^{\alpha}_{mm} + \sum_{\substack{k \\ \neq j,m}} q_k J_{mk} + (q_j - \tfrac{1}{2}) J_{mi} + (q_m + 1) J_{mm} - \sum_{\substack{k \\ \neq m}} q^{\alpha}_k K_{mk} - (q^{\alpha}_m + \tfrac{1}{2}) K_{mm} +$$

$$+ H^{\beta}_{mm} + \sum_{\substack{k \\ \neq j,m}} q_k J_{mk} + (q_j - \tfrac{1}{2}) J_{mi} + (q_m + 1) J_{mm} - \sum_{\substack{k \\ \neq j,m}} q^{\beta}_k K_{mk} - (q^{\beta}_m + \tfrac{1}{2}) K_{mm} - (q^{\beta}_j - \tfrac{1}{2}) K_{mi} -$$

$$- H^{\beta}_{jj} - \sum_{\substack{k \\ \neq j,m}} q_k J_{jk} - (q_j - \tfrac{1}{2}) J_{jj} - (q_m + 1) J_{jm} + \sum_{\substack{k \\ \neq j,m}} q^{\beta}_k K_{jk} + (q^{\beta}_m + \tfrac{1}{2}) K_{jm} + (q^{\beta}_j - \tfrac{1}{2}) K_{jj} +$$

$$+ \left[(q^{\alpha}_m + \tfrac{1}{2}) - \tfrac{1}{2} \right] L_{mm} + \left[(q^{\beta}_m + \tfrac{1}{2}) - \tfrac{1}{2} \right] L_{mm} - \left[(q^{\beta}_j - \tfrac{1}{2}) - \tfrac{1}{2} \right] L_{jj}$$

which by use of eqs. (B.13) and (G.33) simply means

$$(L.46) = {}^{t}\varepsilon^{\alpha}_{m} + {}^{t}\varepsilon^{\beta}_{m} - {}^{t}\varepsilon^{\beta}_{j} \qquad (L.47)$$

where the transition state means that one half electron has been added to ψ^{α}_m and one half has been excited from ψ^{β}_j to ψ^{β}_m.

The shake-up configuration A can interact with a primary attachment configuration B, in which one electron has been attached to ψ_h. The matrix element is

$$G_{BA} = (jm|mh) \qquad (L.48)$$

Finally, there is interaction between A and $C\alpha$ and $C\beta$.

$$G_{A\alpha} = \frac{\sqrt{2}}{2} \left[2(ia|mj) - (ij|ma) \right] \qquad \text{for} \qquad (L.49)$$

interaction $A(j \to m) - C\alpha(i \to a)$

$$G_{A\beta} = -\frac{\sqrt{6}}{2} \cdot (ij|ma) \qquad \text{for} \qquad (L.50)$$

interaction $A(j \to m) - C\beta(i \to a)$

The secular determinant becomes finally

$$\begin{vmatrix} -EA_h - E & G_{B\alpha} & G_{B\beta} & G_{BA} \\ G_{B\alpha} & E^{\circ}_{\alpha} - E & G_{\alpha\beta} & G_{A\alpha} \\ G_{B\beta} & G_{\alpha\beta} & E^{\circ}_{\beta} - E & G_{A\beta} \\ G_{BA} & G_{A\alpha} & G_{A\beta} & E^{\circ}_{A} - E \end{vmatrix} = 0$$

L.6. Shake ups in EA in small molecules: some results

Shake up in EA in cyanogen [2]

The shake-up processes are defined in Sec.L.5.

The EA of orbitals 10 and 11 are given in Sec.J.3. as +0.10 eV and the IP of orbitals 8 and 9 in Sec.H.5. as 13.4 eV. This gives

$$S_{ia} = {}^t\varepsilon_a - {}^t\varepsilon_i + K_{ia} = -0.10 + 6.105 + 1.30 = 7.305 \text{ eV}$$

$$T_{ia} = {}^t\varepsilon_a - {}^t\varepsilon_i - K_{ia} = \qquad = 4.705 \text{ eV}$$

and

$$E^{\circ}_{\alpha} = -0.10 + 7.305 + \tfrac{1}{2}\cdot 0.18 - \tfrac{1}{2}\cdot 0.19 = 7.20 \text{ eV}$$

$$E^{\circ}_{\beta} = -0.10 + 4.705 + \tfrac{3}{2}\cdot 0.18 + \tfrac{1}{2}\cdot 0.19 = 4.97 \text{ eV}$$

$$E^{\circ}_{A} = -0.10 - 0.10 + 6.105 = 5.91 \text{ eV}$$

The matrix elements are taken from the HAM/3 print-out and have been calculated according to Sec.I.2.

The only large quantity which cannot be checked from experiment is therefore the one-determinant eigenvalue of orbital 12. Since according to Sec.H.5. the bonding π orbital $1\pi_u$ is calculated too high in HAM/3 (-15.069 instead of -15.6), we expect that the antibonding orbital $2\pi_g$ is calculated much too low. Instead of 4.713 eV we take the eigenvalue 6.3 eV. This gives the secular determinant

$$\begin{vmatrix} 6.30-E & -1.57 & -0.07 & 1.26 \\ -1.57 & 7.20-E & 0 & -1.17 \\ -0.07 & 0 & 4.97-E & 0.22 \\ 1.26 & -1.17 & 0.22 & 5.91-E \end{vmatrix} = 0$$

Solving the matrix gives the following results:

a) The primary attachment is displaced from 6.30 to 5.08 eV with intensity 93 %.

b) $C\alpha$ is obtained at 9.25 eV with intensity about 100 %.

c) $C\beta$ is obtained at 4.67 eV with intensity 100 %.

d) A is obtained at 5.38 eV with intensity 32 %.

A comparison with experiment [21] gives the following result.

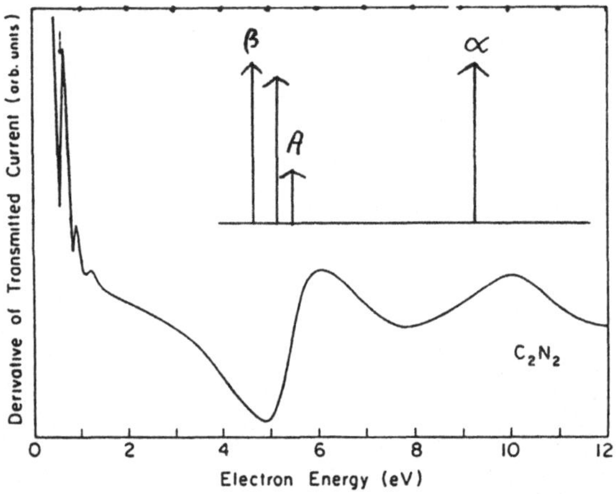

The agreement with experiment indicates that both the band at 5.5 eV and the band at 9.1 eV may be caused by formation of negative ions. Comparison with the UV spectrum, presented in Sec.I.10., indicates that the band at 9.1 eV can not be due to triplet or singlet excitation. For the band at 5.5 eV no such statement can, however, be made.

This is the fifth study of cyanogen. Previously have photoelectron spectrum, UV spectrum, electron affinities and shake up in the photoelectron spectrum been studied.

Shake up in EA in butadiene

The orbitals in butadiene are presented in Sec.H.5. and Sec. J.3. With two π^* orbitals the only shake up of interest is $11 \rightarrow 12$ of Type A with energy $E_A = 2 \cdot 0.502 + 3.607 = 4.611$ eV.

It interacts with the primary attachment in orbital 13:

$$\begin{vmatrix} 3.409 - E & -1.01 \\ -1.01 & 4.611 - E \end{vmatrix} = 0$$

Diagonalization gives: primary attachment at 2.84 eV and A at 5.18 eV. This result is compared with experiment in Sec.J.3.

L.7. The UV spectrum of the naphthalene anion

An important way to study a negative ion is to study its absorption spectrum. The main difficulty is then to produce anions in sufficient quantities.

If naphthalene is dissolved in an organic solvent (2-methyl-tetrahydrofuran) and sodium metal is added, the 3s electron will leave the sodium and go to the naphthalene and enter its lowest unoccupied orbital, 25 [22] . Another method is to irradiate the solution with γ -rays [23].

The UV absorption spectrum of the naphthalene radical anion is shown below [24 - 26].

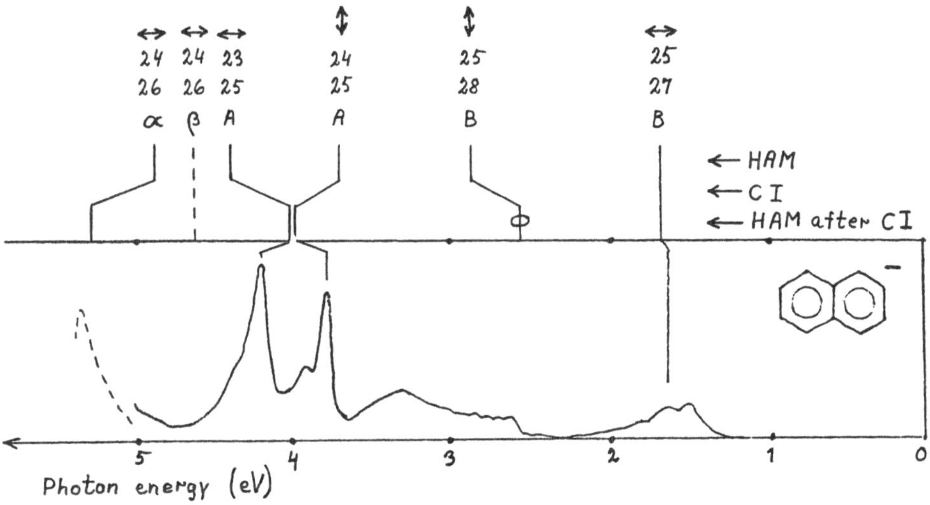

Since naphthalene is a large molecule with many orbitals and complicated symmetries, we will not give a complete presentation here (for a partial discussion of photoelectron spectrum and singlet excitations in UV, see ref. [26]).

In the spectrum above the different peaks have different origin. If a calculation is not accurate enough, the calculated energies will be displaced with respect to each other. This can easily spoil our possibilities to find the correct interpretation.

We can therefore not rely upon the calculated eigenvalues but must correct them first.

In ref. [26] the experimental PES of naphthalene is compared with HAM/3 calculations. It appears that some HAM eigenvalues have to be corrected: ε_{22} by +0.04 eV, ε_{23} by +0.45 eV and ε_{24} by +0.45 eV.

In ref. [26] the experimental UV spectrum is found to agree very closely with the calculated UV spectrum (supposed that the Mataga γ is used). This shows that the unoccupied orbitals have to be corrected by the same amount: ε_{25} by +0.45 eV and ε_{26} by +0.45 eV.

Since we have no information about how to correct ε_{27} , ε_{28} and ε_{29} , we will use the HAM/3 values.

Finally we assume that all matrix elements can be taken from the HAM calculation, partly since they are small, partly since the forms of the orbitals depend mainly upon the symmetries.

We are now prepared to discuss the UV spectrum.

Since the extra electron in the anion is in orbital 25, we find that the first excitation is $25 \rightarrow 27$ with excitation energy $= \varepsilon_{27} - \varepsilon_{25}$. No K_{ia} is involved since 25 is singly occupied. Our use of the HAM/3 value for ε_{27} is supported by the agreement with the spectrum. (The excitation $25 \rightarrow 26$ is forbidden.)

Next, the excitations $25 \rightarrow 28$ and $24 \rightarrow 25$ interact. The excited configuration in the first excitation is of Type B and in the second of Type A. The energy displacements are small, but $24 \rightarrow 25$ takes nearly all intensity.

Third, excitation $24 \rightarrow 26$ gives $C\alpha$ and $C\beta$, which interact with $23 \rightarrow 25$ of Type A. Again, the displacements are small, but $C\beta$ gets a negligible intensity.

The calculated result is plotted on top of the experimental anion spectrum above. We see that the agreement is reasonable.

Since the CI calculations do not influence the energies too much, the explanation for the agreement is that the eigenvalues, which we have used, are good. Our correction procedure, using PES and UV, is thus proven to be reliable.

In previous studies calculated eigenvalues have been used and the explanation for the bands is therefore different in different studies [7, 24, 25, 27]. Our result is most similar to that in ref. [7] .

L.8. Shake ups in EA in larger molecules

Shake ups in EA in naphthalene

The electron affinity spectrum of naphthalene has been recorded in several laboratories [28-31] with very nearly the same results. The first three bands are strong and well defined, the following five are weaker and very broad. The spectrum is shown below.

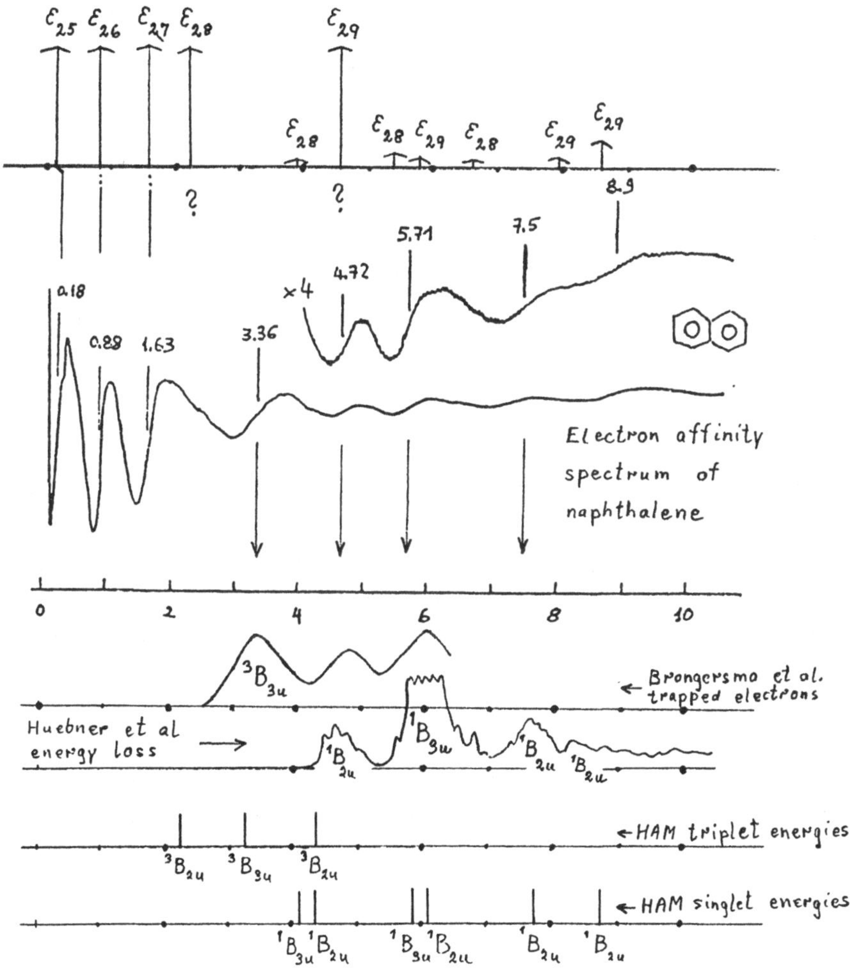

The highest π^* orbital in naphthalene has according to Table J.3. ε_{29} = +5.60 eV. Since excitation energies in naphthalene in some cases are as low as 3 eV, there are many shake ups which can interact with this primary attachment configuration. This gives a determinant, in which many weak shake ups have been omitted.

$$
\begin{array}{ccccccccc}
\text{prim.} & \alpha & \beta & \alpha & \beta & A & \alpha & \beta & A
\end{array}
$$

$$
\begin{vmatrix}
5.60-E & 0.76 & 0.54 & 0.78 & 0.21 & 0.51 & 0.81 & 0.55 & -0.67 \\
0.76 & 6.68-E & 0.74 & 0 & 0 & 0 & 0.45 & 0 & 0 \\
0.54 & 0.74 & 6.90-E & 0 & 0 & 0 & 0 & 0.27 & 0 \\
0.78 & 0 & 0 & 7.80-E & 0.45 & 0 & 0 & 0 & 0 \\
0.21 & 0 & 0 & 0.45 & 7.86-E & 0 & 0 & 0 & 0 \\
0.51 & 0 & 0 & 0 & 0 & 5.66-E & 0 & 0 & 0 \\
0.81 & 0.45 & 0 & 0 & 0 & 0 & 7.63-E & -0.25 & 0 \\
0.55 & 0 & 0.27 & 0 & 0 & 0 & -0.25 & 6.76-E & 0 \\
-0.67 & 0 & 0 & 0 & 0 & 0 & 0 & 0 & 7.38-E
\end{vmatrix} = 0
$$

We observe that all shake ups have higher energy than +5.60 eV. All interactions will therefore push down the primary attachment energy, and after the diagonalization it has got the energy +4.56 eV. In the diagram above this is plotted as an arrow, marked ε_{29}. The shake ups from this determinant, which all obtain their intensities from ε_{29}, are also marked ε_{29}.

In the same way the next highest orbital with ε_{28} = +2.94 eV is pushed down to +2.17 eV, but the following orbitals with ε_{27} = +1.78 and ε_{26} = +0.94 and ε_{25} = +0.09 eV are so far below the shake ups that they are pushed down very little.

When we compare the experimental electron affinity spectrum above with the HAM/3 EA's we see that there is a very good agreement for the first three. But ε_{28} and ε_{29} have no counterpart in the experimental curve. We have varied ε_{29} = +5.60 eV, but no real improvement could be obtained. We must also observe a restraint: the non-degenerate orbital energy in benzene is approximately the average of ε_{28} and ε_{29} (see [32]).

It was stressed in Sec.J.2.a) that the trapped electron method exhibits maxima for EA's and excitation in the same curve. This observation makes it natural to inquire whether perhaps excitation processes are responsible for the higher bands in the electron affinity spectrum.

In the lower part of the figure we have plotted the result from a trapped electron study [33] (electron energy $\approx$ 0 after the collision), which shows mainly triplets, and an electron impact energy loss study [34] (incident electron energy 100 eV), which shows only singlets. The interpretations follow from the HAM/3 calculations at the bottom (see also ref. [26]).

The coincidence of the energies indicates that the features in the electron affinity spectrum at 3.36 eV, 4.72 eV, 5.71 eV, 7.5 eV and perhaps also 8.9 eV are due to triplet and singlet excitation, as already discussed in Sec.J.2.d).

Shake ups in EA in benzene

The electron affinity spectrum of benzene has been recorded by Burrow et al. [35] and others. The band at 1.15 eV exhibits vibrational structure, the band at 4.85 eV is diffuse.

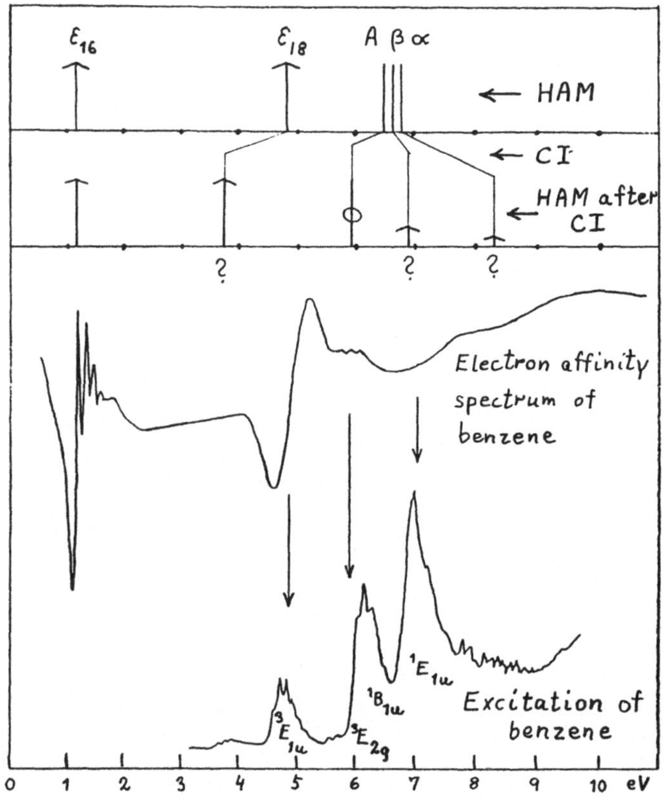

We show first the π orbitals, obtained in an EA calculation.

Orbital	13	14	15	16	17	18
Eigenvalue	-6.676	-4.317	-4.317	+1.107	+1.107	+4.857
Species in C_{2v}	b_1	a_2	b_1	a_2	b_1	b_1
Node planes	|	—	+	X	✳	

We assume that the extra electron in the anion is in orbital 16.

The eigenvalues 14, 15, 16 and 17 are rather good as can be seen from the PES in Sec.H.5. and from the UV spectrum in Sec.I.10. No correction is therefore required.

The electron affinity spectrum exhibits an EA at -1.15 eV, which directly corresponds to our negative eigenvalue ε_{16} or ε_{18} : -1.107 eV.

The primary attachment to orbital 18 interacts strongly with three shake ups:

A: attachment to 16 and excitation 15 → 16: E_A = 6.53 eV

α: attachment to 16 and excitation 14 → 17: E_α = 6.87 eV

β: attachment to 16 and excitation 14 → 17: E_β = 6.67 eV

The determinant

$$\begin{array}{cccc} \text{prim.} & A & \alpha & \beta \end{array}$$

$$\begin{vmatrix} 4.86-E & 1.07 & 1.06 & 0.78 \\ 1.07 & 6.53-E & 0.74 & -0.55 \\ 1.06 & 0.74 & 6.87-E & -0.68 \\ 0.78 & -0.55 & -0.68 & 6.67-E \end{vmatrix} = 0$$

has thus three shake ups of higher energy. The primary attachment energy is therefore pushed down to 3.76 eV as indicated in the figure. We see that this value has no direct counterpart in the experimental spectrum.

Looking for the same explanation as for naphthalene we have therefore inserted the electron impact energy loss spectrum from Sec.I.10. in the diagram. We see that the strongest triplet, $^3E_{1u}$, seems to be responsible for the band at 4.85 eV in the electron affinity spectrum, and that also two singlets can be seen as weak bands in the spectrum. It is remarkable that the observed EA value, 4.85 eV [35], agrees exactly with the vertical energy in $^3E_{1u}$, 4.85 eV [36].

Another way to attain agreement between the calculated and the observed EA would be to assume that the eigenvalue ε_{18} is higher than 4.857 eV. Due to the properties of the CI calculation it is then necessary to have ε_{18} as high as 6.80 eV. Such a high value is in disagreement with chemical experience (π-electron theory) and is improbable.

References.

1. E. Lindholm and L. Åsbrink, J. Electron Spectrosc. 18, 121 (1980).
2. D.P. Chong, L. Åsbrink and E. Lindholm, in preparation.
3. S.P. McGlynn, L.G. Vanquickenborne, M. Kinoshita and D.G. Carroll, Introduction to Applied Quantum Chemistry, Holt, Rinehart and Winston, New York (1972).
4. H.C. Longuet-Higgins and J.A. Pople, Proc.Phys.Soc. A 68, 591 (1955).
5. P. Balk, S. de Bruijn and G.J. Hoytink, Rec.Trav.Chim.Pays-Bas 76, 907 (1957).
6. R. Zahradnik and P. Carsky, J.Phys.Chem. 74, 1235 (1970).
7. A. Ishitani and S. Nagakura, Theor.Chim.Acta 4, 236 (1966).
8. D.A. Lowitz, J.Chem.Phys. 46, 4698 (1967).
9. P. Carsky and R. Zahradnik, Topics in Current Chemistry 43, 1 (1973).
10. R.L. Martin and E.R. Davidson, Chem.Phys.Letters 51, 237 (1977).
11. L.S. Cederbaum, W. Domcke, J. Schirmer, W. von Niessen, G.H.F. Diercksen and W.P. Kraemer, J.Chem.Phys. 69, 1591 (1978).
12. J. Baker, Chem.Phys.Letters 101, 136 (1983).
13. H. Nakatsuji, J.Chem.Phys. 80, 3703 (1984).
14. L.S. Cederbaum, W. Domcke, J. Schirmer and H. Köppel, J.Chem.Phys. 72, 1348 (1980).
15. J. Müller, R. Arneberg, H. Ågren, R. Manne, P.Å. Malmqvist, S. Svensson and U. Gelius, J.Chem.Phys. 77, 4895 (1982).
16. S. Svensson, P.Å. Malmqvist, M.Y. Adam, P. Lablanquie, P. Morin and I. Nenner, Chem.Phys.Letters 111, 574 (1984).
17. L. Åsbrink, C. Fridh and E. Lindholm, Chem.Phys. 27, 159 (1978).
18. D. Demoulin, Chem.Phys. 11, 329 (1975).
19. D.G. Wilden, P.J. Hicks and J. Comor, J.Phys.B: Atom. Molec. Phys. 10, L403 (1977).
20. K.H. Kochem, W. Sohn, K. Jung, H. Erhardt and E.S. Chang, J.Phys.B: Atom. Molec. Phys. 18, 1253 (1985).
21. L. Ng, V. Balaji and K.D. Jordan, Chem.Phys.Letters 101, 171 (1983).
22. P. Balk, G.J. Hoytink and J.W.H. Schreurs, Rec.Trav.Chim. Pays-Bas 76, 813 (1957).

References (cont.)

23. T. Shida and W.H. Hamill, J.Chem.Phys. 44, 2375, 4372 (1966).

24. P. Balk, S. de Bruijn and G.J. Hoytink, Rec.Trav.chim. Pays-Bas 76, 908 (1957).

25. T. Shida and S. Iwata, J.Am.Chem.Soc. 95, 3473 (1973).

26. L. Åsbrink, C. Fridh and E. Lindholm, Z.Naturforsch. 33a, 172 (1977).

27. H.M. Chang, H.H. Jaffé and C.A. Masmanidis, J.Phys.Chem. 79, 1118 (1975).

28. K.D. Jordan and P.D. Burrow, Chem.Phys. 45, 171 (1980).

29. P.D. Burrow, private information.

30. M. Allan, private information.

31. A. Modelli, private information.

32. E. Lindholm, C. Fridh and L. Åsbrink, Faraday Discuss.Chem. Soc. 54, 127 (1972).

33. G.J. Verhaart, P. Brasem and H.H. Brongersma, Chem.Phys. Letters 62, 519 (1979).

34. R.H. Huebner, S.R. Mielczarek and C.E. Kuyatt, Chem.Phys. Letters 16, 464 (1972).

35. P.D. Burrow, J.A. Michejda and K.D. Jordan, J.Am.Chem.Soc. 98, 6392 (1976).

36. J.P. Doering, J.Chem.Phys. 67, 4065 (1977).

M. Total energy

M.1. The total energy of a molecule

The total energy of a molecule is obtained from eqs.(D.22 - 29). The different terms have been discussed in Sec.F.3.

The HAM/3 method can be described as an approximation to the rigorously derived HAM model. In Sec.F.1. most of the approximations are described. It is therefore sufficient here to mention the results from the use of the HAM/3 program.

Table M.1., which is taken from a study of cyclobutane [1], gives the difference between the experimental and theoretical total energies (the "error") for some hydrocarbons using HAM/3. The unit, kcal/mol, is further discussed in Sec.L.2.

Table M.1.

Heats of formation, calculated with HAM/3. The table gives the difference between experimental and theoretical heats of formation (the "error") for the experimental geometries

Molecule	Error (kcal/mol)
methane	0
ethane	−4
propane	−2
n-butane	−11
cyclohexane	−2
norbornane	−12
adamantane	−10
ethylene	1
propene	4
2-butene-cis	3
2-butene-trans	1
cyclopentene	−12
cyclohexene	−5
allene	2
butadiene	8
cyclopentadiene	−12
norbornadiene	−40
acetylene	6
diacetylene	5
benzene	8
naphthalene	−3
fulvene	−22
dimethylfulvene	−23
cyclobutane	−37
cyclopropane	−65
cyclopropene	−86
spiropentane	−144

The table shows that there is generally a reasonable agreement between the HAM/3 total energies and the experimental ones. The explanation is, of course, that the correlation energy has been taken care of in the method.

A comparison the the HAM/3 results with results from other methods shows first of all that Hartree-Fock ab-initio calculations (without CI) give an "error" which is much larger since in such calculations the correlation energy is not taken care of.

The semiempirical methods MINDO/3 and MNDO [2] give "errors", which are smaller by a factor of 2 or 3 compared to the HAM/3 results. Those methods are parametrized especially to give good total energies. It is therefore satisfactory to find that HAM/3, which was parametrized mainly for other properties of the molecule, can be considered as competitive also in this respect.

Regarding the possibilities to improve the HAM results in the future the following can be said.

There is no doubt that eqs.(D.22-29) can represent the total energy of a molecule with good accuracy. A less approximate treatment can therefore be expected to give considerably reduced errors, which would mean an approach towards chemical accuracy.

A difficulty in this prediction is the large errors in Table M.1. for small rings, for which the calculation gives too high total energies (too unstable compounds). This result, which is colloquially called "strain" , seems to be impossible to understand from eqs.(D. 22-29).

M.2. Heat of formation ΔH_f

The total energy of ethylene, which comes out from a HAM/3 calculation, is -2137.54 eV. This number is too large to be convenient in chemistry. Therefore the dissociation energy is calculated instead.

The dissociation energy is defined as the energy, which must be supplied to a molecule to decompose it into atoms.

In ethylene the calculated energy of two carbon atoms and four hydrogen atoms is according to Sec. E.5.

$$- 2\cdot 1029.8749 - 4\cdot 13.606 = - 2114.17 \text{ eV}$$

The calculated dissociation energy is thus

$$- 2114.17 + 2137.54 = 23.37 \text{ eV}.$$

Also this number is too large to be convenient. Therefore, the heat of formation is calculated instead.

The heat of formation is defined as the energy which is released when the molecule is changed into graphite and gas.

When four hydrogen atoms form $2H_2$ $2 \cdot 4.52$ eV are released (4.52 eV is the dissociation energy of H_2) and when two carbon atoms form graphite $2 \cdot 7.41$ eV are released (7.41 eV is the heat of sublimation of graphite). Altogether 23.86 eV are released.

The total release of energy when one ethylene molecule is changed into $2H_2$ + graphite is therefore:

heat of formation = 23.86 - 23.37 =

= + 0.49 eV

= + $0.49 \cdot 23.06$ kcal/mol

= + 11.3 kcal/mol

Experimental value + 12.5 kcal/mol

The heat of formation measures thus the total energy. A very stable molecule has a negative heat of formation (CH_4: -17.9 kcal/mol) and an unstable molecule a large, positive heat of formation.

Since most changes in nature try to diminish the total energy of the molecule, the total energy or the heat of formation govern the geometry of the molecule and its chemical reactions.

In the HAM/3 program the printout gives always the heat of formation, ΔH_f, for the molecule or ion, which is studied in the SCF calculation.

M.3. Check of the transition state method

In Sec.H.1. the ionization energy was calculated by eq. (H.1)

$$IP_i = E(q_i - 1) - E(q_i) \qquad (H.1)$$

This expression was then rearranged and gave eq.(H.6)

$$IP_i = -\,{}^t\varepsilon_i \qquad (H.6)$$

We intend now to verify that these two methods give the same result also when the computer and the HAM/3 program are involved. We choose orbital 6 in ethylene for our test.

In Sec.H.5. the eq.(H.6) calculation gave ε_6 = -10.912 eV.

To perform the eq.(H.1) calculation we remove one whole electron from orbital 6 and obtain the print-out

	1	2	3	4	5	6	7	8	9	10	11	12
	-27.812	-23.144	-19.749	-18.640	-16.493	-14.880	-8.370	12.585	16.274	23.308	27.675	34.404
q_j →	2.000	2.000	2.000	2.000	2.000	1.000	0.0	0.0	0.0	0.0	0.0	0.0
1 C 1	0.482	0.345	0.000	0.049	-0.000	0.000	-0.000	0.668	0.831	0.000	0.015	0.000
2 C 1	-0.070	0.225	0.000	-0.526	0.000	-0.000	-0.000	0.405	-0.405	0.000	0.992	0.000
3 C 1	-0.000	0.000	-0.435	-0.000	-0.496	0.000	-0.000	0.000	-0.000	0.677	0.000	-0.849
4 C 1	-0.000	-0.000	0.000	0.000	-0.000	-0.642	-0.797	0.000	0.000	-0.000	0.000	-0.000
5 C 2	0.482	-0.345	-0.000	0.049	0.000	0.000	0.000	0.668	-0.831	-0.000	-0.015	-0.000
6 C 2	0.070	0.225	-0.000	0.526	-0.000	0.000	0.000	-0.405	-0.405	0.000	0.992	0.000
7 C 2	0.000	0.000	-0.435	-0.000	0.496	-0.000	-0.000	0.000	-0.000	0.677	-0.000	0.849
8 C 2	0.000	0.000	0.000	0.000	-0.000	-0.642	0.797	0.000	0.000	-0.000	-0.000	0.000
9 H 3	0.161	0.246	-0.235	-0.179	-0.291	0.000	-0.000	-0.470	-0.281	-0.640	-0.486	0.657
10 H 4	0.161	0.246	0.235	-0.179	0.291	-0.000	-0.000	-0.470	-0.281	0.640	-0.486	-0.657
11 H 5	0.161	-0.246	-0.235	-0.179	0.291	-0.000	-0.000	-0.470	0.281	-0.640	0.486	-0.657
12 H 6	0.161	-0.246	0.235	-0.179	-0.291	0.000	-0.000	-0.470	0.281	0.640	0.486	0.657

HEAT OF FORMATION= 263.512 KCAL/MOL = 11.427EV

Then we study the neutral molecule and obtain the print-out

	1	2	3	4	5	6	7	8	9	10	11	12
	-20.734	-16.444	-12.858	-11.397	-9.902	-7.062	-0.968	18.268	22.723	27.525	32.079	38.175
1 C 1	0.459	-0.328	-0.000	0.091	0.000	-0.000	-0.000	0.680	0.837	0.000	-0.003	-0.000
2 C 1	-0.051	-0.208	0.000	-0.521	0.000	-0.000	-0.000	0.415	-0.370	-0.000	1.009	0.000
3 C 1	-0.000	0.000	0.414	0.000	0.475	-0.000	0.000	-0.000	-0.000	0.690	0.000	-0.861
4 C 1	0.000	0.000	0.000	-0.000	0.000	0.642	0.797	0.000	-0.000	0.000	0.000	-0.000
5 C 2	0.459	0.328	-0.000	0.091	-0.000	0.000	0.000	0.680	-0.837	-0.000	0.003	-0.000
6 C 2	0.051	-0.208	-0.000	0.521	-0.000	0.000	-0.000	-0.415	-0.370	-0.000	1.009	0.000
7 C 2	0.000	-0.000	0.414	-0.000	-0.475	-0.000	0.000	0.000	-0.000	0.690	0.000	0.861
8 C 2	0.000	-0.000	0.000	-0.000	0.000	0.642	-0.797	-0.000	0.000	-0.000	-0.000	0.000
9 H 3	0.184	-0.263	0.255	-0.180	0.306	-0.000	0.000	-0.461	-0.291	-0.633	-0.472	0.649
10 H 4	0.184	-0.263	-0.255	-0.180	-0.306	0.000	0.000	-0.461	-0.291	0.633	-0.472	-0.649
11 H 5	0.184	0.263	0.255	-0.180	-0.306	-0.000	-0.000	-0.461	0.291	-0.633	0.472	-0.649
12 H 6	0.184	0.263	-0.255	-0.180	0.306	-0.000	0.000	-0.461	0.291	0.633	0.472	0.649

HEAT OF FORMATION= 11.133 KCAL/MOL = 0.483EV

The difference 11.427 - 0.483 = 10.944 eV agrees well with the eq.(H.6) value 10.912 eV above.

M.4. Doubly charged ions [3]

To find the energy of a doubly charged ion in relation to that of the molecule ("appearance potential"), the doubly charged molecule is studied by use of the option CHARGREF 2. We find with N_2 as an example

NITROGEN 4 DOUBLY OCCUPIED MOLECULAR ORBITALS CHARGE= 2.000

			1	2	3	4	5	6	7	8
			-54.940	-34.787	-32.770	-32.770	-32.342	-22.644	-22.644	14.406
1	N	1	-0.584	-0.532	0.0	0.0	0.383	0.0	0.0	0.556
2	N	1	0.0	0.0	0.605	-0.250	-0.000	0.772	0.0	0.0
3	N	1	0.0	0.0	0.250	0.605	0.0	0.0	-0.772	0.0
4	N	1	0.225	-0.416	0.000	0.0	0.595	0.0	0.0	-0.789
5	N	2	-0.586	0.543	0.0	0.0	0.365	0.0	0.0	-0.556
6	N	2	0.0	0.0	0.611	-0.252	-0.000	-0.767	0.0	0.0
7	N	2	0.0	0.0	0.252	0.611	0.0	0.0	0.767	0.0
8	N	2	-0.226	-0.436	-0.000	0.0	-0.581	0.0	0.0	-0.788

HEAT OF FORMATION= 919.775 KCAL/MOL = 39.884EV

For the neutral molecule we obtain

NITROGEN 5 DOUBLY OCCUPIED MOLECULAR ORBITALS CHARGE= 0.000

			1	2	3	4	5	6	7	8
			-31.523	-13.134	-10.892	-10.892	-10.153	-2.462	-2.462	30.933
1	N	1	-0.601	0.529	0.0	0.0	0.349	0.0	0.0	0.564
2	N	1	0.0	0.0	0.658	0.0	0.0	0.0	0.770	0.0
3	N	1	0.0	0.0	0.0	0.658	0.0	0.770	0.0	0.0
4	N	1	0.200	0.438	0.0	0.0	0.597	0.0	0.0	-0.782
5	N	2	-0.601	-0.529	0.0	0.0	0.349	0.0	0.0	-0.564
6	N	2	0.0	0.0	0.658	0.0	0.0	0.0	-0.770	0.0
7	N	2	0.0	0.0	0.0	0.658	0.0	-0.770	0.0	0.0
8	N	2	-0.200	0.438	0.0	0.0	-0.597	0.0	0.0	-0.782

HEAT OF FORMATION= -52.485 KCAL/MOL = -2.276EV

The appearance potential of N_2^{++} is thus 39.884 + 2.276 = = 42.160 eV to compare with the experimental value 42.9 eV [3].

The importance of this study is that the agreement with experiment is reasonable although the correlation energy change is large.

Chong [4] has demonstrated how Auger electron spectra can be obtained by studying the excitation of the doubly charged molecule.

References

1. L. Åsbrink, C. Fridh, E. Lindholm and G. Ahlgren, Chem.Phys. 33, 195 (1978).
2. M.J.S. Dewar and W. Thiel, J.Am.Chem.Soc. 99, 4899, 4907 (1977).
3. L. Åsbrink, C. Fridh and E. Lindholm, Int.J.Mass Spectrom. 32. 93 (1979).
4. D.P. Chong, Chem.Phys.Letters 82, 511 (1981).

N. Dipole moments

N.1. Calculation of dipole moment

The calculation of dipole moments is performed in exactly the same way in the HAM/3 computer program as in other methods (see ref. [1] page 87 (where $\zeta_A = n_\mu \zeta_\mu$)).

In a diatomic molecule $\oplus \overset{}{\underset{R}{\text{—}}} \ominus$ we assume that one atom has the charge $+Q$ (measured in atomic units) and the other $-Q$ and that the internuclear distance is R (measured in Å). The dipole moment μ is then

$$\mu = -4.80 \cdot Q \cdot R \quad \text{(unit: D = Debye)}$$

(since the dipole moment is 10^{-18} D, when Q = 1 electronic charge and R = 1 cm).

For a polyatomic molecule we have

$$\mu_x = -4.80 \cdot \sum_A Q_A \, x_A \tag{N.1}$$

For a molecule with lone-pair orbitals the local dipole moments contribute.

A lone-pair orbital can be written $2s_A + 2px_A$ which can be illustrated  or 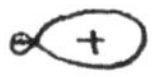 . The electron charge is thus located to the right and forms with the nucleus to the left a local dipole.

To calculate the local dipole moment we calculate

$$\text{average of } e \cdot x = \int \psi_i^* \cdot ex \cdot \psi_i \cdot d\tau \tag{N.2}$$

by inserting a LCAO expansion. This gives a term

$$\mu_x = e \cdot 2P_{sx_A} \cdot \int \phi_{2s} \cdot x \cdot \phi_{2px} \cdot d\tau \tag{N.3}$$

$$\approx P_{sx_A} \cdot \frac{7.34}{\zeta} \tag{N.4}$$

(see Sec. F.2.) with μ in Debye and x in Å.

N.2. Dipole moment of HCN

A HAM/3 calculation for neutral HCN

COORDINATES IN ANGSTROM UNITS

	ATOM	X	Y	Z
1	H	-1.0650	0.0	0.0
2	C	0.0	0.0	0.0
3	N	1.1560	0.0	0.0

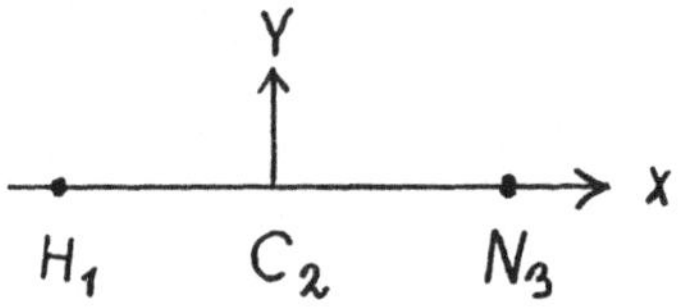

gives orbital energies and orbitals

	1	2	3	4	5	6	7	8	9
	-22.030	-16.260	-10.067	-9.440	-9.440	-0.888	-0.888	14.360	38.292
1 H 1	0.192	0.450	0.132	0.0	0.0	0.0	0.0	0.585	-1.002
2 C 2	0.500	0.356	-0.145	0.0	0.0	0.0	0.0	-0.936	0.228
3 C 2	0.168	-0.437	-0.282	0.0	0.0	0.0	0.0	-0.220	-1.174
4 C 2	0.0	0.0	0.0	0.0	0.617	-0.000	0.813	0.0	0.0
5 C 2	0.0	0.0	0.0	-0.617	-0.000	-0.813	0.0	0.0	0.0
6 N 3	0.643	-0.351	0.498	0.0	0.0	0.0	0.0	0.405	0.506
7 N 3	-0.114	-0.041	0.736	0.0	0.0	0.0	0.0	-0.669	-0.522
8 N 3	0.0	0.0	0.0	-0.000	0.674	0.0	-0.766	0.0	0.0
9 N 3	0.0	0.0	0.0	-0.674	0.0	0.766	0.000	0.0	0.0

We see that orbital 3 has lone-pair character with large $c_{s_N,i}$ and $c_{x_N,i}$. We can then calculate P_{sx_N}

$$P_{sx_N} = 2 \cdot \sum_i c_{s_N i}\, c_{x_N i} \approx 2 \cdot 0.498 \cdot 0.736 = 0.73$$

which gives in eq. (N.4)

$$\mu_x = 0.73 \cdot \frac{7.34}{2} = 2.68$$ Debye or if also the other local dipole moments are included: 1.68 D.

The print-out gives also the atomic charges:

Q on H_1 = + 0.125

Q on C_2 = + 0.047

Q on N_3 = - 0.173

which gives in eq. (N.1)

$$\mu_x = -4.80 \cdot \left[0.125 \cdot (-1.065) + (-0.173) \cdot 1.156\right] = 1.60 \text{ Debye}$$

The total dipole moment is therefore calculated as 1.60 + 1.68 = 3.28 Debye. The experimental dipole moment is 2.99 Debye.

Usually, the calculated dipole moments are less good, and HAM/3 should <u>not</u> be used to calculate dipole moments.

We will now try to illustrate our result by presenting a picture of the charge distribution in the HCN molecule. We must then consider the electronic charge outside the N atom. If we assume that the distance from this charge to the N atom is 0.5 Å, we get the correct local dipole moment with $e = -0.700$ ($\mu_x = 4.8 \cdot 0.700 \cdot 0.5 = 1.68$ D)

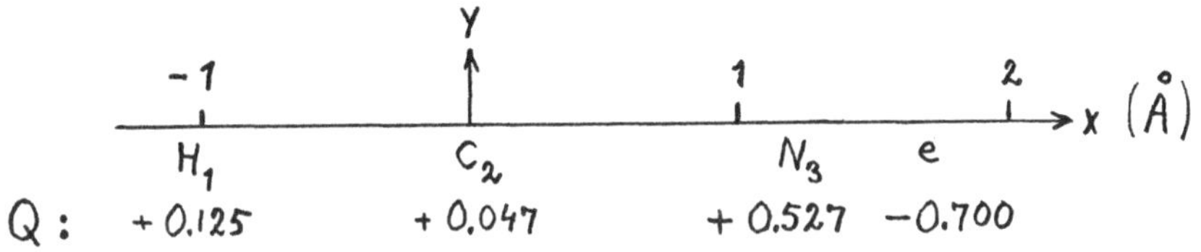

With another assumed distance we get another distribution.

This picture illustrates the statement in Sec. F.2. that in calculation of the electrostatic interaction the expression $\sum Q_A Q_B \gamma_{AB}$ is not sufficient.

Reference

1. J.A. Pople and D.L. Beveridge, Approximate Molecular Orbital Theory, McGraw-Hill, New York (1970).

O. Chemical reactions

O.1. Can a HAM model be used?

In a chemical reaction two molecules approach and form one or two other molecules. In the "transition point" the distance between the two molecules is comparatively large. Only if the theoretical method is reliable also for large internuclear distances R_{AB} can we expect to be able to handle also the "transition state" in a successful way.

It is probable that the HAM model is good enough and that the difficulty therefore concerns only its parametrization. It is obviously necessary to use molecules during the parametrization whose dimensions are comparable to the distance between the two nearest atoms in the transition state. If this distance is of the order of 3 Å, it is sufficient to use some molecules of the size of naphthalene or anthracene. It is therefore probable that a HAM method can be used.

It must be remarked here that it is usually necessary to remove the degeneracy near the transition point by use of a 2×2 CI.

Secondly, we will discuss the formal ways in which a chemical reaction can be studied. There are several possibilities.

a) One can study how the orbitals change during the chemical reaction. Both the form and the energy of an orbital must change in a continuous way during the reaction, and this means that the symmetry does not change. Such studies were performed by Woodward and Hoffmann [1].

A correct study should give the orbital energies of the neutral molecule complex. However, to enable a comparison with photoelectron spectra we present below the ionization energies instead.

b) One can also study the change of the total energy or of the heat of formation during the reaction. This is often done by use of Hartree-Fock ab-initio methods (especially by Pople and coworkers), but since the correlation energies are neglected, these studies are limited to certain types of reactions [2]. It has therefore been claimed that MINDO/3 or MNDO are better. Since HAM/3 gives total energies, which are less accurate than MNDO, the possible advantage of HAM/3 is that the studies of types a) and b) can be combined. This is done in some examples in Sec. O.2. and O.3.

O.2. Dissociation of cyclobutane [3]

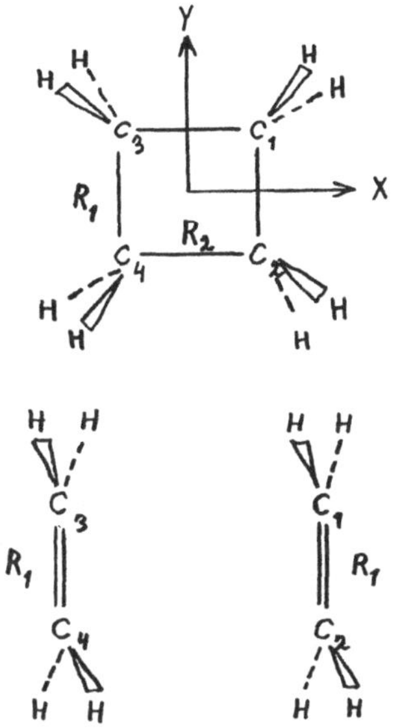

In the square molecule cyclobutane the C-C distances R_1 and R_2 are equal. If energy is supplied, R_2 (the "reaction coordinate") increases and the molecule dissociates into two ethylenes.

This is the first reaction studied by Woodward and Hoffmann in "The Conservation of Orbital Symmetry" [1].

To study the dissociation we perform HAM/3 calculations for many different R_2, from the "initial" distance 1.5 Å up to the "final" distance 4 Å. The results are shown below for R_2 = 1.55 Å and R_2 = 2.70 Å (Table O.2.a. and Table O.2.b.)

We observe that the orbitals have the same form and the same symmetry during the reaction. The reason is that the hamilton operator is symmetric and its solutions must then also be symmetric. We can therefore follow every orbital during the reaction, even if it changes its energy.

The configurational ionization energies of the orbitals are shown in Fig.O.2.c.This figure is quite reliable as to the occupied orbitals for cyclobutane and ethylene and as to the π^* orbitals of ethylene from comparisons with experiment (PES, UV and EA). It is then likely that also the intermediate configurational energies can be relied upon.

We observe from Fig.O.2.c. that most orbitals do not alter their energies. Only the "frontier orbitals", 12 and 13, change their energy with a crossing point at 2.04 Å. This defines the "transition point".

We have now completed the study, denoted as a) in Sec. O.1. and turn over to b), the study of the total energy.

Table O.2.a. Configurational ionization energies and molecular orbitals of planar cyclobutane. The reaction coordinate R_2 = 1.55 Å.

	1	2	3	4	5	7	6	9	11	10	8	13	12	17	14
	-25.837	-21.281	-21.222	-18.266	-16.397	-15.249	-13.902	-13.858	-12.150	-11.418	-11.172	-11.071	5.115	5.616	6.110
C1	0.344	0.270	0.269	0.224	0.0	-0.068	0.0	0.0	0.001	0.0	-0.092	-0.092	-0.332	-0.337	-0.007
C1	-0.033	0.120	-0.020	0.123	0.0	0.222	0.0	0.0	0.298	0.0	0.018	0.431	0.559	-0.064	0.496
C1	-0.033	-0.021	0.121	0.123	0.0	0.223	0.0	0.0	-0.297	0.0	0.430	0.015	-0.066	0.562	-0.482
C1	0.0	0.0	0.0	0.0	0.285	0.0	0.324	0.325	0.0	0.358	0.0	0.0	0.0	0.0	0.0
H1	0.096	0.158	0.159	0.192	0.173	0.153	0.197	0.198	0.000	0.220	0.093	0.092	-0.027	-0.027	0.001
	$2a_g$	$2b_{2u}$	$2b_{3u}$	$2b_{1g}$	$1b_{1u}$	$3a_g$	$1b_{3g}$	$1b_{2g}$	$4a_g$	$1a_u$	$3b_{2u}$	$3b_{3u}$	$4b_{2u}$	$4b_{3u}$	$3b_{1g}$

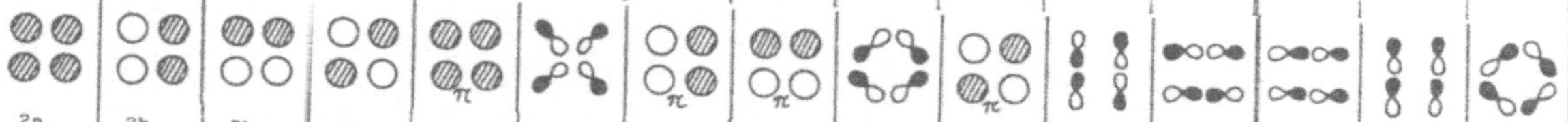

Table O.2.b. Configurational ionization energies and molecular orbitals of a dissociating cyclobutane. The reaction coordinate R_2 = 2.70 Å.

	1	2	3	4	5	6	7	8	9	10	11	12	13	14	15
	-23.674	-23.190	-19.499	-18.822	-15.807	-15.150	-14.366	-13.870	-12.884	-12.222	-10.076	-9.176	-4.165	-3.020	13.308
C1	0.321	0.334	0.226	0.239	0.0	0.0	-0.081	-0.052	0.0	0.0	0.018	-0.023	0.011	-0.013	0.477
C1	-0.007	0.009	-0.007	0.008	0.0	0.0	-0.011	0.015	0.0	0.0	0.441	0.470	0.550	0.574	-0.019
C1	-0.035	-0.039	0.142	0.153	0.0	0.0	0.359	0.373	0.0	0.0	0.013	-0.017	0.014	-0.016	0.300
C1	0.0	0.0	0.0	0.0	0.286	0.304	0.0	0.0	0.327	0.344	0.0	0.0	0.0	0.0	0.0
H1	0.129	0.129	0.185	0.185	0.181	0.179	0.129	0.128	0.217	0.215	0.008	-0.008	0.003	-0.003	-0.309
	$2a_g$	$2b_{2u}$	$2b_{3u}$	$2b_{1g}$	$1b_{1u}$	$1b_{3g}$	$3a_g$	$3b_{2u}$	$1b_{2g}$	$1a_u$	$4a_g$	$4b_{2u}$	$3b_{3u}$	$3b_{1g}$	$5a_g$

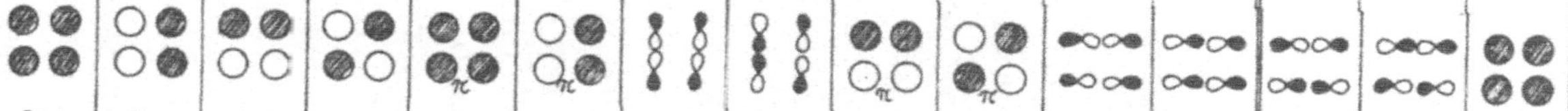

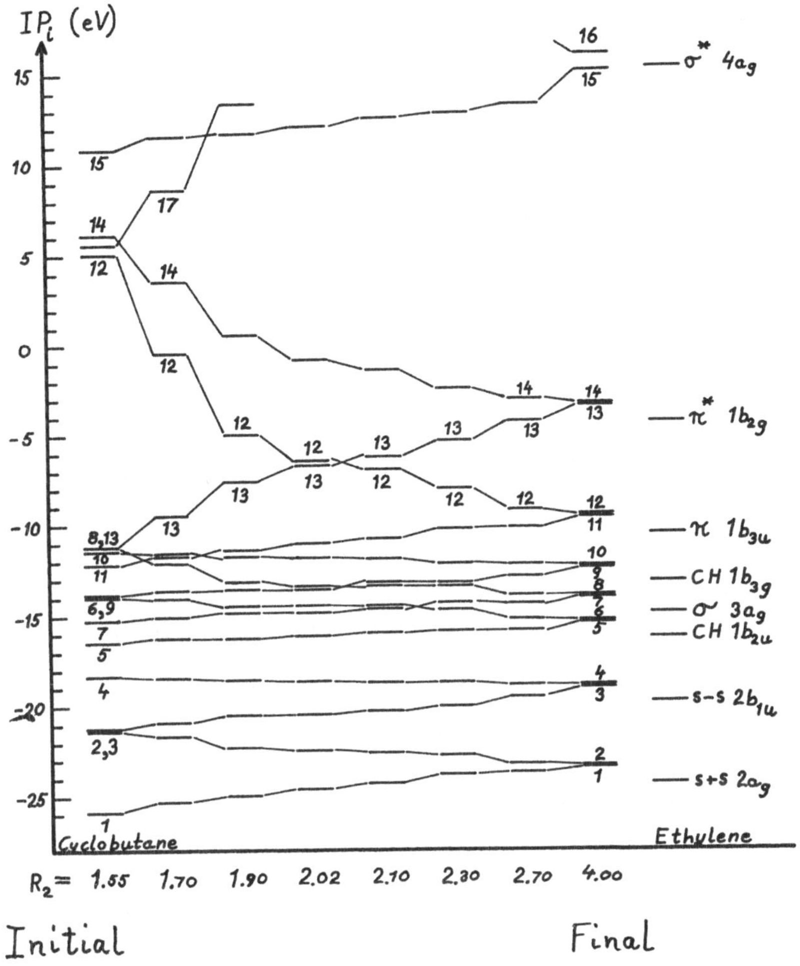

Fig.O.2.c. Configurational ionization energies (eV) of the dissociating cyclobutane at different values of the reaction coordinate R_2. In the initial (cyclobutane) and final (two ethylenes) conformations the calculated ionization energies agree reasonably well with the experimental photoelectron spectra and UV spectra.

In Fig.O.2.d. the total energies from the HAM/3 calculations are plotted for different R_2. We observe that the curves for the energies of cyclobutane and 2 ethylenes cross at 2.04 Å. This defines again the "transition point".

Since the HOMO-LUMO gap near 2.04 Å is small, we must apply the theory presented in Sec. G.8. This means that we have to

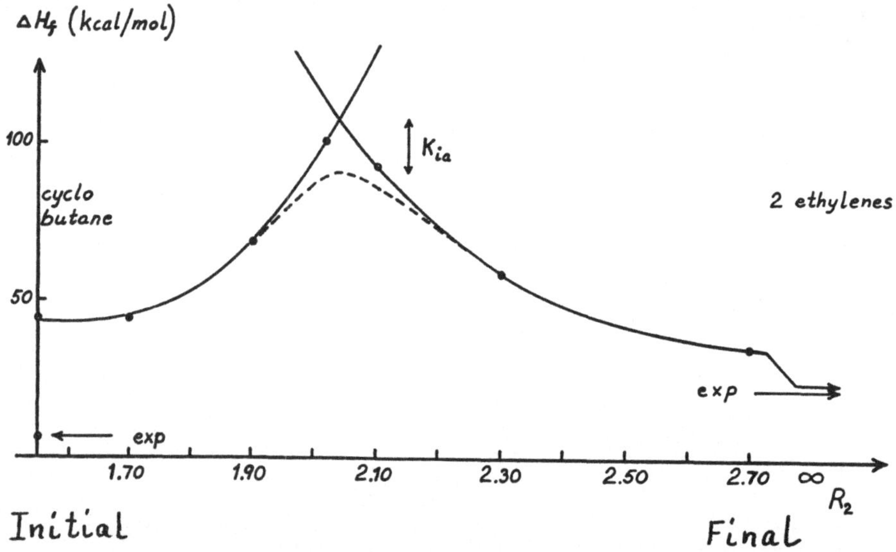

Fig.O.2.d. The total energies along the reaction path, given as heat of formation. The full lines give the configurational energies, to the left for a more or less deformed cyclobutane, to the right for the two approaching ethylenes. The dotted line gives the state energy for the ground state of the dissociating cyclobutane, of which, however, only one point, the highest, has been calculated by use of CI.

subtract (and add, respectively) the CI interaction matrix element K_{ia} from the energy at the crossing point. Around this point the change must be smaller.

The result of the last study is a value for the activation energy in reasonable agreement with the experimental values.

Remark 1: The calculations are performed at several distances but not at 2.04 Å. The reason is that the computer is instructed to put two electrons into every orbital, starting from the lowest energy. Difficulties arise when HOMO and LUMO are too near each other. It is therefore advisable to avoid this critical distance and to rely upon extrapolation instead.

Remark 2: In Fig.O.2.c. the ionization energies are denoted as "configurational". This means that they will be influenced by the 2 x 2 CI, described in Sec. G.8. The influence is important mainly for HOMO, orbital 13 or 12, for which the true IP is obtained by adding part of K_{ia} .

0.3. The internal rotation of ethylene [4]

The internal rotation of ethylene can be understood as a simple chemical reaction.

We start with a planar ethylene, supply energy and twist it until we reach the "transition point" at 90°. Then the twisting continues, energy is released and we obtain finally a "reaction product", a twisted ethylene.

To study this reaction we perform HAM/3 calculations for several twisting angles, of course avoiding the transition point 90°. The print-outs are shown in Table 0.3. and the ionization energies are plotted in Fig. 0.3.a.

We observe that the σ-type orbitals 1, 2 and 4 have about constant energies during the twisting. The pseudo-π orbitals 3 and 5 are strongly affected and are at 85° nearly degenerate.

HOMO and LUMO, 6 and 7, π and π^{*}, are also strongly influenced and are also at 85° nearly degenerate. At 90° they cross.

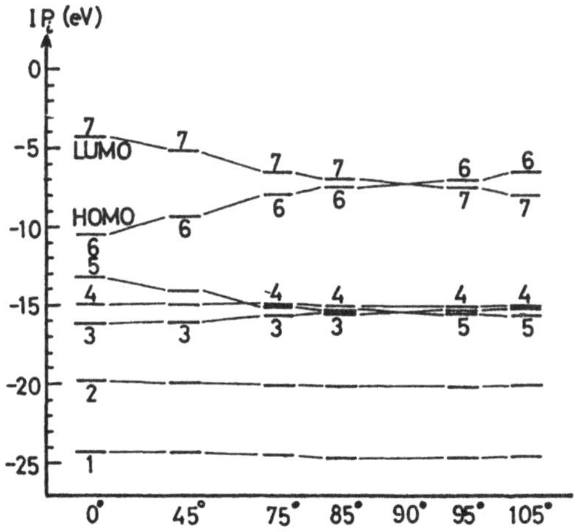

Fig. 0.3.a. Configurational ionization energies of the twisted ethylene at different values of the twist angle, calculated by HAM/3. In the limits at 0° and 180° the ionization energies agree reasonably well with experimental photoelectron and UV spectra. It is seen, that HOMO and LUMO cross during the reaction.

. ONE HALF ELECTRON DIFFUSELY REMOVED. FILLED ORBITALS GIVE IONIZATION ENERGIES

ETHYLENE PLANAR

	1	2	3	4	5	6	7	8
	-24.293	-19.849	-16.256	-14.883	-13.174	-10.539	-4.268	15.421
C 1	-0.466	-0.334	0.0	0.081	0.0	0.0	0.0	0.676
C 1	0.055	-0.213	0.0	-0.522	0.0	0.0	0.0	0.413
C 1	0.0	0.0	-0.420	0.0	-0.482	0.0	0.0	0.0
C 1	0.0	0.0	0.0	0.0	0.0	-0.642	-0.797	0.0
C 2	-0.466	0.334	0.0	0.081	0.0	0.0	0.0	0.676
C 2	-0.055	-0.213	0.0	0.522	0.0	0.0	0.0	-0.413
C 2	0.0	0.0	-0.420	0.0	0.482	0.0	0.0	0.0
C 2	0.0	0.0	0.0	0.0	0.0	-0.642	0.797	0.0
H 3	-0.177	-0.257	-0.249	-0.180	-0.301	0.0	0.0	-0.464
H 4	-0.177	-0.257	0.249	-0.180	0.301	0.0	0.0	-0.464
H 5	-0.177	0.257	-0.249	-0.180	0.301	0.0	0.0	-0.464
H 6	-0.177	0.257	0.249	-0.180	-0.301	0.0	0.0	-0.464

$2a_g$ $2b_{3u}$ $1b_{2u}$ $3a_g$ $1b_{1g}$ $1b_{1u}$ $1b_{2g}$ $4a_g$

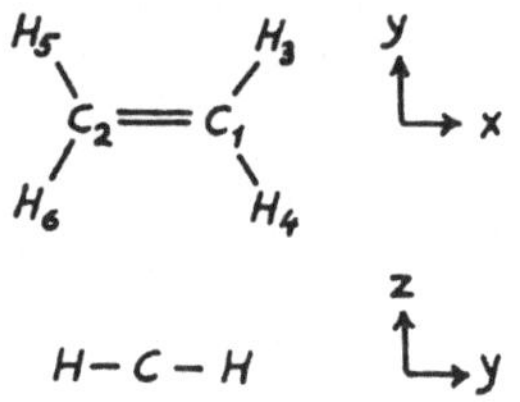

ΔH_f = 11 kcal/mol

ETHYLENE ROTATED 45 DEG

	1	2	3	4	5	6	7	8
	-24.379	-19.931	-16.009	-14.908	-14.115	-9.293	-5.158	15.619
C 1	-0.468	-0.334	0.0	-0.079	0.0	0.0	0.0	-0.677
C 1	0.055	-0.212	0.0	0.522	0.0	0.0	0.0	-0.413
C 1	0.0	0.0	-0.416	0.0	-0.343	-0.396	0.229	0.0
C 1	0.0	0.0	-0.100	0.0	-0.329	0.529	-0.737	0.0
C 2	-0.468	0.334	0.0	-0.079	0.0	0.0	0.0	-0.677
C 2	-0.055	-0.212	0.0	-0.522	0.0	0.0	0.0	0.413
C 2	0.0	0.0	-0.416	0.0	0.343	0.396	0.229	0.0
C 2	0.0	0.0	0.100	0.0	-0.329	0.529	0.737	0.0
H 3	-0.176	-0.257	-0.250	0.180	-0.269	-0.125	-0.076	0.466
H 4	-0.176	-0.257	0.250	0.180	0.269	0.125	0.076	0.466
H 5	-0.176	0.257	-0.250	0.180	0.269	0.125	-0.076	0.466
H 6	-0.176	0.257	0.250	0.180	-0.269	-0.125	0.076	0.466

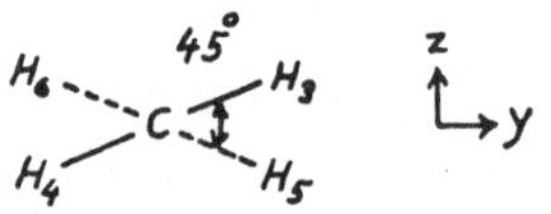

ΔH_f = 42 kcal/mol

ETHYLENE ROTATED 75 CEG

	1	2	3	5	4	6	7	8
	-24.528	-20.056	-15.594	-15.004	-14.962	-7.903	-6.483	15.722
C 1	0.469	0.336	0.0	0.000	0.078	0.0	0.0	-0.677
C 1	-0.055	0.211	0.0	-0.000	-0.523	0.0	0.0	-0.413
C 1	0.0	0.0	-0.405	0.246	-0.000	0.536	-0.351	0.0
C 1	0.0	0.0	-0.173	0.388	-0.000	-0.443	0.647	0.0
C 2	0.469	-0.335	0.0	0.0	0.078	0.0	0.0	-0.677
C 2	0.055	0.211	0.0	0.000	0.523	0.0	0.0	0.413
C 2	0.0	0.0	-0.405	-0.246	0.000	-0.536	-0.351	0.0
C 2	0.0	0.0	0.173	0.388	-0.000	-0.443	-0.647	0.0
H 3	0.175	0.257	-0.253	0.257	-0.180	0.132	0.109	0.469
H 4	0.175	0.257	0.253	-0.257	-0.180	-0.132	-0.109	0.469
H 5	0.175	-0.257	-0.253	-0.256	-0.180	-0.132	0.109	0.469
H 6	0.175	-0.257	0.253	0.256	-0.180	0.132	-0.109	0.469

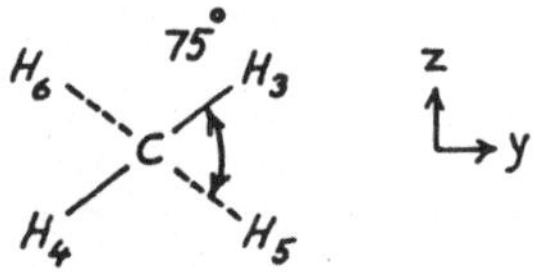

ΔH_f = 92 kcal/mol

ETHYLENE ROTATED 85 DEG.

	1	2	3	5	4	6	7	8
	-24.590	-20.104	-15.420	-15.273	-14.987	-7.430	-6.989	15.694
C 1	0.470	-0.336	0.0	0.0	-0.078	0.0	0.0	-0.676
C 1	-0.055	-0.210	0.0	0.0	0.523	0.0	0.0	-0.413
C 1	0.0	0.0	-0.398	0.220	0.0	-0.575	-0.386	0.0
C 1	0.0	0.0	-0.191	0.399	0.0	0.413	0.612	0.0
C 2	0.470	0.336	0.0	0.0	-0.078	0.0	0.0	-0.677
C 2	0.055	-0.210	0.0	0.0	-0.523	0.0	0.0	0.413
C 2	0.0	0.0	-0.400	-0.215	0.0	0.576	-0.385	0.0
C 2	0.0	0.0	0.202	0.396	0.0	0.413	-0.611	0.0
H 3	0.175	-0.257	-0.253	0.255	0.180	-0.128	0.117	0.469
H 4	0.175	-0.257	0.253	-0.255	0.180	0.128	-0.117	0.469
H 5	0.175	0.257	-0.256	-0.253	0.180	0.128	0.117	0.469
H 6	0.175	0.257	0.256	0.253	0.180	-0.128	-0.117	0.469

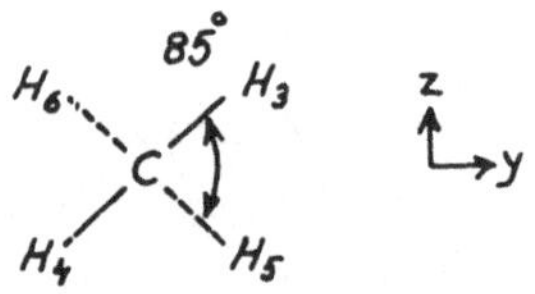

ΔH_f = 112 kcal/mol

Table O.3. HAM/3 calculations for planar and twisted ethylene. The tables give the configurational ionization energies (eV) and the orbitals. The numbers of the orbitals correspond to the order in planar ethylene. To the right the geometry is shown. The heats of formation, ΔH_f , obtained from separate calculations, are also given.

It is also important to study the total energy, which is shown in Fig. O.3.b.

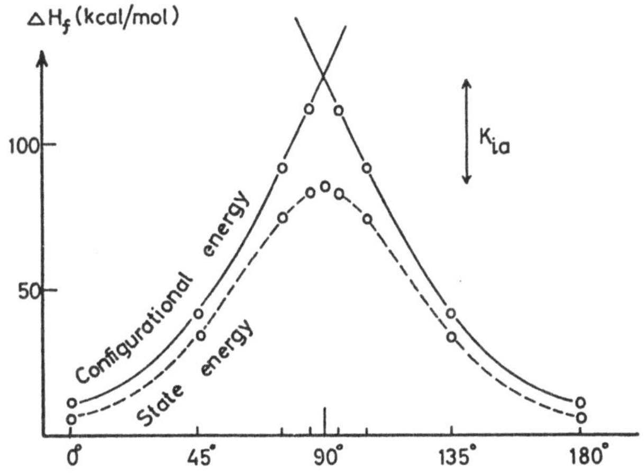

Fig. O.3.b. The total energies during the rotation, given as heat of formation. The full lines give the configurational energies and the dotted line gives the state energy for the ground state of the molecule. Only one point, the highest, is of importance for the determination of the activation energy.

We observe that the configurational energy at 90° is 120 kcal. If now the CI procedure, described in Sec. G.8., is applied at this angle, the energy must be reduced by K_{ia} = = 1.61 eV = 37 kcal/mol, and if it is applied at 0° the energy must be reduced by further 6 kcal/mol. The final computed activation energy is thus 120-37-6 = 77 kcal/mol, in good agreement with the experimental value 65 kcal/mol.

References

1. R.B. Woodward and R. Hoffmann, The Conservation of Orbital Symmetry, Verlag Chemie, Weinheim (1970).
2. P. Carsky and M. Urban, Ab-initio Calculations, Springer, Berlin (1980), p. 74.
3. L. Åsbrink, C. Fridh, E. Lindholm and G. Ahlgren, Chem.Phys. 33, 195 (1978).
4. L. Åsbrink, C. Fridh, E. Lindholm and G. Ahlgren, to be published.

Lecture Notes in Chemistry

Vol. 1: G. H. Wagnière, Introduction to Elementary Molecular Orbital Theory and to Semiempirical Methods. V, 109 pages. 1976.

Vol. 2: E. Clementi, Determination of Liquid Water Structure. VI, 107 pages. 1976.

Vol. 3: S. R. Niketic and K. Rasmussen, The Consistent Force Field. IX, 212 pages. 1977.

Vol. 4: A. Graovac, I. Gutman and N. Trinajstić, Topological Approach to the Chemistry of Conjugated Molecules. IX, 123 pages. 1977.

Vol. 5: R. Carbo and J. M. Riera, A General SCF Theory. XII, 210 pages. 1978.

Vol. 6: I. Hargittai, Sulphone Molecular Structures. VIII, 175 pages. 1978.

Vol. 7: Ion Cyclotron Resonance Spectrometry. Edited by H. Hartmann and K.-P. Wanczek. VI, 326 pages. 1978.

Vol. 8: E. E. Nikitin and L. Zülicke, Selected Topics of the Theory of Chemical Elementary Processes. X, 175 pages. 1978.

Vol. 9: A. Julg, Crystals as Giant Molecules. VII, 135 pages. 1978.

Vol. 10: J. Ulstrup, Charge Transfer Processes in Condensed Media. VII, 419 pages. 1979.

Vol. 11: F. A. Gianturco, The Transfer of Molecular Energies by Collision: Recent Quantum Treatments. VIII, 328 pages. 1979.

Vol. 12: The Permutation Group in Physics and Chemistry. Edited by J. Hinze. VI, 230 pages. 1979.

Vol. 13: G. Del Re et al., Electronic States of Molecules and Atom Clusters. VIII, 177 pages. 1980.

Vol. 14: E. W. Thulstrup, Aspects of the Linear and Magnetic Circular Dichroism of Planar Organic Molecules. VI, 100 pages. 1980.

Vol. 15: A.T. Balaban et al, Steric Fit in Quantitative Structure-Activity Relations. VII, 178 pages. 1980.

Vol. 16: P. Čársky and M. Urban, Ab Initio Calculations. VI, 247 pages. 1980.

Vol. 17: H. G. Hertz, Electrochemistry. X, 254 pages. 1980.

Vol. 18: S. G. Christov, Collision Theory and Statistical Theory of Chemical Reactions. XII, 322 pages. 1980.

Vol. 19: E. Clementi, Computational Aspects for Large Chemical Systems. V, 184 pages. 1980.

Vol. 20: B. Fain, Theory of Rate Processes in Condensed Media. VI, 166 pages. 1980.

Vol. 21: K. Varmuza, Pattern Recognition in Chemistry. XI, 217 pages. 1980.

Vol. 22: The Unitary Group for the Evaluation of Electronic Energy Matrix Elements. Edited by J. Hinze. VI, 371 pages. 1981

Vol. 23: D. Britz, Digital Simulation in Electrochemistry. X, 120 pages. 1981.

Vol. 24: H. Primas, Chemistry, Quantum Mechanics and Reductionism. XII, 451 pages. 1981.

Vol. 25: G. P. Arrighini, Intermolecular Forces and Their Evaluation by Perturbation Theory. IX, 243 pages. 1981.

Vol. 26: S. Califano, V. Schettino and N. Neto, Lattice Dynamics of Molecular Crystals. VI, 309 pages. 1981.

Vol. 27: W. Bruns, I. Motoc, and K. F. O'Driscoll, Monte Carlo Applications in Polymer Science. V, 179 pages. 1982.

Vol. 28: G. S. Ezra, Symmetry Properties of Molecules. VIII, 202 pages. 1982.

Vol. 29: N.D. Epiotis, Unified Valence Bond Theory of Electronic Structure. VIII, 305 pages. 1982.

Vol. 30: R.D. Harcourt, Qualitative Valence-Bond Descriptions of Electron-Rich Molecules: Pauling "3-Electron Bonds" and "Increased-Valence" Theory. X, 260 pages. 1982.

Vol. 31: H. Hartmann, K.-P. Wanczek, Ion Cyclotron Resonance Spectrometry II. XV, 538 pages. 1982.

Vol. 32: H.F. Franzen, Second-Order Phase Transitions and the Irreducible Representation of Space Groups. VI. 98 pages. 1982.

Vol. 33: G.A. Martynov, R.R. Salem, Electrical Double Layer at a Metal-dilute Electrolyte Solution Interface. VI, 170 pages. 1983.

Vol. 34: N.D. Epiotis, Unified Valence Bond Theory of Electronic Structure · Applications. VIII, 585 pages. 1983.

Vol. 35: Wavefunctions and Mechanisms from Electron Scattering Processes. Edited by F.A. Gianturco and G. Stefani. IX, 279 pages. 1984.

Vol. 36: J. Ugi, J. Dugundij, R. Kopp and D. Marquarding, Perspectives in Theoretical Stereochemistry. XVII, 247 pages. 1984.

Vol. 37: K. Rasmussen, Potential Energy Functions in Conformational Analysis. XIII, 231 pages. 1985.

Vol. 38: E. Lindholm, L. Åsbrink, Molecular Orbitals and their Energies, Studied by the Semiempirical HAM Method. X, 288 pages. 1985.